Mineral Processing and Technology International Conference 2017

IOP Conference Series: Materials Science and Engineering
Volume 285

Jakarta, Indonesia
23 – 24 October 2017

ISBN: 978-1-5108-5747-6
ISSN: 1757-8981

Printed from e-media with permission by:

Curran Associates, Inc.
57 Morehouse Lane
Red Hook, NY 12571

Printed by Curran Associates, Inc. (2018)

For permission requests, please contact the Institute of Physics
at the address below.

Institute of Physics
Dirac House, Temple Back
Bristol BS1 6BE UK

Phone: 44 1 17 929 7481
Fax: 44 1 17 920 0979

techtracking@iop.org

Additional copies of this publication are available from:

Curran Associates, Inc.
57 Morehouse Lane
Red Hook, NY 12571 USA
Phone: 845-758-0400
Fax: 845-758-2633
Email: curran@proceedings.com
Web: www.proceedings.com

Table of contents

Volume 285

Mineral Processing and Technology International Conference 2017

23–24 October 2017, Jakarta, Indonesia

Accepted papers received: 14 December 2017
Published online: 30 January 2018

Preface

Papers

IOP Conf. Series: Materials Science and Engineering **285** (2017) 011001 doi:10.1088/1757-899X/285/1/011001

Preface

Mineral Processing and Technology International Conference (mineprocèt) 2017 was a scientific meeting that is technically organized by Research Unit for Mineral Technology in conjunction with Indonesian Science Expo (ISE) 2017 organized by Indonesian Institute of Sciences (LIPI). mineprocèt 2017 was held on 23 - 24 October, 2017 at Balai Kartini, Jakarta, Indonesia.

This symposium aimed to bring together researchers, academicians, and scholars all around the world to share information and their research results. This event was also expected to become a beneficial media in exchanging their innovative ideas and build research collaborations in the future. The symposium covered all scientific and technical activities in mineral processing and technology – from basic to the advanced applications – to pitch ideas, insight and findings for optimal and sustainable use of sub-economic mineral resources especially those related to mineral beneficiation, extractive metallurgy (bio-hydrometallurgy, pyrometallurgy, electrometallurgy), separation and purification, physical metallurgy and foundry, powder and nano technology, mineral industry application, energy and coal technology, mineral economy and policy, phytoextraction and environmental remediation, control and modelling for mineral processing.

We specially invited 4 scientists as keynote speakers from various countries to share their knowledge to all participants. The total participants attending this symposium were 44 persons. This symposium has successfully attracted 39 submitted papers. Papers that have been presented in the mineprocèt 2017 were eligible for publishing in IOP Journal of Material Science and Engineering: Conference Series. The editorial board has decided to accept 33 papers from Indonesia, China, India and South Korea. The accepted papers have passed through strict reviewing process by scientific committee of this symposium. EDAS system has been used to facilitate the submission, reviewing, and revision processes.

We would like to appreciate all of our co-organizers, supporting societies, sponsors, and Research Unit for Mineral Technology staffs who have contributed generous amount of grants and effort to support us during the symposium. We hope this symposium has successfully enabled us to share and communicate our knowledge and skills and to build collaborations among the participants. We are looking forward to welcoming you in Indonesia again. See you in the next mineprocèt!

Dr. Eng. Widi Astuti, ST, MT
Editor-in-Chief
Research Unit for Mineral Technology
Indonesian Institute of Sciences

IOP Conf. Series: Materials Science and Engineering **285** (2017) 011001 doi:10.1088/1757-899X/285/1/011001

Conference Organizers

mineprocèt'17
Mineral Processing and Technology International Conference 2017

Scientific Organizing Committee

Erik Prasetyo, Ph. D.
Conference Chair/Editor
Research Unit for Mineral Technology
Indonesian Institute of Sciences

Dr. Eng. Widi Astuti, ST, MT
Conference Co-chair/Editor-in-Chief
Research Unit for Mineral Technology
Indonesian Institute of Sciences

Fajar Nurjaman, ST, MT
Conference Co-chair/Editor
Research Unit for Mineral Technology
Indonesian Institute of Sciences

Editorial Board

IOP Conf. Series: Materials Science and Engineering **285** (2017) 011001 doi:10.1088/1757-899X/285/1/011001

Under auspice of

Balai Penelitian Teknologi Mineral – Lembaga Ilmu Pengetahuan Indonesia
Research Unit for Mineral Technology – Indonesian Institute of Sciences

Jl. Ir. Sutami km. 15 Tanjung Bintang
Lampung Selatan 35361, Indonesia
Ph. +62721350054 Fax +02721350056
http://bptm.lipi.go.id/

IOP Conf. Series: Materials Science and Engineering **285** (2017) 011001 doi:10.1088/1757-899X/285/1/011001

Keynote Speakers

Tsuyoshi Hirajima

Department of Earth Resource Engineering
Kyushu University

Prof. Tsuyoshi Hirajima is specialist in mineral processing, powder and recycling technology, which research interest is to promote the development of an advanced mineral processing technology with a special focus on low emission application and resources recycling, highly required in the present century, for a sustainable development of modern societies.

Currently, his research group is studying the development of advanced mineral processing technology, resources recycling, waste treatment and utilization, and environmental remediation. He had authored and co-authored more than 200 peer reviewed scientific papers and several books. He is member of Society for Mining, Metallurgy and Exploration, The Japan Institute of Energy, The Mining and Materials Processing Institute of Japan, The Resources Processing Society of Japan, The Society of Powder Technology of Japan, Japan Society of Material Cycles and Waste Management and Energy and Resources Institute of Japan.

M. Akbar Rhamdhani

Deputy Chair Dept Mechanical Product Design Engineering
Swinburne University of Technology

Associate Professor M. Akbar Rhamdhani obtained his PhD from McMaster University Canada in Materials Science and Engineering. Akbar's research focuses on advanced metal/material refining and impurities removal, development of new processes for metal production, thermodynamics and kinetics of high temperature metal and chemical processes, and physical chemistry of interface. Akbar's recent research projects include: Thermodynamic behaviour of valuable trace elements during e-waste processing in copper smelting; Oxidation behaviour of rare-earth elements in permanent magnet scrap; Electrically enhanced silicon refining; Impurities control in electronic conductor grade aluminium; Removal of impurities from Titanium ore. Akbar has been a Visiting Professor at Katholieke Universiteit Leuven Belgium and University of Indonesia. He has been working with metals industries in Europe and Indonesia. Akbar is an author/co-author of more than 140 publications. Recently, Akbar has been awarded a number of international award including the 2015 Mann Redmayne Medal by IOM3 United Kingdom, the 2015 Marcus Grossmann Medal by ASM International USA, and the 2015 MetSoc Award by Metallurgical Society of Canada.

Naoko Okibe
Department of Earth Resource Engineering
Kyushu University

Assoc. Prof. Naoko Okibe specializes in Biohydrometallurgy, Bioremediation, Bioengineering, Environmental Microbiology, which research interest includes bioleaching of waste catalysts, manganese removal from refinery waste waters using Mn(II)-oxidizing bacteria, bioleaching of waste PCBs, nickel laterite and refractory copper sulfides.

She had authored and co-authored more than 35 peer reviewed scientific papers. She is currently member of Japan Society for Environmental Biotechnology, The Society for Biotechnology of Japan and The Mining and Materials Processing Institute of Japan.

Agustinus R. Uria
Faculty of Pharmaceutical Sciences
Hokkaido University

Dr. rer. nat. Agustinus R. Uria studied Biotechnology at Wageningen University, Netherlands, obtaining MSc degree on February 2004 through the Netherlands Fellowship Programmes (NFP) (2002-2004). Under the scheme of DAAD long-term research grant (2007-2011), he conducted his PhD project with Professor Joern Piel, developing metagenomics approaches to study natural product biosynthesis in uncultured symbiotic bacteria.

Upon obtaining his doctoral degree (Dr. rer. nat.) on December 2012 from the University of Bonn, Germany, he moved to Swiss Federal Institute of Technology - ETH Zurich, Switzerland, working as a postdoctoral research associate at Institute of Microbiology from February 2013 to January 2015. Some of his research data were published in *Nature*, *Science*, *Nature Chem Biol* and *ACS Chem Biol*. He is currently a JSPS postdoc research fellow under the scheme of Japan Society for the Promotion of Science at Faculty of Pharmaceutical Sciences, Hokkaido University, Japan.

IOP Conf. Series: Materials Science and Engineering **285** (2017) 011002 doi:10.1088/1757-899X/285/1/011002

Peer review statement

All papers published in this volume of *IOP Conference Series: Materials Science and Engineering* have been peer reviewed through processes administered by the proceedings Editors. Reviews were conducted by expert referees to the professional and scientific standards expected of a proceedings journal published by IOP Publishing.

IOP Conf. Series: Materials Science and Engineering **285** (2017) 012001 doi:10.1088/1757-899X/285/1/012001

Nickel extraction from nickel matte

R Subagja

Research centre for Metallurgy and Material LIPI
Gd. 470, Kawasan Puspiptek, Tangerang Selatan, Indonesia

Email: rudi.subagja@lipi.go.id

Abstract. In present work, the results of research activities to make nickel metal from nickel matte are presented. The research activities were covering **a**) nickel matte characterization using Inductively Couple plasma (ICP), Electron Probe Micro Analyzer (EPMA) and X-Ray Diffraction (XRD), **b**) nickel matte dissolution process to dissolve nickel from nickel matte into the spent electrolyte solutions that contains hydrochloric acid, **c**) purification of nickel chloride leach solution by copper cementation process to remove copper using nickel matte, selective precipitation process to remove iron, solvent extraction using Tri normal octyl amine to separate cobalt from nickel chloride solutions and **d**) Nickel electro winning process to precipitate nickel into the cathode surface from purified nickel chloride solution by using direct current. The research activities created 99, 72 % pure nickel metal as the final product of the process.

1. Introduction

Nickel is an important metal for making stainless steel, high temperature steel, magnetic materials, nickel plating, non ferrous alloys, catalyst, fuel cell and chemicals. Cause of wider application of nickel at Industries, the world nickel production tends to increase. In 2013, the world nickel production increased 10 times compared with the world nickel production in 1950 [1]. The increased of world nickel consumption was stimulated by a new discoveries of variety nickel alloys for industrial application and development of new materials [2, 3].

In line with the progress of Indonesian industries, the nickel demand increases, mostly of nickel are used as raw materials for electroplating industries in Indonesia. However in future, in line with the progress of Indonesian industries, the nickel demand is estimated will increase. Nowadays to full fill nickel demand in Indonesia, nickel have to be imported from foreign countries because there are not industry that can produce pure nickel metals from the available raw material in Indonesia. Referring to the data from the ministry of Industry of the republic of Indonesia, the value of imported nickel to Indonesia is 36 million US $ at 2016 [4]. The high imported nickel product to Indonesia creates an opportunity to make nickel metal from the available nickel resources in Indonesia. There are nickel laterite ores that can be used for nickel making and spreads at several areas in Indonesia such as at South Sulawesi, Maluku and Papua that can be used as raw material for nickel making [5,6]. Some of those nickel laterite ores has already been utilized by P.T Aneka Tambang to make Ferro nickel (25 % nickel and 75 % iron) and by P.T Vale to make nickel matte (73 % nickel and 20 % sulfur), however industry that can produce pure nickel metals has not available yet in Indonesia. Therefore to obtain the prospective process for nickel making from the available nickel resources in Indonesian, the Research centre for metallurgy and material LIPI conducted several laboratory tests to make pure nickel metal from nickel matte. The results of those laboratory tests are presented in this paper.

2. Characteristic of nickel matte

Nickel matte characterizations process was carried out by using Inductively Couple Plasma (ICP), Electron Probe Micro Analyzer (EPMA) and X-Ray Diffraction (XRD) that has already been reported in our paper [7)]. The result of nickel matte analysis using ICP at Table 1 shows that nickel and sulfur becomes the main elements in nickel matte. Another element such as cobalt, iron, copper, arsenic and lead exist as minor elements. The result of microanalysis study using EPMA shows that nickel, sulfur and other elements that available at nickel matte forms a complex compound formation, while result of analysis by using XRD shows that Ni_2S_3 exists as a main phase in nickel matte.

Table 1. The chemical composition of nickel matte [7]

Elements	Ni	Co	Fe	Cu	As	Pb	S
%	73	0.8	0.6	3	0.2	0.003	20

3. Process Flow sheet for nickel making from nickel matte

Result of nickel matte characterization using ICP, EPMA and XRD shows that Ni_2S_3 is the main phase in nickel matte, another elements such as cobalt, iron, arsenic, copper and lead are existing as minor elements that forms a complex compound formation, and therefore a process alternative to separate nickel from another elements in nickel matte are through nickel matte electro refining [8], or through a combination processes of hydro and electrometallurgy as shown in process flow sheet at Figure 1.

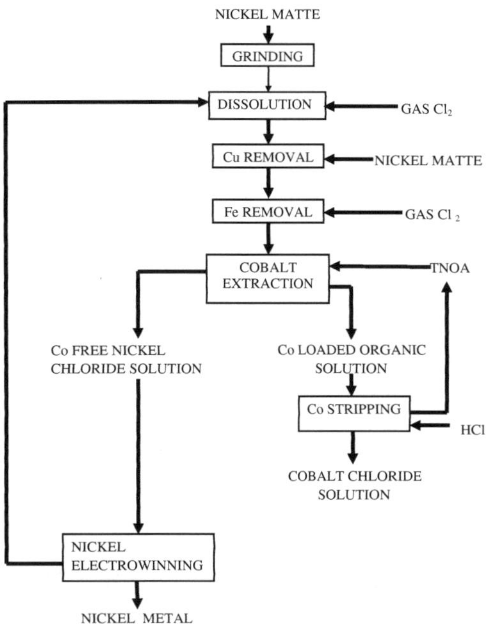

Figure 1. Process flow sheet for nickel making

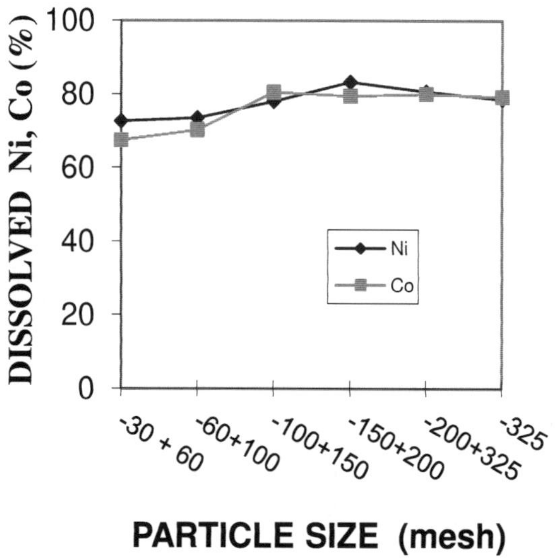

Figure 2. Effect of particle size on dissolution Ni and Co [9]

The process flow sheet at Figure 1 consists of several steps of process i.e. grinding process, nickel matte dissolution process to dissolve nickel from nickel matte into spent electrolyte solutions that contains hydro chloride acid from electro winning process, purification process to remove, copper, iron and cobalt from leached nickel chloride solution. In this purification process, copper are removed from the leach solutions by cementation process using nickel matte, iron are removed from leach solutions by selective precipitation process using chlorine gas and cobalt are separated from nickel chloride leach solutions by solvent extraction process. The final process is nickel electro winning. At this process nickel is deposited from purified nickel chloride solutions at the surface of cathode by using direct current.

4. Grinding process

Nickel matte particle size plays an important role in nickel dissolution process from nickel matte into the acid solutions. The more finer particle size are generally gives better result to the dissolution process because the smaller particle has larger surface area that creates good contact between nickel matte particle with the acid leach solution, and therefore it will stimulate and accelerated the nickel and cobalt dissolution process from nickel matte into acid solutions.

An experiment to investigate effect of nickel matte particle size on nickel and cobalt dissolution from nickel matte in to hydrochloric acid solutions has been investigated by author. At that experiment, the nickel matte particle size were varied from – 30 mesh up to -325 mesh, and dissolution experiment were carried out using 16 % volume aqueous hydrochloric acid solution at temperature 95°C, with chlorine gas flow rate 100 ml/minute, for dissolution time 1 hours. The result at Figure 2 shows that the finer nickel matte particle size increased the nickel and cobalt dissolution from nickel matte. For ground nickel matte that has particle size - 60 mesh, only 73% of nickel and 68 % of cobalt were dissolved from nickel matte. However if the nickel matte particle size was ground up to -100 mesh, the dissolved nickel and cobalt increased respectively to 78 % for nickel and 80 % for cobalt [9].

5. Nickel matte dissolution process into spent electrolyte contains hydrochloric acid solutions

The simplified nickel matte dissolution reaction in hydrochloric acid solution under chlorine gas addition is taking place according to the following reactions:

$$Ni_3S_2 + Cl_2 \rightarrow 3\,Ni^{2+} + 2\,S^{o} + 2\,Cl^{-} \quad (1)$$

That dissolution reaction is affected by dissolution temperature, hydrochloric acid concentration in leach solutions, copper and chlorine gas addition into the leach solution [7)]. The effect of temperature on nickel and cobalt dissolution from nickel matte into the 16 % volume hydrochloric acid solutions at temperature from 50 °C to 95 °C, and chlorine gas flow rate 60 ml/minute for 90 minute dissolution time at Figure 3 shows that the increase of dissolution temperature from 50 °C to 80 °C did not give any significant effect to the dissolution of nickel and cobalt from nickel matte. However if the dissolution temperature were increased from 80 °C to 95 °C, the dissolved nickel and cobalt were respectively increased to 100 % for nickel and 98 % for cobalt.

The effect of hydrochloric acid concentration on nickel dissolution has also been observed using spent electrolyte as leach solutions and the concentration of hydrochloric acid in spent electrolyte were varied from 1 % up to 16 % volume. The result of experiment at Figure 4 shows that the increase of hydrochloric acid concentration in spent electrolyte solutions from 1 % to 16 % increased the dissolved nickel from 25 % to 90 % and dissolved cobalt from 48 % to 100 % [7].

Copper in leach solutions influences the nickel dissolution process. Copper gives a catalytic effect to the nickel dissolution process. Our result of experiment at Figure 5 shows that the increase of $CuCl_2$ concentration from 0 gram/liter to 40 gram/liter in the leach solutions lead to increase the dissolved nickel from 61 % to 90 % and dissolved cobalt from 77 % to 93 % [7]. The same result of copper catalytic effect during nickel dissolution process was also reported by Wladyslawa Mulak [10].

IOP Conf. Series: Materials Science and Engineering **285** (2017) 012001 doi:10.1088/1757-899X/285/1/012001

The catalytic effect of copper during nickel matte dissolution process is taking place according to the following reactions:

$$2\,Cu^{+} + Cl_2 \rightarrow 2\,Cu^{2+} + 2\,Cl^{-} \quad (2)$$
$$Ni_3S_2 + 2\,Cu^{2+} \rightarrow 2\,NiS + Ni^{2+} + 2\,Cu^{+} \quad (3)$$
$$NiS + 2\,Cu^{2+} \rightarrow Ni^{2+} + 2\,Cu^{+} + S^{o} \quad (4)$$
$$Cu_2S + S^{o} \rightarrow 2\,CuS \quad (5)$$

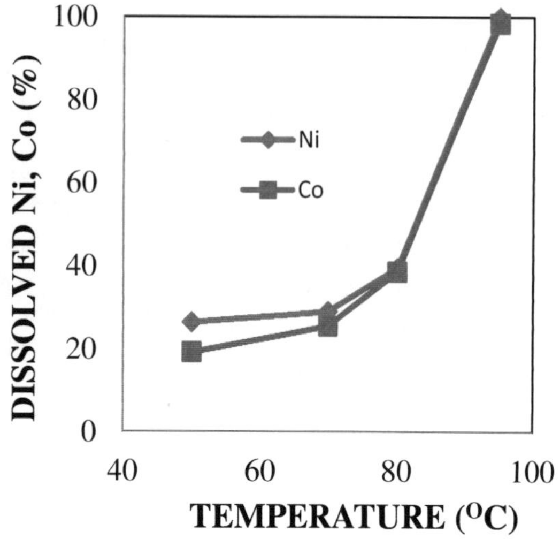

Figure 3. Effect of Temperature on dissolution Ni and Co

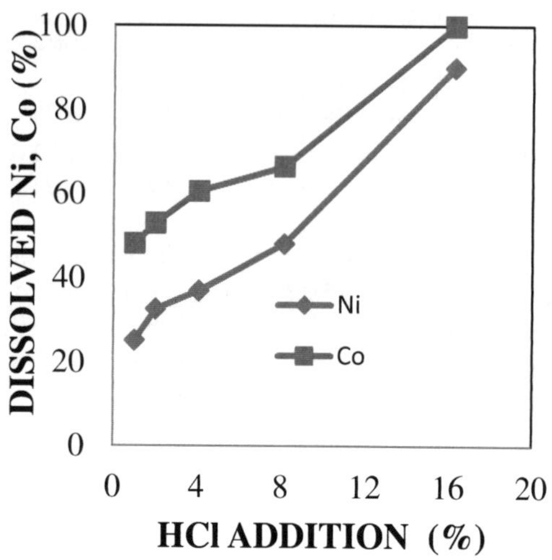

Figure 4. Effect of HCl addition on dissolution Ni and Co

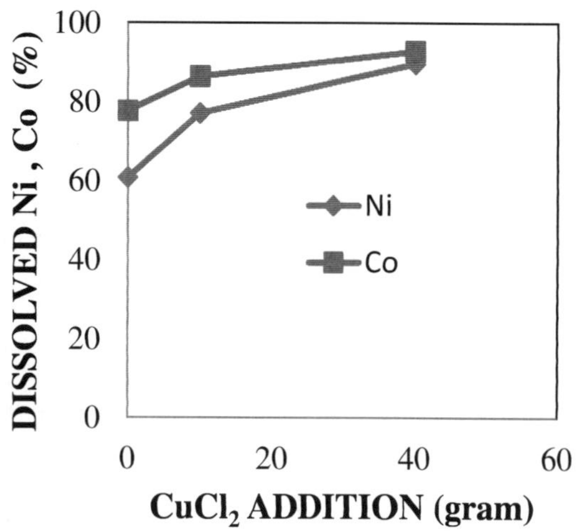

Figure 5. Effect of $CuCl_2$ addition on dissolution Ni and Co [7]

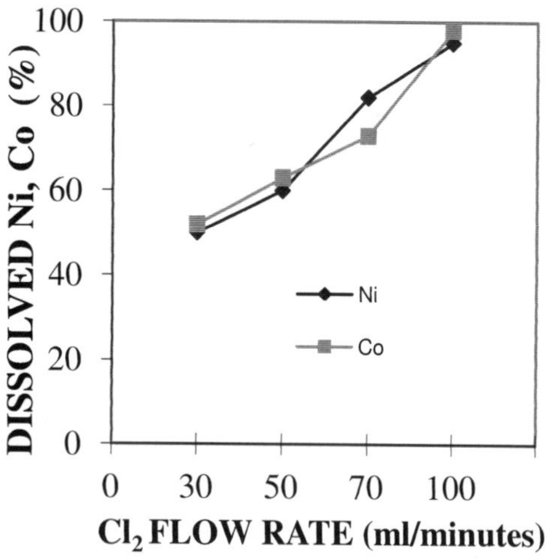

Figure 6. Effect of Cl_2 addition on dissolution Ni and Co [7]

In case of chlorine gas effect on nickel dissolution process, our result of experiment at Figure 6 shows that during nickel matte dissolution into 16 % volume hydrochloric solution at temperature 95 ^{o} C, the increase of chlorine gas flow rate from 30 ml/minute to 100 ml/minute lead to increase the dissolved nickel from 50 % to 95 %, and dissolved Cobalt from 52 % to 98 % [7].

6. Copper removal from nickel chloride leached solutions

The produced nickel chloride leached solutions from the nickel matte dissolution process were containing impurities elements such as Cu 0.15 gram/liter, Fe 0.33 gram/liter and cobalt 0.013 gram/liter. Those impurities elements have to be removed from the nickel chloride solutions because it will influence purity of the final product nickel metal. There are many alternative process to remove ionic copper from the solutions such as by precipitation of ionic copper using activated sulfur powder [11], copper cementation using Aluminum canning sheet [12], copper cementation using nickel plate steel [13], the cementation of copper from copper sulfate on zinc using organic compound sodium diethyldithiocarbamate as additive [14], copper cementation from copper sulfate solutions using iron powder [15], copper cementation from spent copper-pickle sulfate solution by zinc ash [16], and copper removal from the nickel chloride solutions using nickel matte that was investigated by author [17]. Comparing of those copper removing processes, the copper cementation using nickel matte is most prospective ways to remove ionic copper from solutions because nickel matte is raw materials for nickel making process, and the successfully of this latest process is affected by the quantities of nickel matte added, temperature and copper precipitation time. Result of our experiment to remove ionic copper from nickel chloride solutions that contains copper 0.15 gram/liter at temperature 75 o C and cementation time 60 minutes, at Figure 7 shows that the increase of nickel matte addition from 0.14 gram to 0.84 gram lead to increase the percentage of precipitated copper from 28 % to 94 %, and when the cementation temperature were increased from 30 0 C to 90 o C, the result at Figure 8 shows that the cemented copper increased, almost 100 % of ionic copper can be cemented from nickel chloride solutions at temperature 90 0 C using nickel matte [17].

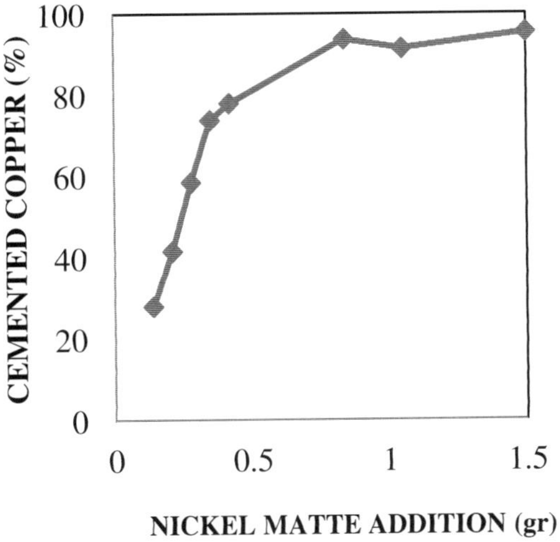

Figure 7. Effect of nickel matte addition on cemented copper

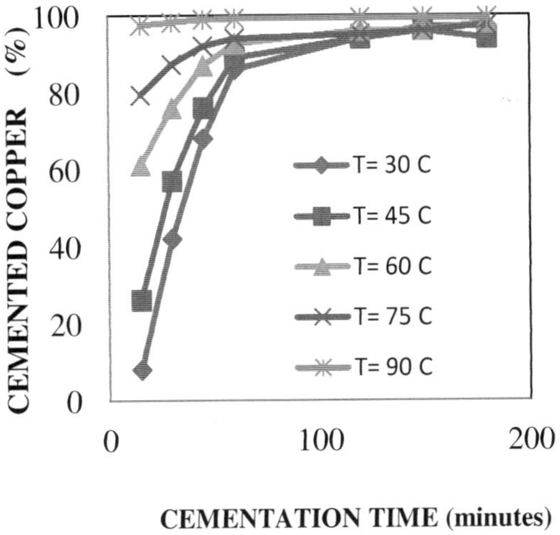

Figure 8. Effect of temperature on cemented copper

7. Iron removal from nickel chloride leached solutions

Nickel chloride leached solution produced from nickel matte dissolution process contains 0.33 gram/liter iron. That ionic iron should be removed from nickel chloride leached solutions because it influences the purity of the final product nickel metal. There are many alternative process to remove ionic iron from the leach solutions such as by using solvent extraction [18], oxidation of ionic iron available in the solution using dissolved oxygen [19], oxidation using mixed SO_2 and oxygen [20], iron removal with the iron phosphate precipitation process [21], magnetic precipitation from nickel rich solution by slow oxidation at pH 2.0-2.2 at temperature 90 -100° C [22], and in our experiment, iron was removed from nickel chloride solutions by oxidation of ionic Ferro to ionic ferric in the solution using chlorine gas, and controlling the solution potential & pH to precipitate ionic ferric from the solutions[23]. Among those methods, the iron removal by using chlorine gas and pH controlling to precipitate iron from the solutions is the most potential ways to make a purest nickel metal. This method is influenced by temperature and solutions pH. Our results of iron precipitation experiments at Figure 9 and Figure 10 show that the precipitated iron at solution pH = 3.25 increased if the precipitation temperature were increased from 40° C to 50° C. The increased of the precipitated iron with the temperature increase is caused by kinetic and thermodynamic aspect of the precipitation reactions. From kinetic aspect, the rate of precipitation will increase if the solution temperatures are increased and from thermodynamic aspect the increase of solution temperature shift the equilibrium curve of iron with its oxide to the left side and therefore at higher temperature iron will easily be precipitated [24]. Another experimental result at Figure 10 shows that for temperature 50° C and precipitation time 60 minutes, the precipitated iron increased from 85 % to 100 %. if the solution pH were increased from pH 2 to pH 3.25.

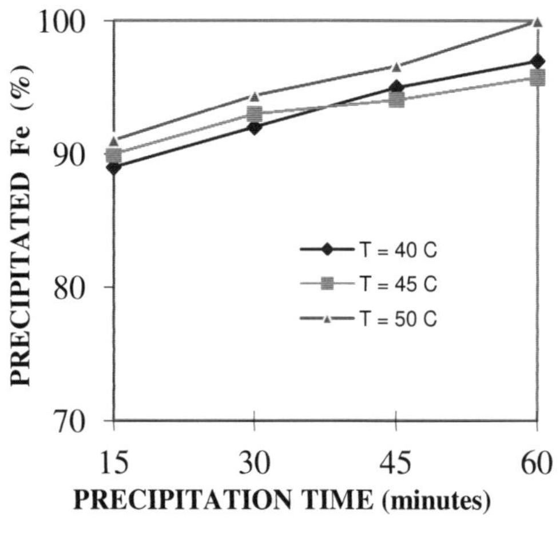

Figure 9. Effect of temperature on precipitation Fe

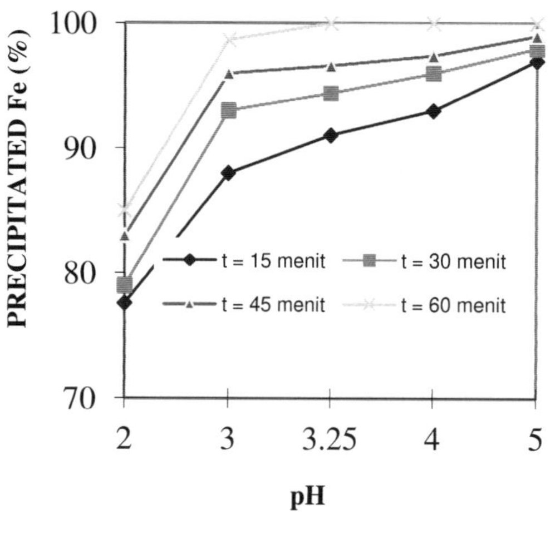

Figure 10. Effect of solution pH on precipitation Fe [23]

8. Cobalt Separation from nickel chloride solutions by solvent extraction

Cobalt separation from nickel chloride leached solutions is a final step in purification of nickel chloride leached solutions. There are many alternative processes to separate cobalt from the aqueous solution such as by selective precipitation and solvent extraction [25]. The separation of cobalt from nickel by solvent extraction will depend on the characteristic of their mother solutions [26]. Therefore, there are many commercial organic materials to separate cobalt from nickel by solvent extraction. The choice of those organic materials for solvent extraction process is depending on the characteristic of aqueous solution in which cobalt and nickel exist together in the solutions. Amine compound with varieties trade name are generally used to separate cobalt from nickel chloride solutions. Those amine compound are *Alamine® 336, Alamine® 308, and Adogen® 381,* while for the aqueous sulfate solutions, versatic 10, D2EHPA (di(2-etylheksyl) phosphoric acid, organo phosphoric acid with trade name PC-88, and Cyanex 272 are generally used to separate cobalt by solvent extraction. In ammoniac solution *ketoximes* with trade name *LIX® 87QN, LIX® 8* are generally used to separate cobalt. Although there were organic solvent for solvent extraction process to separate nickel and cobalt, there are still problems in application of those solvent, and therefore Cheng conducted research to separate nickel and cobalt containing in the solution from nickel laterite processing which contains Mn, Mg and Ca. Their research result shows that mixed organic compound containing Versatic 10, LIX 63 and Tri butyl phosphate can be used to separate nickel and cobalt from solutions that containing Mn, Mg and Ca [27, 28], and mixed organic solution Versatic 10 and Acorga CLX 50 can be used to recover nickel and cobalt from leach solutions of nickel laterites by solvent extraction [29]. Those result of solvent extraction experiments need additional study to separate cobalt from nickel containing solutions. In our study to separate cobalt from nickel chloride leach solutions, the solvent extraction process were choose using organic compound Tri Normal octyl Amine (TNOA) as solvent and Shell soll AB as diluents [30]. The result of our solvent extraction experiment at Figure 11 shows that cobalt extraction slightly decreased if the solution pH were increased from pH 3 to pH 4, that is caused by characteristic of octyl amine that generally more active to extract cobalt at lower solution pH [31]. Result of our experiment shows that almost 98 % of cobalt can be extracted from nickel chloride solutions by TNOA at aqueous to organic phase (A/O) ratio = 1, room temperature, solution pH = 3 and extraction time 30 minutes [30]. The cobalt extraction by TNOA is effected by interaction between cobalt with chloride ion in solutions as shown at Figure 12. In aqueous phase ionic cobalt reacts with chloric ions to forms a chloro complex compound ($CoCl_4^{2-}$). This compound then move from aqueous phase to interface phase. At same time TNOA that have already reacted with chloride ions ($R_3NH^+Cl^-$) move from organic phase to interface phase and react with ($CoCl_4^{2-}$) to form $(R_3NH^+)_2CoCl_4^{2-}$ and chloride ion Cl^-. The phase $(R_3NH^+)_2CoCl_4^{2-}$ then goes to organic phase and chloride ion goes to aqueous phase [32]. The data at Figure 11 shows that cobalt extraction increases if A/O ratio decreases. At lower A/O, the TNOA concentration is high that lead to increase the cobalt extraction. However, if the A/O ratio increases, the TNOA concentration decreases and cobalt extraction decreases.

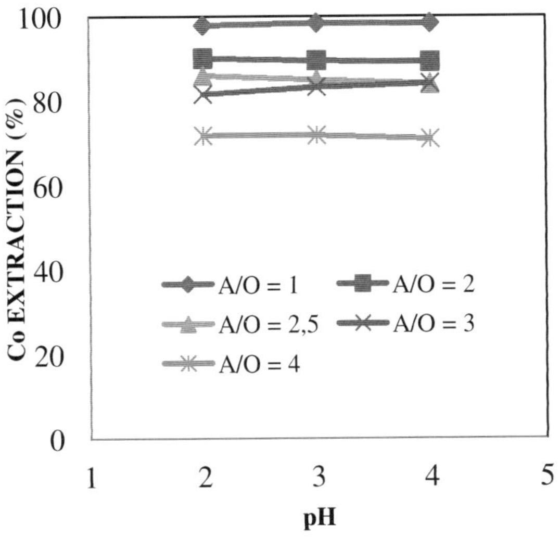

Figure 11. Effect of pH on cobalt extraction [30]

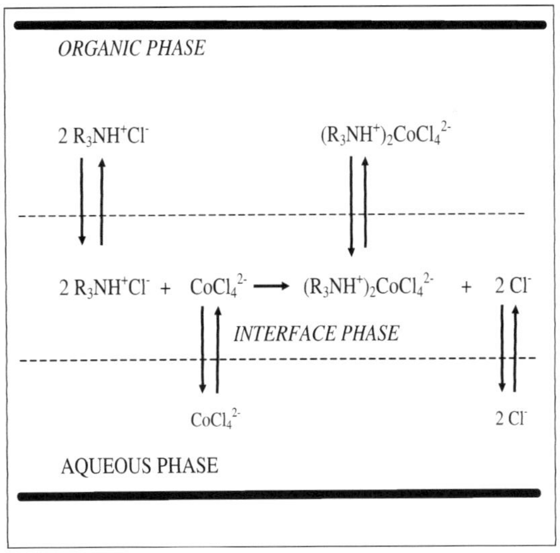

Figure 12. Some equilibrium involved in the extraction of cobalt from a chloride system using

tertiary amine [32]

9. Nickel electro winning process

The schematic apparatus for nickel electro winning process is presented at Figure 13. The electro winning cell consists of a bath that is containing of nickel chloride electrolyte solutions, positive electrode and negative electrode. The negative electrode (cathode) was made from titanium plate and the positive electrode (Anode) was made from Ruthenium oxide coated titanium plate material. The main reaction for nickel electro winning process in chloride solution takes place according to the equations 6 until 9. Nickel is deposited at the surface of cathode and hydrogen gas is generated according to the following reactions:

$$Ni^{2+} + 2e^- \rightarrow Ni^o \quad (6)$$

$$2H^+ + 2e- \rightarrow H_2 \quad (7)$$

At anode, chlorine gas and oxygen gas will be generated according to the following equation:

$$2Cl^- \rightarrow Cl_2 + 2e \quad (8)$$

$$H_2O \rightarrow 2H^+ + \tfrac{1}{2}\,O_2 + 2\,e- \quad (9)$$

The chlorine, oxygen and hydrogen gases are produced by side reaction of nickel electro winning process, they consumes energy and reduced the current efficiency of nickel deposition process at cathode. The nickel current efficiency is ratio of actual nickel deposited at cathode with theoretical weight of nickel deposited at cathode that is calculated by equations 10.

$$\text{Current efficiency} = \frac{W_A}{W_T} \times 100\% \quad (10)$$

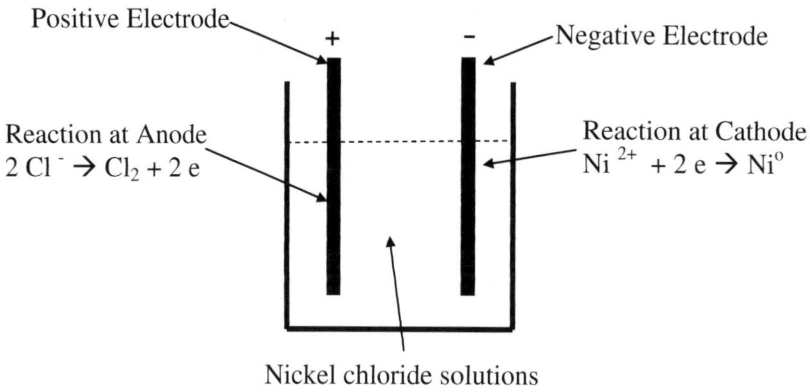

Figure 13. Nickel electro winning apparatus

Theoretical weight of nickel deposited at cathode (W_T) was calculated by Faraday law [33] according to equation 11.

$$W_T = i\, e_{Ni}\, t \quad (11)$$

Whereas:
W_A = weight of deposited nickel at cathode
W_T = theoretical weight of nickel deposited at cathode calculated by Faraday law
i = Current (Ampere)

e_{Ni} = equivalent weight for nickel
t = deposition time (second)

The result of our nickel electro winning experiment using nickel chloride electrolyte solutions contains 55 gram/liter nickel, at current density 2 Ampere per square decimeter (ASD), at temperature 50°C and solution pH = 1 show that almost 97 % of current efficiency was obtained and the final nickel metal deposited at cathode side has chemical composition as shows at Table 2.

Table 2. The chemical analysis of nickel deposited at cathode.

Element	Ni	Co	Zn	Cu	Fe
%	99,72	0,16	0,04	0,04	0,04

10. Conclusions

- Nickel metal with purity 99, 72 % can be made from nickel matte by using Hydro and electrometallurgy processes that consist of a) nickel matte dissolution process using spent electrolyte that containing hydrochloric acid solutions to dissolve nickel and cobalt from nickel matte, b) solution purification to remove copper, iron and cobalt from nickel chloride leached solutions and c) nickel electro winning process to deposited nickel from nickel chloride solutions.
- At dissolution process, almost 90 % of nickel and 100 % of cobalt were dissolved from nickel mate into the spent electrolyte that containing 16 % volume aqueous hydrochloric acid solutions at temperature 95 °C and dissolution time 180 minutes.
- At purification process, Almost 100 % of ionic copper can be removed from the nickel chloride solutions by cementation using nickel matte at temperature 90 °C and almost 100 % of iron was removed from the nickel chloride solution by chlorine gas addition at temperature 50 °C and solution pH 3,25. For Cobalt removal from nickel chloride solutions, the solvent extraction process to separate cobalt from nickel chloride using Tri normal octyl amine able to extract 100 % of cobalt into TNOA solutions at A/O (Aqueous to organic) ratio 1 and solution pH 3.
- The nickel electro winning process to deposit nickel from nickel chloride solutions that containing 55 gram/liter nickel, at current density 2 Ampere per square decimeter (ASD), temperature 50 °C and pH = 1 created current efficiency 97 % and nickel metal with purity 99,72 %.

References

[1] Oxley A, Barcza N 2013 *Int. J. Miner. Eng.* **54** 2-13
[2] Shailesh 2006 *Journal of Metal* 18-20
[3] Shoemaker L E, Smith G D 2006 *Journal of Metal* 22-6
[4] www.kepemperin.go.id.
[5] Departemen Pertambangan dan Energi 1998 *Potensi dan Prospek Investasi di Sektor Pertambangna dan Energi 1998-1999.*
[6] Dit. Sumber daya Mineral Republik Indonesia 1997 *Peta Ekonomi Mineral*
[7] Subagja R 2006 *Metalurgi* **21** 39-46
[8] Subagja R 2008 *Majalah Teknologi Indonesia* **3** 11-9
[9] Subagja R 1999 *Prosiding pemaparan hasil Litbang Ilmu Pengetahuan Teknik, LIPI, Bandung* pp 302-9
[10] Mulak W 1987 Hydromet. **17** 201-14
[11] Kobayashi M, Kametami H, Yamada K 1989 *Metallurgical review of MMIJ* **6** 54-67
[12] Mac Kinnon D J, Ingraham T R 1971 *Canadian Metallurgical Quarterly* **10** 197-201
[13] Popov B N, White R E, Ivshin J V 1991 *Plating and Surface Finishing* 61-5
[14] El-Saharty A A, El-Hammamy N H, El-Araby H A 2015 *Egyptian Journal of Aquatic Research*

41 289–93
[15] Jhajharia R, Jain D, Sengar A, Goyal A, Soni P R 2016 *Powder Technology* **301** 10–5
[16] Ahmed I M, El-Nadi Y A, Daoud J A 2011 Hydromet. **110** 62–6
[17] Subagja R 2003 *Metalurgi* **18** 23-8
[18] Ritcey G M, Lucas B H 1974 *CIM Bulletin* 87-92
[19] Iwai M, Majima H, Awakura Y 1982 *Metallurgical transaction B* **13** 311-8
[20] Tiwari B L, Kobe J, Hayden H W 1979 *Metalurgical trasaction B* **10** 607-12
[21] Masambi S., Dorfling Ch, Bradshaw S 2016 *Miner. Eng.* 98 14-21
[22] Han H, Sun W, Hu Y, Cao X, Tang H, Liu R, Yue T 2016 *Hydromet.* 165 318-22
[23] Subagja R 2006 *Metalurgi* **21** 12-17
[24] Yazawa A 1997 *Metallurgical transaction B* **10** 307-21
[25] Cheng C Y, Urbani M D 2003 *Yazawa International Symposium on Metallurgical and Materials Processing: Principals and Technologies*, TMS Annual Meeting, San Diego, USA pp 251-65
[26] Mihaylov I 2003 *Journal of Metal* **55** 38-42
[27] Cheng C Y, Boddy G, Zhang W, Godfrey M, Robinson D J, Pranolo Y, Zhu Z, Wang W 2010 *Hydromet.* **104** 45–52
[28] Cheng C Y, Boddy G, Zhang W, Godfrey M, Robinson D J, Pranolo Y, Zhu Z, Zeng L, Wang W 2010 *Hydromet.* **104** 53-60
[29] Cheng C Y, Urbani M D, Davies M G, Pranolo Y, Zhu Z 2015 *Minerals Engineering* **77** 17-24
[30] Subagja R 2011 *Teknologi Indonesia* **34** 102-10
[31] Biswas B 1998 *J. Radioanalytic. Nucl. Chem.* **240** 387-90
[32] Hayes P 1985 *Process selection in extractive Metallurgy* Brisbane: Hayes publishing co, Australia
[33] Lantropov 1977 *Theoretical electrochemistry* Moscow: Mir Publisher

IOP Conf. Series: Materials Science and Engineering **285** (2017) 012002 doi:10.1088/1757-899X/285/1/012002

A green preparation of Mn-based product with high purity from low-grade rhodochrosite

F Lian, L Ma, Z Chenli and L Mao

School of Material Science and Engineering, University of Science and Technology Beijing, Beijing 100083, China

Email : lianfang@mater.ustb.edu.cn

Abstract. The low-grade rhodochrosite, the main resources for exploitation and applications in China, contains multiple elements such as iron, silicon, calcium and magnesium. So the conventional preparation of manganese sulphate and manganese oxide with high purity from electrolytic product is characterized by long production-cycle, high-resource input and high-pollution discharge. In our work, a sustainable preparation approach of high pure $MnSO_4$ solution and Mn_3O_4 was studied by employing low-grade rhodochrosite (13.86%) as raw material. The repeated leaching of rhodochrosite with sulphuric acid was proposed in view of the same ion effect, in order to improve the solubility of Mn^{2+} and inhibit the dissolution of the impurities Ca^{2+} and Mg^{2+}. With the aid of theoretical calculation, BaF_2 was chosen to remove Ca^{2+} and Mg^{2+} completely in the process of purifying. The results showed that the impurities such as Ca^{2+}, Mg^{2+}, Na^+ were decreased to less than 20ppm, and the Ni- and Fe- impurities were decreased to less than 1ppm, which meets the standards of high pure reagent for energy and electronic materials. The extraction ratio and the recovery ratio of manganese reached 94.3% and 92.7%, respectively. Moreover, the high pure Mn_3O_4 was one-step synthesized via the oxidation of $MnSO_4$ solution with the ratios of $OH^-/Mn^{2+}=2$ and $Mn^{2+}/H_2O_2=1.03$, and the recovery rate of manganese reaches 99%.

1. Introduction

Low-grade manganese carbonate ores (≤14%) becomes gradually the main resources for exploitation and applications in China according to national mineral resources survey [1, 2]. Therefore, the electrolytic deposition technique of rhodochrosite faces great challenges including low electrolysis efficiency, high-resource input and high-pollution discharge [3-5]. Recently, more and more amount of manganese series with high or super high purity such as $MnSO_4$, MnO_2, Mn_3O_4 and MnO is demanded with the development of energy and electronic materials. Due to low-grade rhodochrosite containing multiple elements such as iron, silicon, calcium and magnesium, conventional preparation process of manganese oxide such as Mn_3O_4 with high purity from electrolytic manganese metal is characterized by long production-cycle, high-pollution discharge and high energy consumption [6, 7]. By contrast, hydrometallurgical method combined with thermal decomposition, hydrothermal synthesis and sol-gel processes has been widely studied because of its high efficiency, low consumption and low cost [8]. But the key problem in direct preparation of manganese series from low-grade rhodochrosite is to remove completely the multiple impurities such as iron, silicon, calcium and magnesium [9]. So it is urgent to develop a sustainable alternative approach of high pure Mn-based products with characters of short process, high efficiency and environmental friendly.

In our work, the preparation technique of manganese sulphate with high purity directly from low-grade rhodochrosite ore (containing 13.86% Mn) was studied and improved intensively. In detail, the repeated leaching of rhodochrosite with sulfuric acid was proposed herein to improve the solubility of Mn^{2+} and inhibit the dissolution of the impurities Ca^{2+} and Mg^{2+}, simultaneously. Additionally, the short purifying process was also explored with the aid of theoretical calculation. The recovery ratio of manganese and removal ratio of the impurity were evaluated via chemical titration and inductively coupled plasma (ICP) analysis, respectively. Furthermore, the high pure Mn_3O_4 was synthesized via the oxidation of $MnSO_4$ solution. Finally, the proposed green preparation directly from low-grade rhodochrosite has also been compared herein with the conventional technique from electrolytic manganese from the point of view of energy-efficiency and environmental friendly.

2. Experimental

The rhodochrosite samples were provided by BaoJingTianhe Manganese Industry Company, Hunan province, China. The samples were ground to powder with ≤2 mm in diameter. Firstly, the repeated leaching approach for rhodochrosite was performed herein, 10 g rhodochrosite powder, 3.6 ml sulfuric acid solution (98 %) and deionized water were added into a 150 ml conical flask with solid-liquid ratio of 1:7 in weight and reacted for 2 h at 85 °C under vigorous stirring (300 r/min) [10-12]. Then 10 g rhodochrosite powder together with X ml sulfuric acid solution (98%) and deionized water were added into the initial leaching solution with stirring. Then the residues were filtered to get the first repeated leaching solution. And the aforementioned procedures were repeated 0-4 times, so the total leaching times disperse in the range of 1-5. Moreover, the amount of sulfuric acid solution for each addition was calculated as the following formula:

$$X=0.0779a-ab/38.25 \quad (1)$$

where X is the volume of sulfuric acid solution (ml), a is the total volume of the leaching solution (ml) and b is the concentration of H^+ in the leaching solution (mol/l).

In the purification process, the coarse manganese sulphate solution was adjusted to pH=4.0 using $Ba(OH)_2$ solution. After standing for 24 h, the precipitate was filtered. Subsequently, BaF_2 powder was introduced in the solution and stirred at 25 °C for 3 h. Then the precipitate was separated and the filtrate was adjusted to pH=8.0 using $Ba(OH)_2$ solution. Finally, the precipitate was filtered repeatedly and pure $MnSO_4$ solution was expected to be obtained.

In the synthesis of Mn_3O_4 process, 10ml pure $MnSO_4$ solution were put in 150ml conical flask, and a certain amount of KOH solution (1 mol/l) was added with the continuous stirring. Then the certain amount of H_2O_2 solution (30 wt.%) was added dropwise. After stirring for 10 minutes, the solution was filtered and the precipitation was washed repeatedly with deionized water. Mn_3O_4 powder was obtained after dried in 120 °C.

The chemical composition of the rhodochrosite samples was chemically analyzed via chemical titration. While, the mineral composition of the samples was investigated by X-ray diffraction (XRD) on a Rigaku D/max (Japan) diffractometer. The data were collected at a scan speed of 5°/min on the 2θ range from 10° to 100°. The recovery ratio of manganese was evaluated via chemical titration with ammonium ferrous sulfate. The concentration of impurities such as calcium and magnesium ions was determined by ICP analysis (Spectroflame ICP, 2.5 kW, 27 MHz).

3. Results and Discussions

3.1. The composition of rhodochrosite samples

The chemical compositions of rhodochrosite samples are listed in table 1. The concentration of manganese is only 13.86 %, which is typical low-grade manganese carbonate ores. In addition, the rhodochrosite samples are mainly composed of $MnCO_3$, SiO_2, $CaMg(CO_3)_2$ and $FeSO_3$ et al as shown in figure 1.

Table 1. The major chemical compositions of the rhodochrosite sample (wt.%).

Mn	Ca	K	Mg	Ni	Zn	Na	Fe	Si
13.86	3.80	1.80	1.32	<0.005	0.023	0.86	3.10	16.63

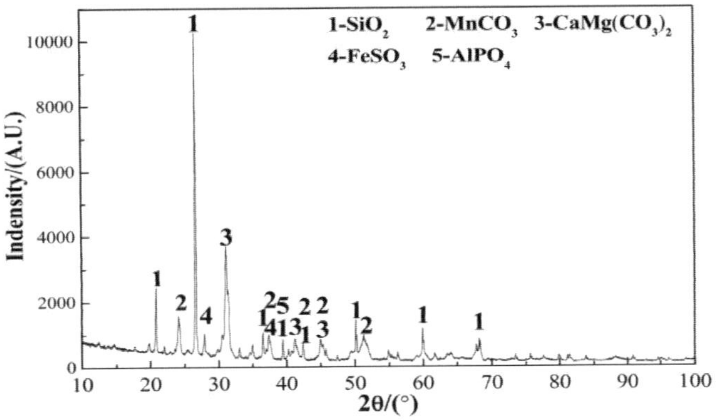

Figure 1. XRD patterns of the low-grade rhodochrosite.

3.2. The preparation of high pure $MnSO_4$ solution

In the repeated leaching process, the accumulative content of impurity in the leaching solution and the extraction ratio of Mn were characterized by ICP. The total recovery ratio of manganese is 94.3% after 5 leaching times. The standard deviation for the four parallel experiment results is below 0.73%, and the average concentration values are given in table 2.

Figure 2 demonstrates that the concentration of manganese is enhanced significantly and the accumulative content of Ca^{2+} is reduced obviously with the increase of the repeated leaching times. Though the absolute concentration of Mg^{2+} still increases, its extraction ratio is reduced in the repeated leaching process. Based on the same ion effect, the repeated leaching approach not only improves the concentration of Mn^{2+} but also inhibits the dissolution of the impurities Ca^{2+} and Mg^{2+}.

The main impurities in the coarse manganese sulphate solution obtained from the leaching process consist of Al^{3+}, Fe^{2+}, Fe^{3+}, Mg^{2+}, Ca^{2+} and a small quantity of heavy metals. Among them, a large proportion of Ca^{2+} ions could be precipitated with SO_4^{2-} and remove with insoluble silica. Moreover, Ca^{2+} and Mg^{2+} cations are removed further by adding BaF_2 as the precipitant at pH=4.0. Finally, the iron, aluminium and heavy metals could be removed in hydrolysis via adjusting pH=8.0 with $Ba(OH)_2$.

According to ICP analysis of the manganese sulphate solution in Table 3, the removal rates of most impurities are above 99%. The Fe element and heavy metal impurities including Ni have been decreased to less than 1 ppm, and the contents of calcium, magnesium and sodium have been decreased to less than 20 ppm, which meets the standards of high pure reagent for energy and electronic materials. Mn content in the manganese sulphate solution is determined by titration with ammonium ferrous sulphate. It is confirmed that the recovery ratio of manganese reaches 92.7% in the purifying process.

Table 2. The average value of ICP analysis of ionic concentration in the repeated leaching solution.

Total leaching times	1	2	3	4	5
Mn^{2+} (g/L)	19.42	38.49	56.02	71.44	84.05
Ca^{2+} (mg/L)	540.2	533.7	522.2	504.0	486.0
Mg^{2+} (mg/L)	1027.3	1165.0	1277.6	1425.0	1687.2

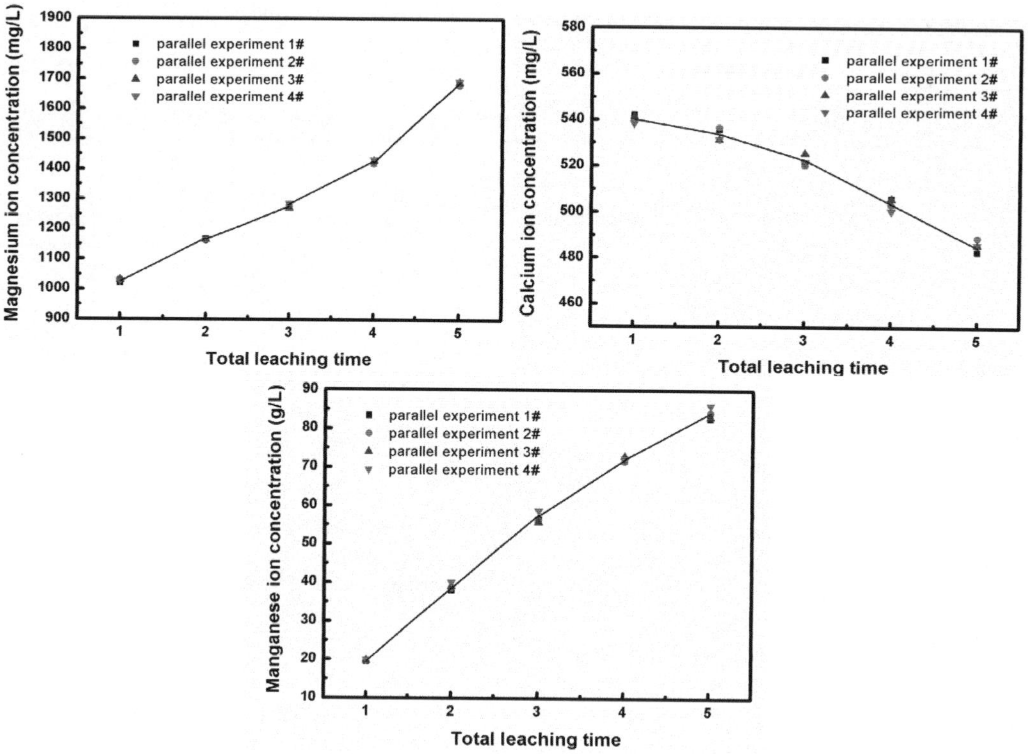

Figure 2. Effect of the leaching times on the ionic concentration.

Table 3. ICP analysis of the manganese sulphate solution.

Elements	Mn	Ca	Mg	Ni	Fe	Al	Na
Before purifying/ppm	84050	486	1687	188	4060	3210	650
After purifying/ppm	77910	2.599	15.82	0.187	0.083	0.006	15.82
Recovery ratio/%	92.7	0.5348	0.9378	0.0995	0.0020	0.0002	2.434

In compare with conventional technique from electrolytic manganese, the power consumption has been reduced to ≤2000 kWh/t(product) in our short-process approach. Moreover, comprehensive recovery rate of manganese increases significantly. As known, the quality of wastewater occurred in the process of electrolytic deposition is complex, containing toxic substances and high content of ammonia/nitrogen. By contrast, the wastewater generated from our short-process approach shows low content of ammonia/nitrogen about 500 mg/t(product) and no toxic emissions. Our study contributes to the sustainable application of low-grade rhodochrosite.

3.3. The preparation of high pure Mn_3O_4

Mn_3O_4 was synthesized from the oxidation of $MnSO_4$ by employing H_2O_2 as oxidant. The one step preparation process at room temperature could significantly reduce the energy consumption in the comparison of the traditional preparation of Mn_3O_4 from electrolytic manganese. The reaction mechanism of Mn_3O_4 in the alkaline solution is shown as follow. The effects of OH^- and oxidant are analysed as seen in figure 3 and 4.

$$3Mn^{2+}+H_2O_2+6OH^- = Mn_3O_4+4H_2O \tag{2}$$

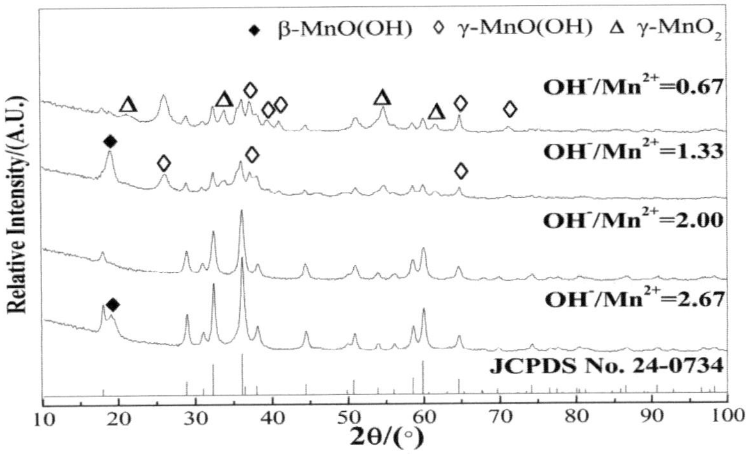

Figure 3. XRD patterns of samples with different rations of OH^-/Mn^{2+}.

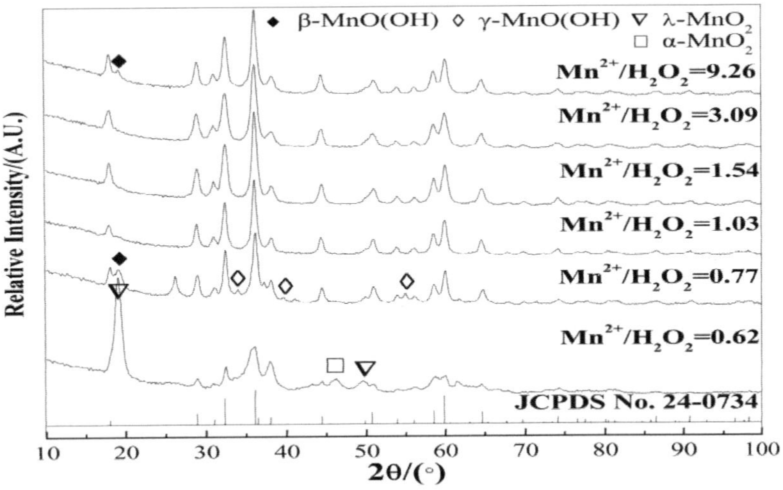

Figure 4. XRD patterns of oxidative products with different rations of Mn^{2+}/H_2O_2.

Figure 3 shows that the oxidative products depends on the concentration of OH^- in the solution. In detail, pure Mn_3O_4 is obtained when OH^-/Mn^{2+} ratio is adjusted to be 2.00. While, MnO(OH) and γ-MnO_2 become the major phases with the decrease of OH^- concentration. By contrast, β-MnO(OH) instead of Mn_3O_4 appears at the much higher pH value.

With a certain concentration of Mn^{2+} and OH^- in the solution, the content of H_2O_2 is responsible for the oxidative products as seen in figure 4. In view of the self-decomposition of H_2O_2 (in formula 3), additional H_2O_2 is introduced to guarantee the complete synthesis of Mn_3O_4.

$$H_2O_2= H_2O+1/2O_2 \tag{3}$$

According to Figure 3 and 4, γ-Mn_3O_4 (JCPDS No. 24-0734) is obtained with high purity when the ratios of OH^-/Mn^{2+} and Mn^{2+}/H_2O_2 are controlled to be 2.00 and 1.03, respectively. In the short process of Mn_3O_4 preparation, the recovery rate of manganese increases to 99 %. Unlike the conventional preparation of Mn_3O_4 from electrolytic manganese, the synthesis of Mn_3O_4 directly from low-grade rhodochrosite is a green technique with character of short process, low energy consumption and environmental friendly.

4. Conclusions

The preparation of high pure $MnSO_4$ and Mn_3O_4 with the sustainable, short and effective process from low-grade rhodochrosite ore (13.86%) has been developed in the paper. The results demonstrate that the extraction ratio and the recovery ratio of manganese in purifying process reach 94.3% and 92.7%, respectively. The concentration of manganese is accumulated in the repeated leaching solution, and simultaneously the extraction ratio of impurities of Ca^{2+} and Mg^{2+} is reduced significantly. The repeated leaching method provides an approach to the effective use of the low-grade rhodochrosite and the reduction of the wastewater discharge. In addition, BaF_2 is chosen as precipitant to remove Ca^{2+} and Mg^{2+} completely. The purification process is combined with adjustment of the pH value of manganese sulfate solution using $Ba(OH)_2$, which contributes to an effective remove of impurities and a short process for purification. Based on the preparation of the high pure $MnSO_4$ solution, Mn_3O_4 is one-step synthesized at room temperature via the oxidation of Mn^{2+} by employing H_2O_2 as oxidant, which is more efficient and less energy consumed than the conventional preparation of Mn_3O_4 from electrolytic manganese. Our study contributes to the sustainable application of low-grade manganese ore.

Acknowledgements

This work was financially supported by the National Science and Technology Pillar Program (2015BAB01B02).

References

[1] Tan Z Z, Mei G G, and Li W J 2004 *Manganese Metallurgy*(Changsha-Central South University Press)
[2] Singh V and Biswas A 2017 *International Journal of Mineral Processing* **158** 35 – 44
[3] Fu J G, He Z X, Wang H, Liang W and Guo C 2010 *Mining Science and Technology* **20** 877 – 81
[4] Michelis I D, Ferella F, Beolchini F, Olivieri A and Vegliò F 2009 *Journal Hazardous Materials* **162** 1285-91
[5] Zhang Y B, You Z X, Li G H and Jiang T 2013 *Hydrometallurgy* **133** 126-32
[6] Subrat M, Kumar M B, Nilima Dand Srinivas R D 2011 *Mining Science and Technology* **21** 803 – 8
[7] Mpho M, Samson B and Ayo A 2013 *International Joural Mining Science and Technology* **23** 537-41
[8] Chen N X and Liao Z W 2008 *Manganese Industry in China* **26** 5 – 8
[9] LiuY C, Lin Q Q and Li L F 2014 *Journal of Mining Science and Technology* **24** 567 – 71
[10] Xie C, Xu L J, Peng T F, Chen K and Zhao J 2013 *Chemical Journal of Geochemistry* **32** 222 – 6
[11] Liu L 2010 *Shanxi Metallurgy* **33** 4 – 6
[12] Zefang C L, Ma L J, Mao L and Lian F 2017 *Pro. of the 8th Int. Conf. on Sustainable Development in the Minerals Industry* (Beijing: China/Canamaple Academia Services) (accepted).

Leaching of silicon from ferronickel (FeNi) smelting slag with sodium hydroxide solution at atmospheric pressure

F R Mufakhir[1,2], **M Z Mubarok**[1], **Z T Ichlas**[1,3]

[1] Department of Metallurgical Engineering, Faculty of Mining and Petroleum Engineering, Institut Teknologi Bandung, Bandung 40132, Indonesia
[2] Research Unit for Mineral Technology, Indonesian Institute of Sciences (LIPI), Lampung 35361, Indonesia
[3] Department of Mining Engineering and Metallurgical Engineering, Western Australian School of Mines, Curtin University, Kalgoorlie 6430, Australia

Email : fika001@lipi.go.id

Abstract. The present paper reports the leaching behavior of silicon from ferronickel slag under atmospheric pressure using sodium hydroxide solution. The effect of several experimental variables, namely concentration of leaching agent, operating temperature, stirring speed, and slurry density was investigated. The leaching kinetic was also investigated by using shrinking core model. It was determined that leaching of silicon from the slag was controlled by diffusion through product layer, although the activation energy was found to be 85.84 kJ/mol, which was unusually high for such a diffusion-controlled process.

1. Introduction

The production of each ton of ferronickel generates about 8 tons of slags. In contrast to the slag produced by copper, iron or steel smelters that are widely consumed in the cement industry, most of ferronickel smelting slag are only disposed at landfills. This may lead to undesirable environmental impacts considering the immense quantity of its production. It is therefore important to valorize these slags to improve the sustainability of the ferronickel industry and to optimize the value of natural resources utilization.

There have been several attempts to provide such alternative usages of ferronickel slags such as for the production of construction materials [1, 2], cements [3, 4], or geopolymers [5, 6], but there has been no attempt to recover the remaining valuable metals within the slags as pure products to date. This is probably because of the fact that the contained valuable metals such as silicon, magnesium, iron, and aluminum are existing primarily as silicate compounds making them difficult to be dissolved. Theoretically, however, the recovery of these metals would be possible by dissolving the silicates with a strong alkaline solution such as NaOH to expose the metals within the slag for the further recovery process.

This approach has been investigated for leaching of steel slags, which also contain significant amounts of silicates, both at elevated and atmospheric pressure. Xiao, et al. (2010) studied the dissolution of silica using either concentrated NaOH or KOH solutions in a Teflon-lined stainless-steel autoclave at 140□240 °C, and reported that 71% of silica can be dissolved in 2 hours at 240 °C [7]. Nikoli□, et al. (2016) reported that leaching of steel slag using concentrated NaOH solutions at 25□60 °C resulted in low silica dissolution with the highest dissolution of only 26% in 2 hours at 60 °C [8]. Despite its higher leaching rate, however, operating at elevated pressure is deemed to be unattractive for the processing of such secondary resources due to the need of an expensive autoclave.

The present study focused to investigate the leaching behavior of silicon from the ferronickel slag with concentrated NaOH solution under atmospheric conditions particularly at near boiling temperatures of the solution and longer leaching duration. The dissolution kinetics of silicon from the slag is also investigated to provide theoretical guidance to elucidate the leaching process.

2. Materials and Methods

2.1. Materials

The slag used in this study was obtained from the ferronickel plant of PT ANTAM (Persero) Tbk in Pomalaa, Indonesia. Molten FeNi slag was immediately granulated by water atomization method by introducing high-pressure water to the molten slag in a water pond and hence resulting in the glassy characteristic of the slag. The slag sample taken from the plant was ground using a rotary ball mill to reduce the particle sizes to between 44 and 74 µm for maximizing the exposure of the minerals within the slag to the leaching agent. Its chemical composition was analyzed using a sequential X-ray fluorescence spectrometer (XRF, ADVANT XP Thermo ARL9900) and the results are presented in table 1. The phase analysis of the slag was performed using X-ray diffraction (XRD, PANalytical X'Pert PRO PW3040/x0) and the results show that the slag comprises of enstatite ($MgSiO_3$, 69.3%), forsterite (Mg_2SiO_4, 27.6%) and magnesium-iron-silicate (3.1%). The used leaching agent was prepared from an analytical grade of sodium hydroxide pellets (Merck) dissolved in deionized water.

Table 1. Chemical composition of ferronickel slag

Element	Weight (%)	Component	Weight (%)
Si	23.06	SiO_2	52.10
Mg	16.90	MgO	27.06
Fe	7.88	Fe_2O_3	11.26
Al	3.43	Al_2O_3	6.49
Cr	0.84	Cr_2O_3	1.22
Ca	0.56	CaO	0.79
Mn	0.48	MnO	0.61
Ni	0.07	NiO	0.09
Na	0.00	Na_2O	0.00
LOI	0.00	0.00	0.00

2.2. Methods

The experiments were performed in batch mode. The leaching tests were carried out in a 5-neck 1000 ml cylindrical Teflon reactor. In order to avoid contamination, polypropylene tubes were used for sampling. The slurry was heated and stirred using a hot plate equipped with temperature control system and magnetic stirrer. Evaporation loss was minimized by installing a condenser that circulates vaporized water back to the reactor. The slag was added into the reactor only after the sodium hydroxide solution has reached the desired temperature. Samples were taken at selected intervals (i.e. after 0.25, 0.5, 1, 2, 4, 8, and 12 hours) and filtered. The filtrate was analyzed using atomic absorption spectrophotometer (AAS, Shimadzu AA-6300), while the residue was washed with deionized water and analyzed using scanning electron microscopy (SEM, JEOL-6510A)) equipped with energy dispersive spectroscopy (EDS). The variations made in leaching experiment are shown in table 2.

Table 2. Variations of experimental variables of FeNi slag leaching

Parameter	Values
NaOH concentration (M*)	6, 8, 10, 12, 14

Agitation speed (rpm)	200, 300, 400, 500, 600
Solid percentage (%)	5, 10, 15, 20, 25
Reaction temperature (±2 °C)	25, 50, 70, 90, 110

*) M= molarity unit (mol/l)

3. Results and discussions

3.1. Leaching chemistry

Figure 1 shows the stability fields for silicon-containing species in sodium hydroxide solutions at temperature of 90 °C and pressure of 1 bar. The figure shows that the dissolution of silica may occur through the formation of soluble sodium silicates (Na_2SiO_3) at pH about 10.5–13.5 at relatively oxidizing environment that is slightly above the water stability region. Based on the XRD result that showed that the ferronickel slag was comprised of forsterite (Mg_2SiO_4), enstatite ($MgSiO_3$), and iron magnesium silicate ($FeMgSiO_4$), the leaching reaction may occur through the following reactions:

$$SiO_{2(s)} + 2\ NaOH_{(aq)} \square\ Na_2SiO_{3(aq)} + H_2O_{(l)} \quad (1)$$
$$MgSiO_{3(s)} + 2\ NaOH_{(aq)} \square Na_2SiO_{3(aq)} + Mg(OH)_{2(aq)} \quad (2)$$
$$Mg_2SiO_{4(s)} + 2\ NaOH_{(aq)} + H_2O_{(l)} \square Na_2SiO_{3(aq)} + 2Mg(OH)_{2(aq)} \quad (3)$$
$$Fe_2Mg_{18}Si_{10}O_{40(s)} + 20NaOH_{(aq)} + 10\ H_2O_{(l)} \square 10Na_2SiO_{3(aq)} + 18\ Mg(OH)_{2(aq)} + 2\ Fe(OH)_{2(aq)} \quad (4)$$

Clearly, these reactions are oversimplified but they show that divalent metals such as magnesium and iron would be dissolved, it given the solution alkalinity, they would be quickly precipitated afterward. The overall process would then result in relatively fine solids that are rich in magnesium and iron for the further recovery process.

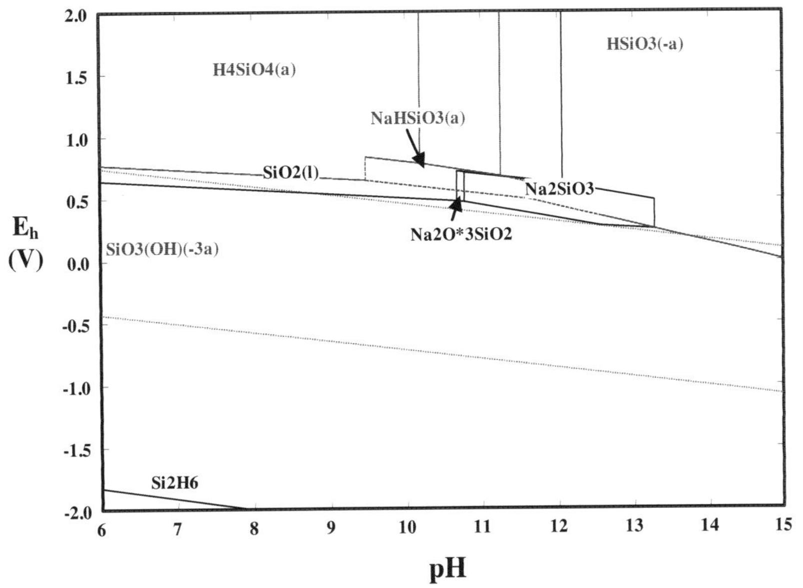

Figure 1. Eh-pH diagram of Si-Na-H_2O system as constructed by HSC software version 5.11 at 90°C, 1 bar, Si (5 M), Na (10 M)

3.2. Effect of sodium hydroxide concentration

The effect of the reagent concentration on the silicon leaching rate was investigated by varying the initial concentration of NaOH solution from 6 to 14 M at operating temperature of 90 °C, solid percentage of 5%, agitation speed of 300 rpm, and leaching duration of 12 hours. The results, as summarized in figure 2, show that the leaching rates are relatively constant at the studied NaOH concentrations except at 10 M in which significant increase in silicon dissolution was observed. A similar trend, wherein silicon dissolution from a ferronickel slag was significantly increased with the increase NaOH concentration up to a certain concentration and decreased significantly afterward, was also observed by Maragkos et al. (2009) [6]. They attributed this behavior to the formation of secondary aluminosilicate precipitates that hinders further dissolution of the silicon at higher NaOH concentrations. This is also seemed to be the case for the present results but in addition to the secondary aluminosilicates, there may also be some significant amounts of Na_2SiO_3 that are precipitated at the surface of the reacted particles due to the lack of water at the higher concentrations of leaching agent.

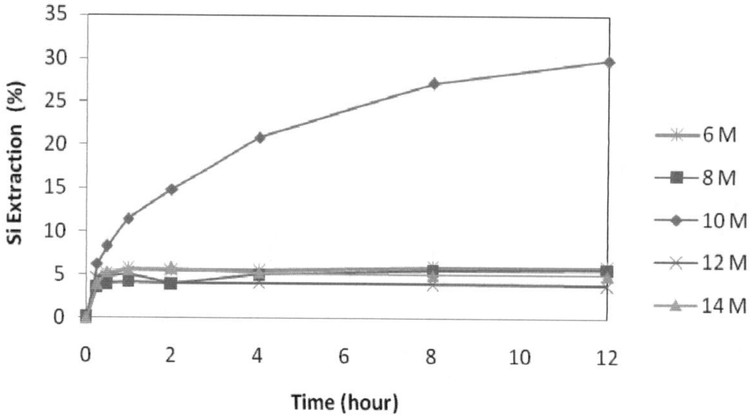

Figure 2. Effect of the NaOH concentration on the leaching rate of silicon from ferronickel slag

3.3. Effect of stirring speed

The effect of stirring speed on the leaching rate was studied to examine the effect of diffusion through liquid film surrounding the solid particles on Si extraction by varying the speed of the magnetic stirrer from 200 to 600 rpm at temperature of 110 °C, NaOH concentration of 10 M, solid percentage of 5%, and leaching duration of 12 hours. The results, as shown in figure 3, show that stirring speed has a notable effect on the leaching rate. Interestingly, the leaching rate of silicon is decreased with the increase in stirring speed suggesting that the reaction is not likely to be controlled by diffusion through the liquid film surrounding the particle. In all the succeeding leaching experiment, therefore, 200 rpm was selected to ensure effective suspension of the particles while minimizing the effect of liquid film diffusion.

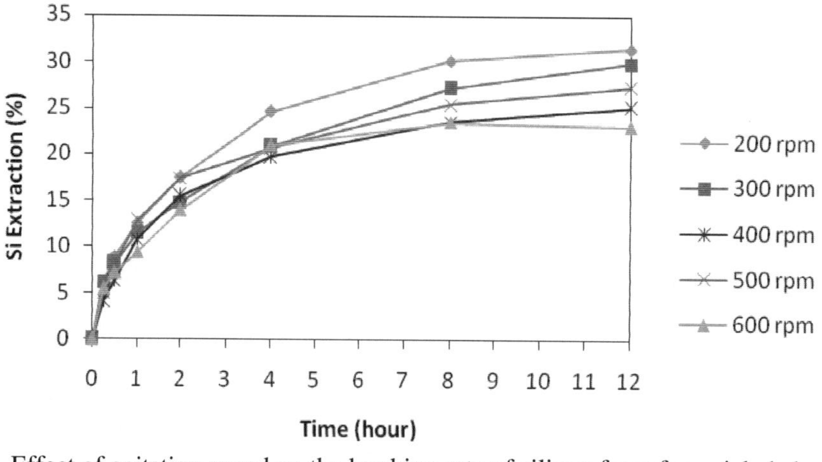

Figure 3. Effect of agitation speed on the leaching rate of silicon from ferronickel slag.

3.4. Effect of reaction temperature

The effect of temperature on the leaching rate of silicon from ferronickel slag was carried out by varying the operating temperature in the range of 25 to 110 °C using NaOH concentration of 10M, agitation speed of 200 rpm, solid percentage of 5%, and leaching duration of 12 hours. The results, as summarized in Figure 4, show that the leaching rate of silicon was highly dependent on temperature. This can be attributed to the higher solubility of sodium silicate coupled with the wider stability of its formation at higher temperature. In addition, this may also be the result of the decrease in the slurry viscosity at the higher temperature improving the mass transfer associated with the acceleration of the leaching reactions [9].

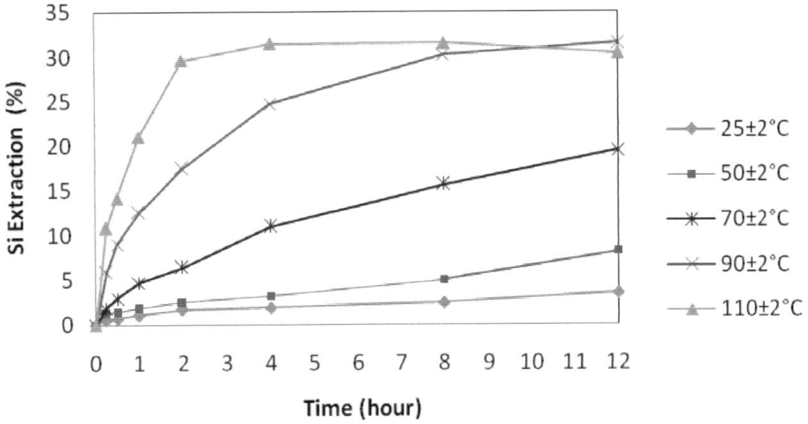

Figure 4. Effect of temperature on the leaching rate of silicon from ferronickel slag.

It is worth to be noted that the experimental result at 110 °C suggests the formation of a product layer on the surface of the leached particles that inhibit further dissolution of silicon. It can be seen that silicon dissolution was relatively fast within the first 4 hours, which was up to 31.48%, but no further dissolution was seemed to occur afterward.

3.5. Effect of slurry density

The effect of slurry density on the leaching rate of silicon from ferronickel slag was conducted by varying the solid percentage from 2 to 25% at temperature of 110 °C, NaOH concentrations of 10 M, stirring speed of 200 rpm, and leaching duration of 12 hours. The results, as shown in figure 5, show that the increase in the slurry density, decrease the leaching rate of silicon to a point where the change is no longer apparent. This is because of the lower the ratio of solid to liquid, the higher the amounts of the free NaOH for the reaction and the lower the viscosity of the slurry.

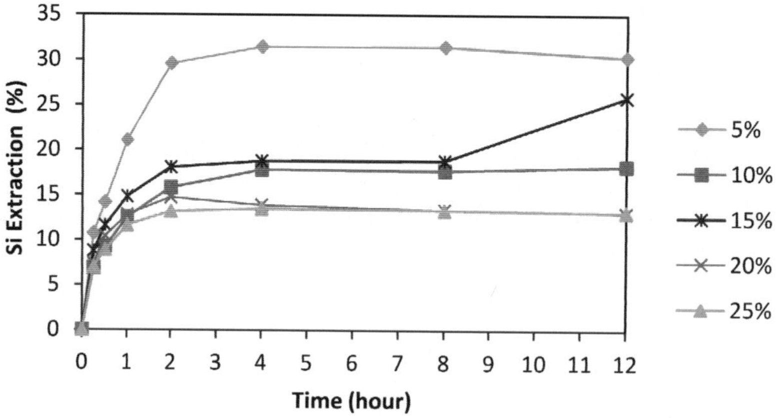

Figure 5. Effect of solid percentage on the leaching rate of silicon from ferronickel slag

3.6. Analysis of leaching kinetics

Visual observations from the experiments suggest that the leaching process of the slag follows the shrinking core model. The leaching process, therefore, can generally be controlled whether by chemical reaction at the surface of the core of un-reacted particle, diffusion of the reactant through liquid film surrounding the particle, or diffusion of reactants through the product layer [10]. Assuming that the geometry of the slag particle is spherical, the following conversion-time equations represent the dissolution kinetics of the system when surface chemical reaction (Eq. (5)), diffusion through liquid film (Eq. (6)), or diffusion through the solid product layer (Eq. (7)) is the rate-controlling step:

$$t/\tau = 1-(1-X_B)^{1/3},\ \tau = \frac{\rho_\beta R_0}{bk"C_A} \qquad (5)$$

$$t/\tau = X_B\ ,\ \tau = \frac{\rho_\beta R}{3bk_f C_{Af}} \qquad (6)$$

$$t/\tau = 1-3(1-X_B)^{2/3}+2(1-X_B),\ \tau = \frac{\rho_\beta R^2}{6b\mathcal{D}_e C_{Af}} \qquad (7)$$

where X_B is the fraction of B in solid that is converted to product at time t, τ is the time for complete conversion of B by reactant A, k" is chemical reaction rate constant, k_f is mass transfer coefficient of the fluid film, b is number of moles of B consumed per mole of reacted A, C_{Af} is the concentration of reactant A in the bulk of solution, ρ_β is the density of the solid, R_0 is the initial radius of the solid, and D_e is the diffusion coefficient through the product layer. By plotting the values of Eq. (5), (6), and (7) using the fractions of the dissolved silicon that were obtained in the previous experiments at various temperatures, it was found that Eq. (7) exhibited the best fit among the three equations based on it corresponding values of R^2 (0.9868-0.9996) as shown in figure 6. The results thus suggest that the leaching process is controlled by diffusion through the product layer.

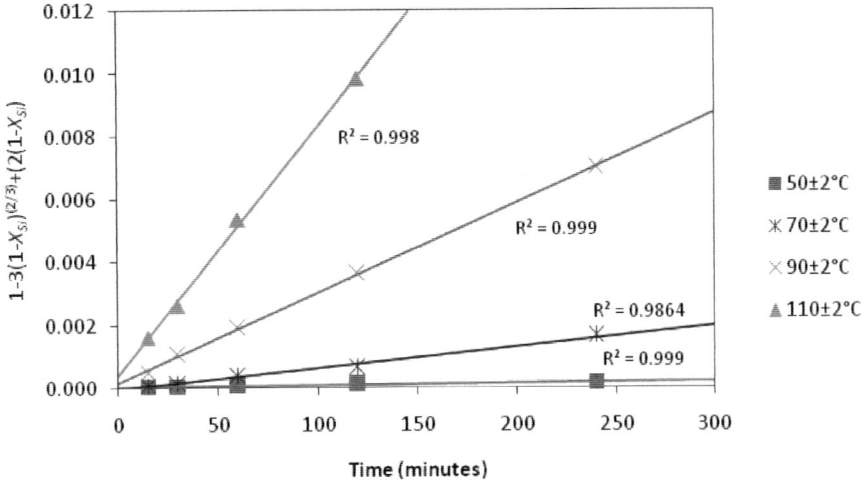

Figure 6. The plot of $1-3(1-X_B)^{2/3}+2(1-X_B)$ versus time at various leaching temperature

The activation energy can then be determined by plotting the dependence of the apparent rate constant that was obtained from Eq. (7) versus 1/T based on Arrhenius equation:

$$k = k_o e^{-Ea/RT} \quad (8)$$

where k is the overall rate/kinetic constant, k_o is the frequency factor, E_a is the apparent activation energy, R is the ideal gas constant (8.314 J/(K.mol), and T is the leaching temperature. The slope of Arrhenius plot in figure 7 reveals that the activation energy for the leaching process is 85.84 kJ/mol. This is unusual because such high value of activation energy usually refers to a chemical reaction controlled process, but previous investigators have reported that some diffusion controlled reactions can also have unusually high activation energy [11, 12, 13, 8].

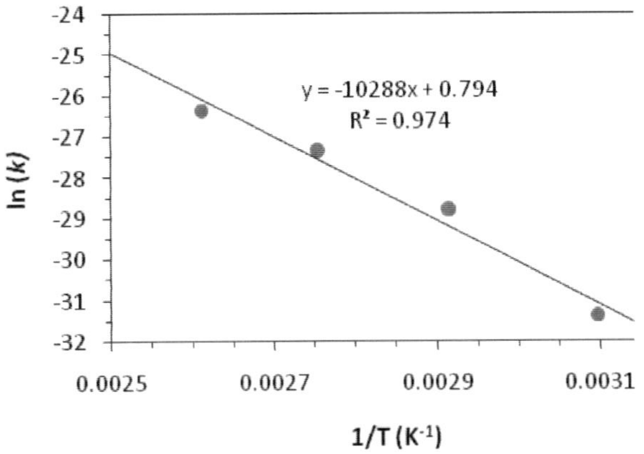

Figure 7. Arrhenius plot for the dissolution of silicon from ferronickel slag

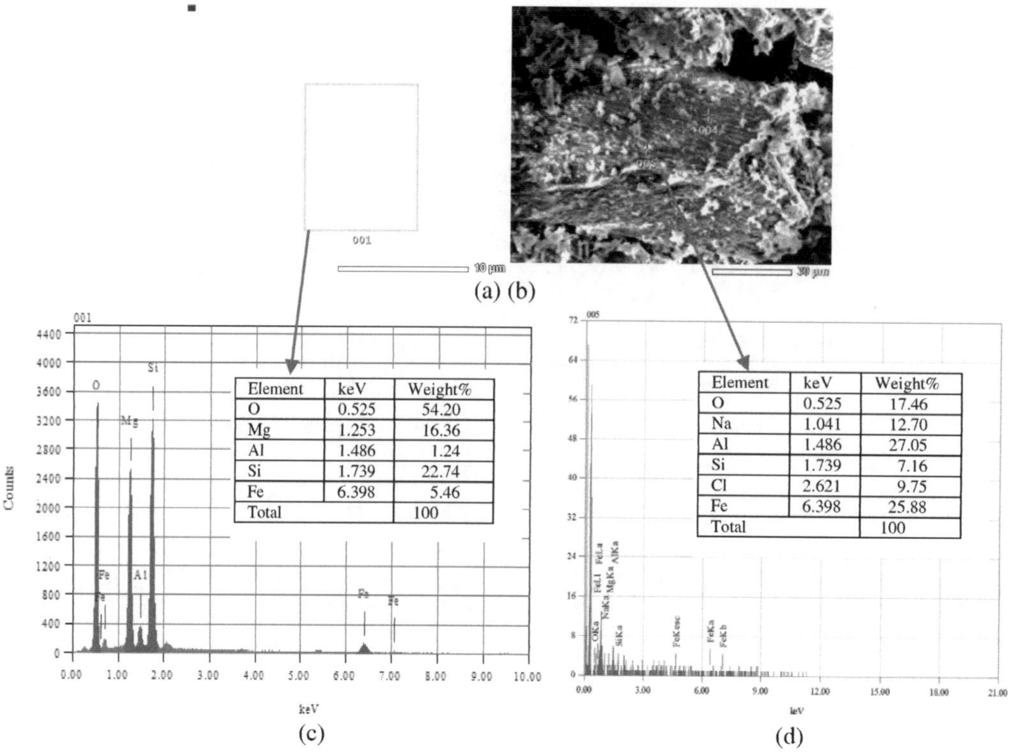

Figure 8. SEM images and EDS analyses of the original ferronickel slag (a and c) and residue (b and d).

The analysis of the leaching residue that was obtained under the optimum leaching conditions (i.e. at NaOH concentration of 10 M, temperature of 110 °C, solid percentage of 5%, and leaching duration of 12 hours) supports that the leaching process is controlled by diffusion through the product layer (figure 8). The morphology of the slag before (figure 8 (a)) and after leaching (figure 8 (b)) shows progressive oxidation on the surface of the particles which is reflected by their rough and porous appearance. The formation of the product layer as aluminosilicates and sodium silicates was indicated by the EDS analyses that show a significant increase of aluminum and sodium concentrations on the surface of the leached particles (figure 8 (c), and (d)).

4. Conclusions

The present study investigated the effect of several operating parameters on silicon dissolution from ferronickel slag under atmospheric pressure and the kinetics of the dissolution process. It was found that concentration of the leaching agent, stirring speed, temperature, and slurry density have the significant effect on the leaching rate of silicon from ferronickel slag under atmospheric pressure. It was determined that the dissolution of silica from the slag was controlled by diffusion through solid product layer, although the apparent activation energy, which was 85.84 kJ/mol, was unusually high for a diffusion-controlled process. The formation of the product layer was apparent from the result of the leaching test at the most optimum conditions found in the present study, namely at NaOH concentration of 10 M, temperature of 110 °C, stirring speed of 200 rpm, and solid percentage of 5%, that showed relatively fast dissolution of silicon up to 31.48% during the first 4 hours, but no further dissolution was observed afterward. This was then supported by dissolution kinetics that was found to fit well to the shrinking core model with diffusion through product layer as the rate-controlling step, and the analysis of the solid residue that indicated the formation of aluminosilicates and sodium silicates precipitates on the surface of the leached particle.

Acknowledgement
The authors would like to thank Ministry of Research, Technology and Higher Education of the Republic of Indonesia through Penelitian Unggulan Perguruan Tinggi (PUPT) scheme for funding the research and PT ANTAM (Persero) Tbk for supplying the ferronickel slag samples.

References

[1] Fidancevska E, Vassilev V, Milosevski M, Parvanov S, Milosevski D and Aljihmani L 2007 *Journal of the University of Chemical Technology and Metallurgy* **42(3)** 285 – 90
[2] Vassilev V, Fidancevska E, Milosevski M, Parvanov S, Milosevski D and Hristova V T 2007 *Journal of the University of Chemical Technology and Metallurgy* **42(4)** 369 – 76
[3] Dourdounis E, Stivanakis V, Angelopoulos GN, Chaniotakis E, Frogoudakis E, Papanastasiou D and Papamantellos D C 2004 *Cement and Concrete Research* **34(6)** 941 – 7
[4] Lemonis N, Tsakiridis P E, Katsiotis N S, Antiohos S, Papageorgiou D, Katsiotis M S and Beazikatsioti M 2015 *Cement and Concrete Research* **81** 130 – 9
[5] Komnitsas K, Zaharaki D and Perdikatsis V 2007 *Journal of Materials Science* **42** 3073 – 82
[6] Maragkos I, Giannopoulou I P and Panias D 2009 *Minerals Engineering* **22** 196 – 203
[7] Xiao Q, Chen Y, Gao Y, Xu H and Zhang Y 2010 *Hydrometallurgy* **104(2)** 216 – 21
[8] Nikoli□ I, Drin□i□ A, Djurovi□ D, Karanovi□ L, Radmilovi V V and Radmilovi□e V R 2016 *Construction and Building Materials* **108** 1 – 9
[9] Chen G, Wang J, Wang X, Zheng S L, Du H and Zhang Y 2013 *Hydrometallurgy* **139** 46 – 53
[10] Levenspiel O 1999 *Chemical Reaction Engineering 3rd ed.* John Wiley & Sons, Inc
[11] Tsuchida H, Narita E, Takeuchi H, Adachi M and Okabe T 1982 *Bulleting of Chemical Society of Japan* **55(6)** 1934 – 38
[12] Olanipekun E 1999 *Hydrometallurgy* **53** 1 – 10
[13] Gbor P K, Ahmed I B and Jia, C Q 2000 *Hydrometallurgy* **57** 13 - 22

IOP Conf. Series: Materials Science and Engineering **285** (2017) 012004 doi:10.1088/1757-899X/285/1/012004

Effect of pulp density and particle size on indirect bioleaching of Pomalaa nickel laterite using metabolic citric acid

H B T M Petrus[1], K C Wanta[2], H Setiawan[1], I Perdana[1] and W Astuti[3]

[1]Advanced Material and Sustainable Mineral Processing Research Group, Chemical Engineering Department, University of Gadjah Mada, Yogyakarta, Indonesia
[2] Chemical Engineering Department, Parahyangan Catholic University, Bandung, Indonesia
[3]Research Unit for Mineral Technology, Indonesian Institute of Sciences, Lampung Selatan, Indonesia

Email: bayupetrus@ugm.ac.id

Abstract. Nickel laterite ore contains oxide of iron, aluminum or both with nickel, cobalt and chromium which can be leached out using hydrometallurgical process. For the purpose of meeting the world's increasing demand of nickel, there is a need to invent environmentally friendly process to efficiently leach nickel. This experiment used nickel laterite ore obtained from Pomalaa, South Sulawesi. The leaching agent is metabolic citric acid produced by *Aspergillus niger* under optimum condition. Leaching process was done in three-necked flask in atmospheric temperature and constant stirring speed of 200 rpm. The variable examined in the experiment was pulp density and particle size of nickel laterite ore. Samples were taken at 3, 7, 10, 14, and 17 minutes and then filtered and diluted to be analyzed using ICP-AES. The result of the experiment showed the maximum recovery of metals increase with the decrease of the pulp density. The maximum recovery for varying pulp density were at 5% solid/liquid ratio and the recovery were Ni at 1.63%, Al at 0.47%, Fe at 0.23% and Mg at 1.09%. The effect of particle size on leaching process showed that the leaching process follows the shrinking core model. The maximum recovery of metals at particle size were at 100-120 mesh with Ni at 1.37%, Fe at 0.10%, Al at 0.72% and Mg at 0.62%.

1. Introduction

Nickel as one of the vital industrial metal has been mined since the nineteenth century. Today, the world's demand mostly relies on the sulfide ores as the main source of nickel. Laterite ores which contain about 72% of the world's nickel deposit have not been sufficiently exploited as it only account for only 40% of the world's nickel production. With increasing nickel demand over the years, there is a need to increase the production of nickel based on laterite ore [1]. Indonesia has in its land one of the largest nickel laterite reserves in the world which were spread mainly on Sulawesi and Papua islands. The processing of those ore though has not been efficient. Only high grade saprolite is being processed for nickel production while the low-grade saprolite and limonite are removed to become waste in the mining areas [2].

While used as a resource for nickel, laterite itself contain mainly the oxide of iron, aluminum or both with nickel, cobalt and chromium as minor compound [3]. Leaching method of laterite ore is very dependent of its content. For ore with high iron content, the reduction method is used ore if the nickel content is high enough, it is better for hydrometallurgical method using organic acid [4], [5].

Within a wide range of organic acids that can be used in the leaching of nickel laterite, citric acid gives the highest percentage of nickel recovery and in addition shows good selectivity of magnesium. Citric acid which can form soluble ligand complex using chelating mechanism can be obtained by fermentation of sucrose using *Aspergillus niger* fungi [6]. The citric acid produced by the excretion process of *Aspergillus niger* fungi is called metabolic citric acid (MCA).

The leaching behavior and kinetics of nickel laterite is affected by parameters such as particle size, pulp density, temperature and the acidity of leaching solution as observed by previous work [2]. Citric acid is a carboxylic acid with three steps of disassociation based on that premise, the possible reaction on the leaching of nickel laterite can be hypothesized as follows [7], [8]:

1. Acid disassociation

$$C_6H_8O_7 \; f \; \left(C_6H_8O_7\right)^{3-} + 3H^+ \quad (1)$$

2. Proton attack

$$NiO + 2H^+ \rightarrow Ni^{2+} + H_2O \quad (2)$$

$$MCO_3 + 2H^+ \rightarrow M^{2+} + H_2O + CO_2 \quad (3)$$

With, M = Metals

3. Complexation/chelation

$$2\left(C_6H_5O_7\right)^{3-} + 3Ni^{2+} \; f \; Ni_3\left(C_6H_5O_7\right)_2 \quad (4)$$

The kinetic itself can be characterized as shrinking core model (SCM) which reaction rate is affected by (1) diffusion through film on the particle surface, (2) diffusion through ash layer in the particle and (3) surface reaction on unreacted core of particle [9]. Previous work have assessed that for the leaching process of nickel laterite, diffusion is the rate-controlling step in the process[2], [5]. This work is focused on the effect of pulp density and particle size of laterite ore on the leaching process of Pomalaa nickel laterite.

2. Materials and Method

2.1 Materials

In this experiment, nickel laterite of limonite type was mined from Pomalaa in South Sulawesi Province of Indonesia in the form of powder and was crushed to obtain the desired distribution of particle size

2.2 Citric Acid Production

Citric acid was produced using *Aspergillus niger* fungi. The culture medium of fungi was made from 3.9 gram of potato dextrose agar (PDA) that was solved in 100 mL aquadest and sterilized using autoclave. Fungi spore was then added to the media and cultured for 7 days.

Substrate was made of 0.25% $(NH_4)_2CO_3$, 0.025% $MgSO_4.7H_2O$, 0.25% K_2HPO_4, 0.06 mg/L $ZnCl_2$ and 1.3 mg/L $Fe_2(SO_4)_3$ which were solved in aquadest with pH of initial medium of 3.0 to 4.0. and was sterilized using autoclave.

The fermentation process was conducted by adding cultured Aspergillus Niger fungi spore to substrate solution with 50 g/L corn starch as carbon source and 3% (V/V) methanol as catalyst. The solution was then incubated for 5 days with temperature of 30-35° C in rotary shaker.

The resulting yield was 230 g / 1 kg corn starch with concentration of citric acid produced in medium 0.05 M. Citric acid produced was then separated using centrifugation to obtain metabolic citric acid (MCA) solution.

2.3 Nickel Laterite Leaching

300 mL of metabolic citric acid and 60 gram of Pomalaa nickel laterite with particle size of 100-120 mesh (pulp density 20%w/v) was combined in a three-necked flask and stirred at 200 rpm on

atmospheric temperature (30-35°C). Samples were taken after 3, 7, 10, 14, and 17 minutes. The experiment was then repeated for pulp density of 5% and 10% and particle size of 60-70 mesh and >200 mesh

Samples taken were first separated from its solid content using centrifugation at 1,000 rpm for 10 minutes. Liquid samples obtained were then filtrated using syringe filter and then diluted to be analyzed for its metal content using ICP-AES (Optima 8300, Perkin Elmer, USA).

3. Results and Discussion

3.1 Characterization

The nickel laterite was of limonite type [(Fe,Ni)O(OH).nH_2O] mined from Pomalaa, South Sulawesi. For initial measurement, the ore was characterized using X-ray Fluorescence (XRF) for its initial metal content. The result of analysis is on Table 1.

Table 1. Metal content of limonite type nickel laterite from Pomalaa

Metal	Content, %wt
Fe	26.04
Si	15.46
Mg	9.78
Ni	2.73
Al	2.54
Cr	1.01
Mn	0.51
Co	0.07

From the analysis result in Table 1 it is shown that this nickel laterite ore is dominated by Fe (26.04%) while Ni only comprised 2.73% of the ore. This makes the leaching process can be approached in 2 ways, by reducing unneeded metals such as Fe and Mn to obtain Ni [10] or leaching all of metal contained and purifying it in later stage. This research was focused in the latter method to examine the viability of this method using organic acid.

3.2 Leaching process with various pulp density

Figure 1 shows the recovery of Fe, Ni, Al and Mn over time with varying pulp density. It is shown in Figure 1 for all metals; the maximum recovery is decreasing with increasing pulp density. This is due to the amount of liquid reactant, in this case metabolic citric acid is insufficient to leach metals in nickel laterite ore [11]. The recovered amount of pulp density that was 5% for Ni (1.63%), Al (0.47%), Fe (0.23%) and Mg (1.09%) still falls short than synthetic citric acid that could leach about 90% of Ni [6]. The result shows that the leaching process using metabolic citric acid need to be modified, whether by using $H_2O_2$to reduce metals such as Fe or Mn or refining the metabolic citric acid production process to produce acid with higher concentration.

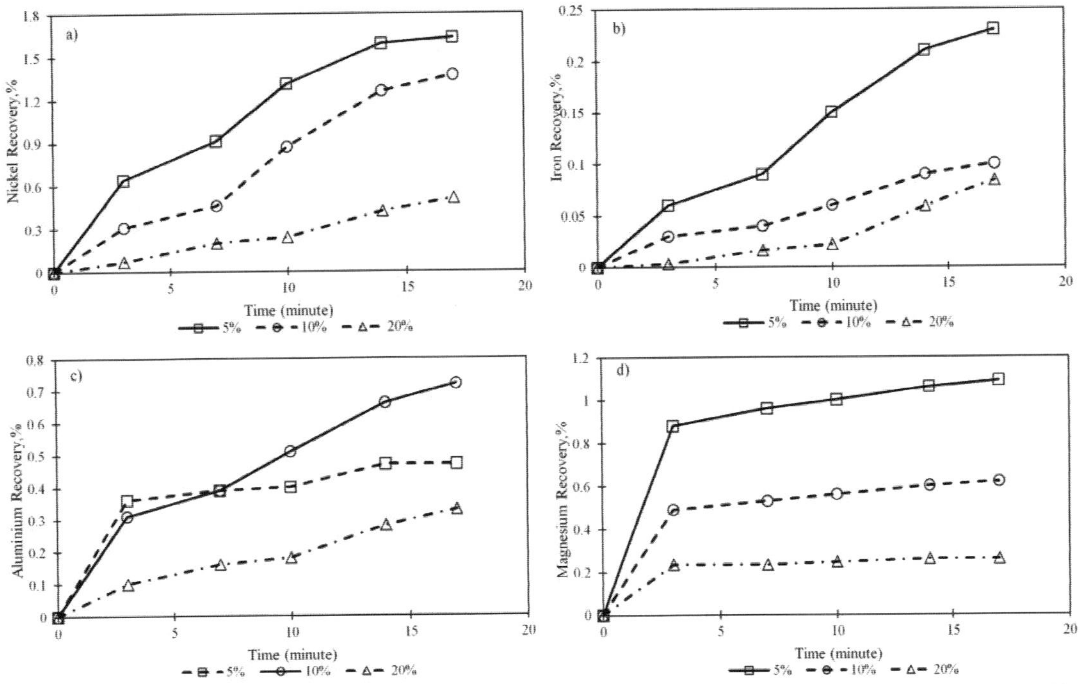

Figure 1. Effect of pulp density on the recovery of (a) Ni, (b) Fe, (c) Al and (d) Mn in leaching process using metabolic citric acid.

3.3 Leaching process with various particle size

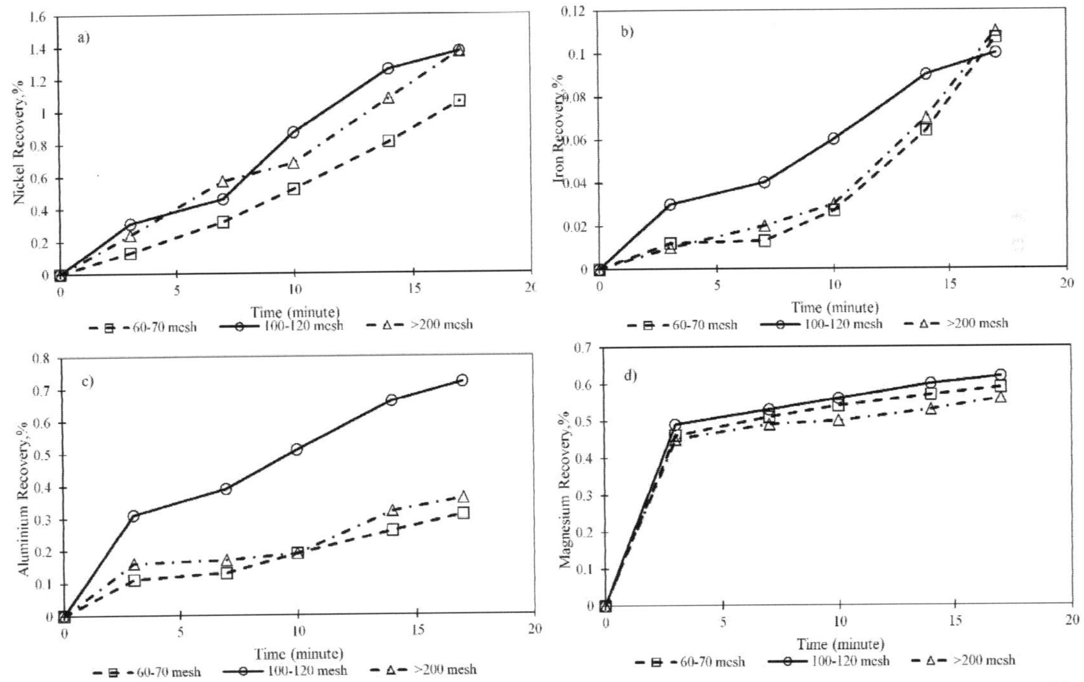

Figure 2. Effect of particle size on the recovery of (a) Ni, (b) Fe, (c) Al and (d) Mn in leaching process using metabolic citric acid

The experiment result in Figure 2 shows the result of recovery of all metals in the leaching process with varying nickel laterite ore size. From the result, it is shown that the maximum recovery of all metals was achieved on ore size of 100-120 mesh with the maximum recovery of Ni at 1.37%, Fe at 0.10%, Al at 0.72% and Mg at 0.62%. The kinetic of leaching process of nickel laterite ore itself followed shrinking core model (SCM) with reactant diffusion through the ash layer as the rate-controlling step [2, 5].

For larger particle size (>200 mesh) it is shown that the recovery rate was slower than from 100-120 mesh of particle size, this was due to wider radius of ash layer formed for larger particle size which hinder the absorption of the reactant to the reaction zone and the desorption of product to the liquid body. For smaller particle size, which means it had larger surface area that supposedly increase the reaction rate. But with increased reaction rate and the complex metal molecules formed on the reaction zone surface caused steric hindrance to the desorption of product. For Mg, particle size did not play role in the recovery rate because of the small size of its molecule which clearly could bypass the hindrance caused by larger product molecules.

4. Conclusion

Experiment results shows that for varying pulp density, the maximum recovery of metals increased with the decrease of the pulp density which was caused by the lack of metabolic citric acid in the solution which also cause the low recovery on this leaching experiment. The maximum recovery of metal at 5% pulp density were at 1.63% for Ni, 0.47% for Al, 0.23% for Fe and 1.09% for Mg.

For the effect of particle size on leaching process, the result shows that the kinetic leaching process followed the SCM with diffusion through the ash layer as the rate-controlling step. The smaller particle size caused lower maximum recovery and recovery rate which was caused by steric hindrance between product molecules, while larger molecule size would also cause the same effect on recovery because the larger ash layer. Mg was exempt from this trend because of its smaller molecule size that will cause it to be able to pass through ash layer and other molecule with less difficulty. The maximum recovery of metals at particle size of 100-120 mesh were Ni at 1.37%, Fe at 0.10%, Al at 0.72% and Mg at 0.62%.

Acknowledgement

The authors acknowledge the support provided by Department of Chemical Engineering, University of Gadjah Mada for the equipment for this experiment. Acknowledgment is also made for materials support provided by Indonesian Institute of Sciences (LIPI).

References

[1] Dalvi A D, Bacon W G, Osborne R C 2004 *The Past and the Future of Nickel Laterites* PDAC Int. Conv. pp 1-27
[2] Astuti W, Hirajima T, Sasaki K, Okibe N 2015 *Miner. Metall. Process.* **32** 176-85
[3] Olanipekun E O 2000 *Int. J. Miner. Process.* **60** 9-14
[4] Petrus H T B M, Rhamdani A R, Putera A D P, Warmada I W, Yuliansyad A T, Perdana I 2016 *IOP Conf. Ser. Mater. Sci. Eng.* **162** pp 12019
[5] Wanta K C, Perdana I, Petrus H T B M 2016 *IOP Conf. Ser. Mater. Sci. Eng.* **162** pp 12018
[6] Mubarok M, Astuti W, Chaerun S 2012 *The XIIIth International Mineral Processing Symposium*
[7] Simate G S, Ndlovu S, Walubita L F 2010 *Hydrometal.* **103** 150-157
[8] Behera S K, Panda P P, Singh S, Pradhan N, Sukla L B, Mishra B K 2011 *Int. Biodeterior. Biodegrad.* **65** 1035-1042
[9] Levenspiel O 1999 *Ind. Eng. Chem. Res.* **38** 4140-4143.
[10] Senanayake G, Childs J, Akerstrom B D, Pugaev D 2011 *Hydrometal.* **110** 13-32
[11] Coto O, Galizia F, Hernández I, Marrero J, Donati E 2008 *Hydrometal.* **94** 18-22

IOP Conf. Series: Materials Science and Engineering **285** (2017) 012005 doi:10.1088/1757-899X/285/1/012005

Decomposition of banten ilmenite by caustic fusion process for TiO_2 photocatalytic applications

Y Aristanti[1], Y I Supriyatna[2], N P Masduki[1] and S Soepriyanto[3]

[1]Department of Metallurgical and Material Engineering, Bandung Institute of Technology and Science, Bekasi 17530, Indonesia
[2]Research Unit for Mineral Technology, Indonesian Institute of Sciences, Lampung 35361, Indonesia
[3]Department of Metallurgical Engineering, Bandung Institute of Technology, Bandung 40132, Indonesia

Email: y.aristant@gmail.com; yayat_iman@yahoo.com

Abstract. Decomposition of Banten ilmenite by caustic fusion process for TiO_2 photocatalytic applications has been done. Caustic fusion process using NaOH to obtain sodium titanate compound which is soluble in sulfuric acid (H_2SO_4) to produces $TiOSO_4$ as a precursor. Synthesis of TiO_2 from $TiOSO_4$ precursors by variations of pH hydrolysis are 1.0 (TiO_2 A), 1.5 (TiO_2 B) and 2.0 (TiO_2 C). XRD pattern identified TiO_2 structures crystals are anatase phase and traces α-Fe_2O_3 as an impurity. Presence of Fe_2O_3 as an impurities give positive effect on TiO_2 photocatalytic activity that is to narrower the band gap energy thus facilitates of electrons excitation from valence band to conduction band and enlarge the specific surface area thus reaction between Rhodamin B solution and TiO_2 surface can be faster. TiO_2 A, TiO_2 B and TiO_2 C was compared to TiO_2 M (commercial TiO_2) in Rhodamin B solution for the photocatalytic activity where the maximum TiO_2 degradation efficiency was obtained at TiO_2 C 80.0 % while TiO_2 M 59.8 %.

1. Introduction

Ilmenite is a titanium-iron oxide mineral which magnetic weak properties, usually black or gray color and its chemical formulation is $FeTiO_3$ [1] with molecular weight is 151.73 grams. Ilmenite in Indonesia were founded as a one of the cassiterite associated minerals which produces tin metal at Bangka island, along southern coastal areas of Java island in form of Iron sands and associations of zircon sand mineral ($ZrSiO_4$) at Central Kalimantan [2-4]. Ilmenite can be used as a raw material to making the titanium (Ti), iron (Fe) metal and also titanium dioxide (TiO_2). TiO_2 acquired by several synthesis methods such as sulfate and chloride process, co-precipitation, hydrothermal, sol-gel and kroll [1, 5-8].

Commercial TiO_2 on the market are currently produced by sulfate or chloride process but both of these processes use acid concentrated in large quantities. Therefore, caustic fusion used alkali such as sodium hydroxide (NaOH) and potassium hydroxide (KOH) to decompose of ilmenite to form of Na_2TiO_3 and K_2TiO_2 as the titanate compound were more soluble in acid [9-11], at the end amounts of acid were used is fewer than two of those processes. Besides of alkaline that used, hydrolysis pH also have an important part of synthesis process [7, 12].

TiO_2 is a semiconductor which has wide applications in engineering such as sensors, solar cells, photocatalysts, pigments, anti-fogging, ceramics and filler [5]. Besides, TiO_2 is non-toxic and has reduction-oxidation properties which can oxidize organic pollutants and reduce the amount of metal ions in solution [13]. These applications of TiO_2 semiconductor has extended since Fujishima and Honda (1972) published the "Electochemical Photolysis of Water at a Semiconductor Electrode" which uses TiO_2 electrodes for H_2O splitting reaction to be ½ O_2 and H_2 under UV light [14].

Inside synthesize of TiO_2 from ilmenite raw materials, Fe in Fe_2O_3 form which is dominant impurity in ilmenite particularly is avoided due to will be contaminate of TiO_2 product [15] but it has several positive affect that is to narrower the band gap energy of TiO_2 [16-18], slow down the TiO_2 transformation phases of anatase to rutile [16], to prevent recombination of electrons from conduction band to valence band [19-22]. Consequently, photo catalytic activity of TiO_2 increases. Therefore, synthesis of TiO_2 from ilmenite which decomposed by caustic fusion process to obtain $TiOSO_4$ precursor followed by sol-gel method with variation of pH hydrolysis is to studied the TiO_2 photo catalytic properties for water treatment applications.

2. Experimental Procedures

2.1. Decomposition of Ilmenite and Synthesis of TiO_2

Caustic fusion process is used to decompose of ilmenite raw materials from Banten, Java Island, Indonesia. Ratio of ilmenite and NaOH (Merck) is 1 : 2, fusion temperature is 750 °C and fusion time is 30 min. Frit is caustic fusion product, leached by water with ratio 1 : 5 (w/v) [15], let it dry. Then leach using H_2SO_4 (Merck) 42.5 % to obtain $TiOSO_4$ solution as a precursor followed by sol-gel method to synthesis of TiO_2 with addition of ammonium hydroxide (NH_4OH) 10 M to adjust of pH hydrolysis variation are 1.0, 1.5 and 2.0. TiO_2 was calcined at temperature of 500 °C for 1 h.

2.2. Samples Code

TiO_2 synthesized which hydrolysis at pH 1.0, 1.5 and 2.0 are coded TiO_2 A, TiO_2 B and TiO_2 C, respectively while commercial TiO_2 is use Merck was coded TiO_2 M.

2.3. TiO_2 Photo catalytic Activity

TiO_2 synthesized was applied as a photo catalyst by using Rhodamin B (RhB) ($C_{28}H_{31}ClN_2O_3$), as a compare was used commercial TiO_2 (TiO_2 M). RhB is an organic compound which difficult to decomposed by bacteria or heat treatment process, its molecular structure can be seen in Figure 1. Each 0.1 g of TiO_2 A, TiO_2 B, TiO_2 C and TiO_2 M inserted to 100 mL RhB solution (concentration 10 ppm) were under solar radiation and no solar radiation treatment for 3 h.

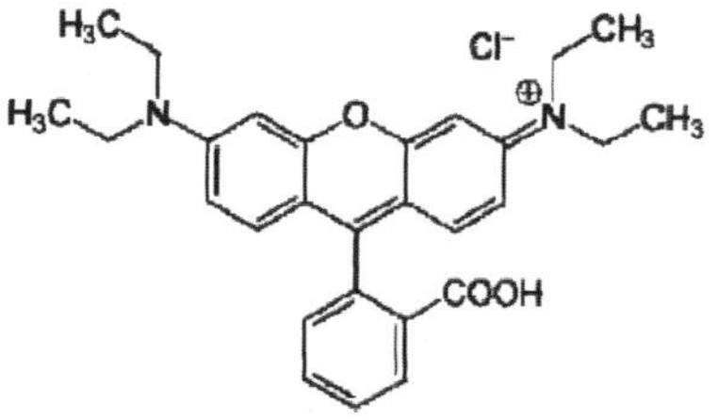

Figure 1. Molecules structure of RhB [23]

2.4. Instrumentations

XRF Thermo Scientific ARL 9900 is used to determine chemical composition of ilmenite, XRD Bruker D8 Advance to identified phases formed and calculated the crystallite size using Debye-Scherrer equation [$D = k\lambda / \beta Cos\theta$] [24], Quantachrome Nova 3200e to determine the specific surface area using BET model and UV-Vis Shimadzu Spectrophotometer to measure the absorbance of RhB and band gap energy of TiO_2 was calculated from these data using Tauc Plot methods.

3. Results and Discussion

3.1. Characteristics of Banten ilmenite

The chemical composition of Banten ilmenite using X-Ray Fluorescence (XRF) can be seen in Table 1 which shows Fe_2O_3 and TiO_2 as the dominant compound. Another compounds in ilmenite that are SiO_2, Al_2O_3, MnO, MgO, V_2O_5, CaO and P_2O_5 have quantities are smaller than Fe_2O_3 and TiO_2. These results are confirmed by XRD pattern (Figure 2) to identify of presence phases which shows $FeTiO_3$ and Fe_2O_3 as the dominant phases, however others phases cannot be detected due to intensities are too small as shown as Table 1.

Table 1. Chemical composition of Banten ilmenite [15]

Phase	Amount (%)
Fe_2O_3	67.46
TiO_2	29.30
SiO_2	1.34
Al_2O_3	0.97
MnO	0.94
MgO	0.91
V_2O_5	0.39
CaO	0.38
P_2O_5	0.15

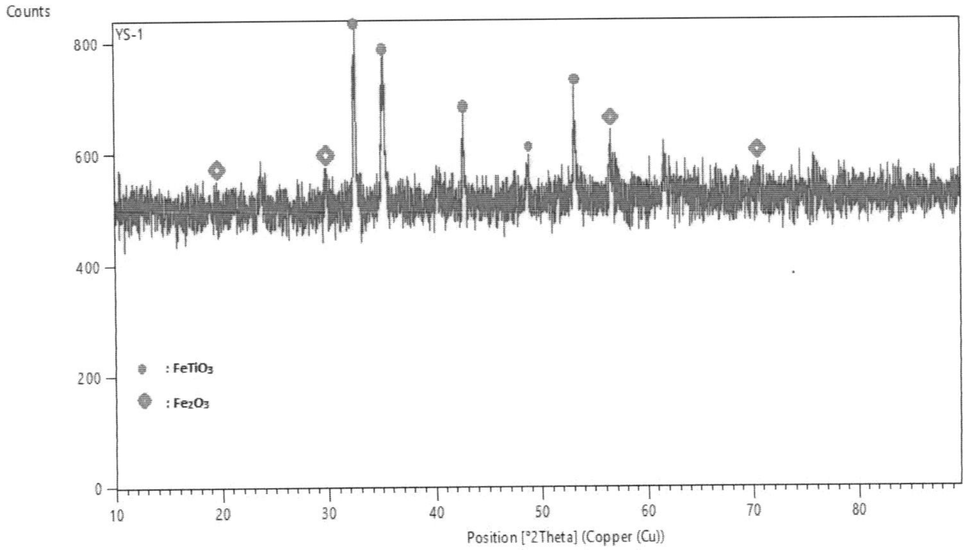

Figure 2. XRD pattern of Banten ilmenite

3.2. Characteristics of TiO_2 synthesized from Banten ilmenite

XRD analysis is used to phases identification formed by pH hydrolysis variation that are TiO_2 A, TiO_2 B and TiO_2 C as shown as Figure 3. Structure crystal phases of TiO_2 synthesized from Banten ilmenite by caustic fusion process are anatase and traces α-Fe_2O_3 as an impurity. Maximum intensities were obtained at pH hydrolysis 1.0 is 25.32°. However when pH hydrolysis are increasing, intensities to be decrease and 2θ angle undergo shifted to be 25.28° and 23.39° for pH hydrolysis 1.5 and 2.0, respectively. These probably caused by increasing of α-Fe_2O_3 amounts as an impurities in TiO_2 synthesized.

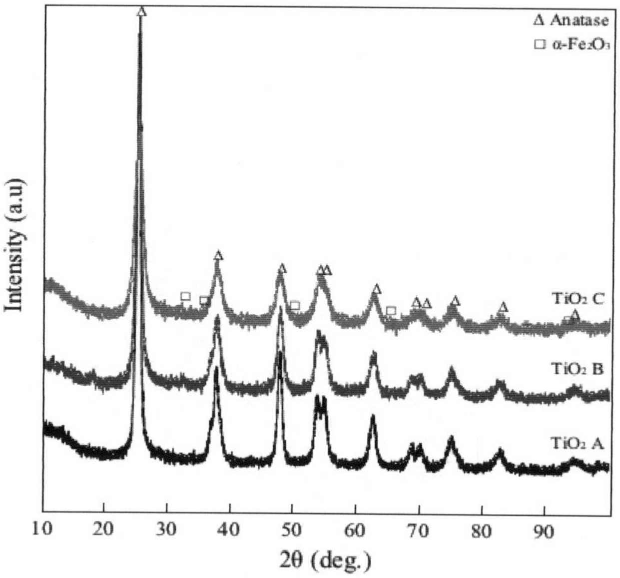

Figure 3. XRD pattern of synthesized TiO_2

Table 2. Phase identified, crystallites size, concentration of RhB and band gap energy of TiO_2.

Sample	Phase	Crystallite Size (nm)	Band Gap Energy (eV)	Concentration (ppm)	
				No Solar Radiation	Solar Radiation
TiO_2 M	A*	11.61	3.22	9.94	4.02
TiO_2 A	A* + α-Fe_2O_3	19.60	2.96	9.76	2.01
TiO_2 B	A* + α-Fe_2O_3	13.36	2.92	9.84	2.12
TiO_2 C	A* + α-Fe_2O_3	7.42	2.88	9.59	2.00

* anatase

Crystallite size can be determined by XRD pattern diffractogram using Debye-Scherrer equation [24]. pH hydrolysis variation was used is affects to crystallites size as shown as Table 2, crystallites size to be decrease when pH hydrolysis are increasing. These caused by intensities of diffraction peaks are getting lower makes width of peaks to be more broadening and its decrease of degree of crystallinity. These results are accordance with the Scherrer method which describes of diffraction peaks width and crystallites size relevance whereby widening of diffraction peaks resulted crystallites size are smaller or vice versa [24].

The crystallites size was calculated has a correlation to specific surface area. Figure 4 shows the specific surface area is increasing when pH hydrolysis also increase. It is based on crystallites size to be smaller caused by increasing of Fe_2O_3 impurities amounts, so the specific surface area to be larger. These means are specific surface area inversely of crystallites size. The α-Fe_2O_3 as impurities were formed by pH hydrolysis is affecting of TiO_2 band gap energy values as shown as Figure 5. Trend of TiO_2 band gap energy was resulted is narrower when pH hydrolysis increase, its means that pH hydrolysis indirectly contribute to band gap energy values. Increasing of pH hydrolysis makes Fe_2O_3 impurities also increase, so the band gap energy of TiO_2 synthesized to be narrower. It is based on intensities of diffraction peaks were lower due to increase of α-Fe_2O_3 as an impurities which can be seen in Figure 3.

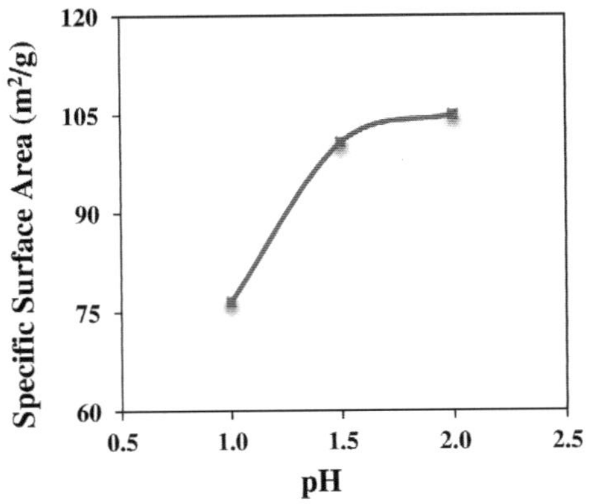

Figure 4. The spesific surface area of TiO_2 synthesized from ilmenite

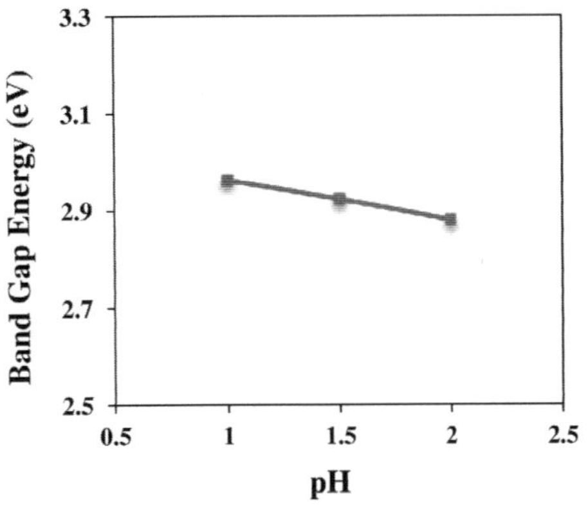

Figure 5. Trend of TiO_2 band gap energy due to Fe_2O_3 impurities

3.3. Photo catalytic activity of TiO_2

On the photocatalytic process, color changes occurs in RhB solution as shown as Figure 6 and Figure 7. At no solar radiation treatment for 3 h, TiO_2 A, TiO_2 B, TiO_2 C and TiO_2 M in RhB solution did not significant discoloration due to absence of UV radiation thus TiO_2 surface are non polar (hydrophobic) [15]. Its confirmed by concentration of RhB solution is tend to stagnate in Table 2. Concentration of RhB solution was determined by measurement of absorbance using UV-Vis Spectrophotometer. However, under solar radiation treatment for 3 h, most significant of color changes was occured at TiO_2 C. This is caused of TiO_2 C has a band gap energy smaller than others to facilitate of electrons excitation from valence band to conduction band and largest specific surface area thus reaction between RhB solution and TiO_2 surface can be faster. It is also confirmed by concentration of RhB solution is down to be 2.00 ppm at TiO_2 C where initial concentration is 10 ppm. Concentration range of TiO_2 synthesized are 2.00-2.12 ppm while TiO_2 M is 4.02 ppm. These results shows the TiO_2 A, TiO_2 B and TiO_2 C are synthesized from ilmenite by caustic fusion process tends to be more effective than TiO_2 M to degrade RhB solution due to Fe_2O_3 as an impurities.

Degradation efficiency (%) of TiO_2 A, TiO_2 B, TiO_2 C and TiO_2 M as shown as Figure 8 can be determined from concentration of RhB solution. Maximum and minimum of degradation efficiency was obtained at TiO_2 C (80.0 %) under solar radiation and TiO_2 M (0.6 %) with no solar radiation treatment, respectively. These caused by presence of Fe_2O_3 as an impurities in TiO_2 synthesis product that are TiO_2 A, TiO_2 B and TiO_2 C which makes the band gap energy is narrower thus facilitate of electrons excitation from valence band to conduction band and its supported by spesific surface area relatively larger than TiO_2 M (13.64 m^2/g).

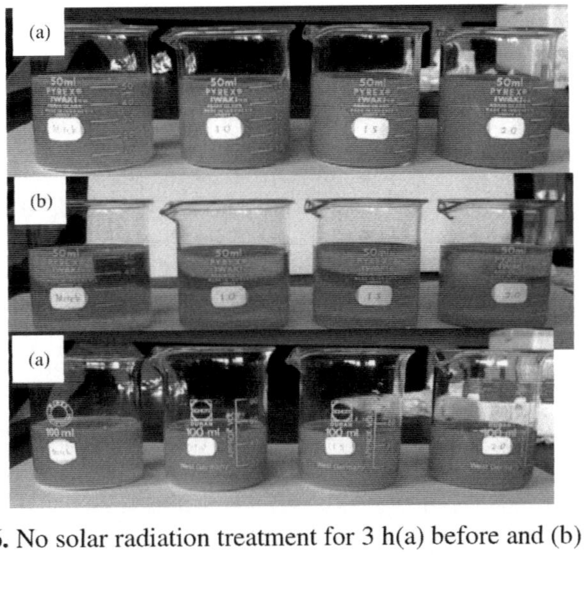

Figure 6. No solar radiation treatment for 3 h(a) before and (b) after

Figure 7. Solar radiation treatment for 3 h (a) before and (b) after

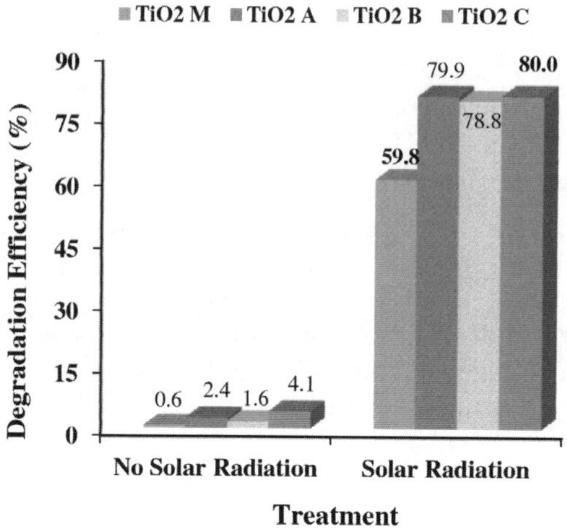

Figure 8. Degradation efficiency of TiO_2 in RhB solution.

4. Conclusion

TiO_2 was obtained from decomposition ilmenite by caustic fusion process followed by sol-gel method to synthesize. Structures crystal phase formed by pH hydrolysis are anatase and traces α-Fe_2O_3. Fe_2O_3 as an impurity in TiO_2 A, TiO_2 B and TiO_2 C give positive effects on photo catalytic activity that are enlarge the specific surface area and to narrower the band gap energy of TiO_2 when pH hydrolysis are increasing. In the case of RhB degradation, TiO_2 synthesized (maximum at TiO_2 C ie 80.0 %) more effective compared to TiO_2 M (59.8 %).

Acknowledgment

The authors would like to thank the Solid Oxide System Laboratory, ITSB and LIPI Lampung for facilitating this research.

References

[1] Chatterjee K K 2007 *Uses of Metals and Metallic Minerals* (New Delhi: New Age International (P) Ltd Publishers)

[2] Annual Report of PT. Timah 2011 *Go Offshore, Go Deeper* (Bangka)

[3] Sardjono P, Adi W A dan Sudirman 2013 *Indonesian Journal Materials Sciences* **14** 209

[4] Mohar M T, Fatmawati D, Sasongko S B 2013 *Jurnal Teknologi Kimia dan Industri* **2** 110

[5] Ahmad A, Awan G H, Aziz S 2007 *70th Annual Session Proceedings* (Pakistan Engineering Congress) 403-12

[6] Kavitha M, Gopinathan C and Pandi P 2013 International Journal of Advancement in Research & Technology **2** 102

[7] Santana-Aranda M A, Moran-Pineda M, Hernandez J, Castillo S, Gomez R 2005 *Superficies y Vacio* **18** 46

[8] Subagja R, Firdiyono F, Setiawan I dan Sufiandi D 2010 Laporan Akhir Program Insentif

Peneliti dan Perekayasa LIPI Tahun 2010
[9] Xue T, Wang L, Qi T, Chu J, Qu J, Liu C 2009 *Hydrometallurgy* **95** 22
[10] Subagja R, Andriyah L and Lalasari L H 2013 *Asian Transactions on Basic and Applied Sciences* **3** 59
[11] Manhique A J 2012 *Titania Recovery from Low-grade Titaniferrous Minerals* (Philosophiae Doctor in Chemistry: University of Pretoria).
[12] Lalasari L H 2014 Sintesis Material Mesopori Fe_2O_3-TiO_2 dari Mineral Ilmenit Bangka sebagai Fotokatalis (Dissertation: Universitas Indonesia).
[13] Slamet, Arbianti R dan Marliana E 2007 *Reaktor* **11** 78
[14] Fujishima A & Honda K 1972 *Nature* **238** 37
[15] Soepriyanto S and Aristanti Y 2016 *Journal of The Australian Ceramic Society* **52** 131
[16] Riyas S and Das P N M 2004 *British Ceramic Transactions* **103** 23
[17] Banisharif A, Khodadadi A A, Mortazavi Y, Firooz A A, Beheshtian J, Agah S, Menbari S 2015 *Applied Catalysis B: Environmental* **165** 209
[18] Ganesh I, Kumar P P, Gupta A K, Sekhar P S C, Radha K, Padmanabham G, Sundararajan G 2012 *Processing and Application of Ceramics* **6** 21
[19] Linsebigler A L, Lu G and Yates J T Jr 1995 *Chem. Rev.* **95** 735
[20] Gnaser H, Huber B, Ziegler C 2004 Encyclopedia of Nanoscience and Nanotechnology **6** 505
[21] Nowotny M K, Sheppard L R, Bak T and Nowotny J 2008 *J. Phys. Chem. C* **112** 5275
[22] Nowotny J 2012 Oxide Semiconductors for Solar Energy Conversion Titanium Dioxide (Boca Raton: CRC Press).
[23] Fikry M, Omar M M & Lotfi Z I 2009 *J. Fluores.* **19** 741
[24] Abdullah M dan Khairurrijal 2009 *Jurnal Nanosains & Nanoteknologi* **2** 1

IOP Conf. Series: Materials Science and Engineering **285** (2017) 012006 doi:10.1088/1757-899X/285/1/012006

The effect of pre-treatments to the nickel limonite leaching using dissolved gaseous SO_2-air

W Wulandari, T H Soerawidjaja, D Alifiani and D A Rangga
Deparment of Chemical Engineering, Faculty of Industrial Technology, Institut Teknologi Bandung, Ganesha 10, Bandung 40132, Indonesia

Email : winny@che.itb.ac.id

Abstract. Nickel limonite leaching has been subjected to a number of studies, one of the method is by using dissolved gaseous SO_2-air. The selectivity of nickel over iron extracted from leaching using dissolved gaseous SO_2-air is advantageous, however the nickel that can be recovered is limited. This paper studies pre-treatments that is applied to the nickel ore prior leaching in order to increase the recovery of dissolved nickel from nickel limonite ore. There two pre-treatments that were carried out in this research, roasting and alkali-roasting using Na_2CO_3. The extraction was carried out for 180 min with pH 2, 3, 4, and 5 and temperature 30, 55, and 80 °C. It is found that the highest yield is achieved at pH 2 and 80 °C with nickel recovery of 61.39%. At pH 2, for alkali-roasting pre-treatment, the nickel yield raised from 28.17% to 100% and for roasting pre-treatment the nickel yield increased from 20.42% to 61.39%. However, at pH 2, the nickel to iron selectivity decreased from 96272 to 534 for roasting pre-treatment and from 1.8 to 1 for alkali-roasting pre-treatment.

1. Introduction

Nickel is a type of the metal that has various uses and application in industry. There are two types of nickel ore, nickel sulphide and nickel laterite. The latter is more abundant, as 70% of nickel ores in the world is categorized as nickel laterite. Currently, 60% of nickel used in industry is originated from nickel sulphide. The limited supply of nickel sulphide forces metallurgy industry to begin using nickel laterite as raw material. Based on the depth zone, nickel laterite is categorized into nickel limonite (low nickel content, <1.5% w/w) and nickel saprolite. Nickel limonite ore reserves in Indonesia reached 464 million wet metric tons while nickel saprolite ore reserves were only about 361.3 million wet metric tons. The common methods to extract nickel from nickel laterite ore is by using hydrometallurgy route, for example high pressure acid leaching (HPAL). This method extracted nickel by using sulphuric acid as leaching agent at 250 °C and 50 bar. Therefore, HPAL is not economical as it required high investment and operating cost.

Another method that has been developed to extract nickel laterite is by using dissolved gaseous SO_2-air. This method is known to provide a high nickel to iron selectivity [1], as well as rapid reaction rates at low temperatures and mild pressure [2]. Some factors that affect extraction results using SO_2 are the concentrations of solvent and operating temperature [3]. The higher the operating temperature as well as the more concentrated of solvent used, the more nickel is obtained. Previous research by the authors has obtained decent nickel selectivity but the nickel recovery is still low [4]. It is found that the good condition was obtained at pH 3, temperature of 80 °C, and 180 min leaching time, with nickel leaching efficiency of 21% from original nickel content. The selectivity nickel towards iron, defined by the extracted nickel (%) divided by extracted iron (%), has been found up to 2743 [4].

Pre-treatment is required prior to extraction to increase the nickel recovery rate. Of many pre-treatments that are available in the literatures, roasting and alkali-roasting are shown to increase the recovery of nickel. The presence of heating before extraction can increase the porosity of the nickel laterite ore and will expose the main mineral, goethite, thus accelerating the interaction between the reducing gas and the nickel species during the leaching process [5]. The pre-treatment of alkali-roasting is done by adding Na_2CO_3 into limonite nickel ore. This alkali-roasting pre-treatment aims to get rid of Cr and Al from nickel ore.

This study is a further development of work of the authors to determine the effect of pre-treatment on the nickel recovery from limonite nickel ore extraction and to understand which of the pre-treatments give the highest nickel recovery.

2. Methodology

This study used nickel laterite ores originated from PT Bhumi Karya Utama located in South East Sulawesi, Indonesia. The composition of the ores is shown in Table 1. The nickel content in the ore is 1.33 %, with Fe content of 15.53 %. The schematic experimental setup is shown in figure 1. The experiments were conducted in three steps: pre-treatment, extraction, and analysis. There are two types of pre-treatment, roasting and alkali-roasting. Roasting pre-treatment was carried out by preparing limonite nickel ore into corundum crucibles then roasted at 650 °C in 46100 muffle furnace for 1 hour.

Table 1. Nickel limonite sample composition from PT Bhumi Karya Utama (in wt%)

Ni (%)	Co (%)	Fe (%)	SiO_2 (%)	CaO (%)	MgO (%)	MnO (%)	Cr_2O_3 (%)	Al_2O_3 (%)	P_2O_5 (%)	TiO_2 (%)	SO_3 (%)
1.33	0.03	15.53	43.19	0.55	19.49	0.23	0.99	2.07	0.02	0.02	<0.01

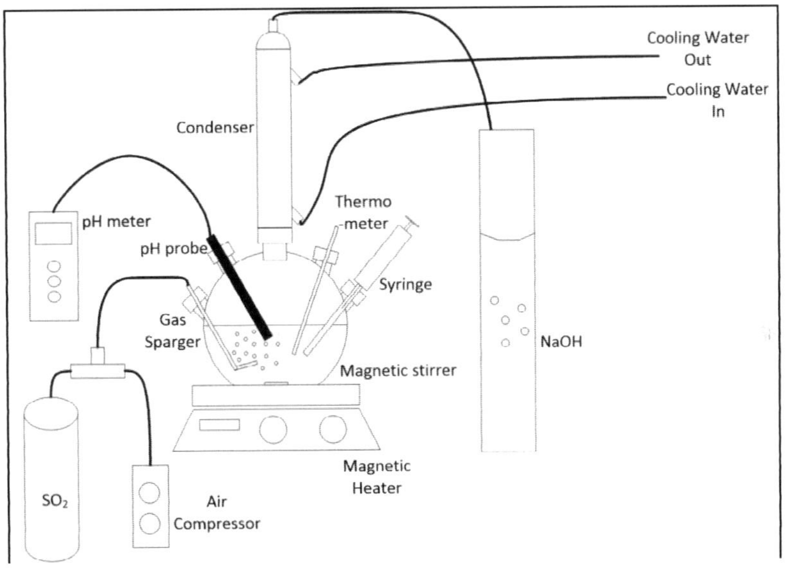

Figure 1. Schematic experimental setup [4]

Alkali roasting pre-treatment was carried out by preparing a mixture of nickel limonite and Na_2CO_3 in corundum crucibles with Na_2CO_3 60% w/w. The mixture is then heated at 940 °C for 2 hours. It was then extracted using water at 95 °C for 30 minutes. The extraction results are then filtered using a filter paper and a Buchner funnel. The residue (unfiltered solids) will be processed at the later stage. The experimental procedure is similar to previous work [4]. The extraction process began by adding 1000 mL deionized water into the reactor and setting the temperature at 30, 55, or 80 °C. Then, 8 gram of pre-treated nickel limonite ore were fed into the reactor. Samples, 10 mL each, were withdrawn from the reactor at specified time range and then filtered using a syringe filter. Its metal content was then analyzed by AAS (atomic absorption spectrometer) analysis. At the end of the extraction (180 minutes), the residue was separated from the solution using a Buchner funnel. The dried residue was then analyzed by using XRD (X-ray diffraction) and TGA-DSC (thermogravimetric analysis – differential scanning calorimetry). The nickel to iron selectivity was calculated using equation 1.

$$nickel\ selectivity = \frac{\left(\frac{nickel}{iron}\right) in\ solution}{\left(\frac{nickel}{iron}\right) in\ ores} \tag{1}$$

3. Results and Discussions

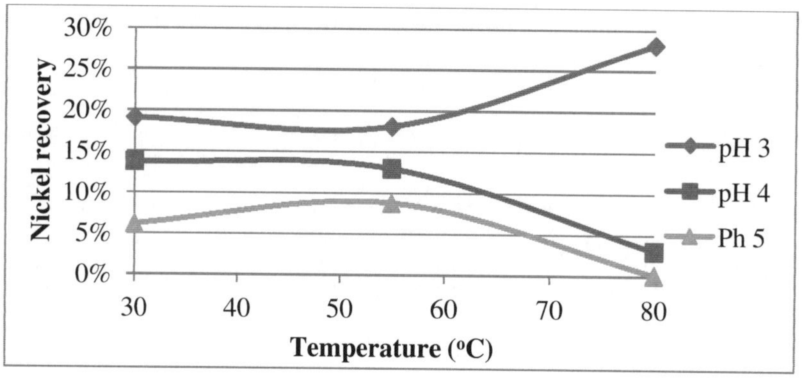

Figure 2. Nickel recovery at various pH and temperature.

3.1. Roasting pre-treatment

Figure 2 shows the nickel recovery at various pH and temperature. In addition to the nickel recovery, the nickel selectivity to iron from limonite nickel extraction is also evaluated. Table 2 shows nickel recovery for roasting pre-treatment and extraction for 180 minutes. Iron recovery from extraction with roasting pre-treatment is much lower compared to nickel recovery. This is due to the use of dissolved gaseous SO_2-air in the extraction process which has higher selectivity rate for nickel than iron [1]. Thus, the use of dissolved gaseous SO_2-air for limonite nickel extraction is considered suitable as the iron in the nickel is not extracted and no further iron separation is needed. At pH 2, the nickel recovery is 61.39%, much higher compared to other process condition; however, iron also starts to dissolve significantly, with a recovery of 3.5% and leads to a selectivity of 17.5.

3.2. Alkali roasting pre-treatment with Na_2CO_3

In alkali-roasting pre-treatment using Na_2CO_3, a TGA-DSC analysis is needed to determine the temperature at which the reaction between nickel limonite and Na_2CO_3 (mass ratio 0.6:1) occur, the result is shown in Figure 3. There are three temperature ranges at mass changes occurs: 30°C to 160°C, 200°C to 330°C, and 730°C to 940°C.

This result was then compared to the work of Astuti *et al* which analyze saprolitic ore using DSC [6]. The largest mass changes (10.3%) occurred when the temperature range 730°C to 940°C. First, mass change was occured at temperatures of 30°C to 160°C followed by endothermic peaks on DSC results, this mass change is caused by the evaporation of water absorbed in nickel ore. Subsequent mass changes was caused by the presence of de-hydroxylation of goethite. This is indicated by the temperature of the mass changes that are in the vicinity of goethite de-hydroxylation temperature [6]. Usually goethite-hematite transformation occurs at temperature 385°C for goethite which has high crystalline structure, whereas goethite with low crystallinity and fine grains will undergo transformation at lower temperature [7]. The reaction between Na_2CO_3 and nickel limonite begins with the decomposition of Na_2CO_3 to Na_2O and then reacts again with chromium(III) oxide (Cr_2O_3), alumina (Al_2O_3), silica (Si_2O_3) and iron(III) oxide (Fe_2O_3). This reaction starts to occur at 850°C.

Table 2. Nickel and iron recovery at various pH and temperature.

pH	Temperature (°C)	Nickel recovery	Iron recovery	Selectivity
2	80	61.39%	3.5%	17.5
3	30	19.08%	0.4203 %	45.4
3	55	18.09%	0.0003 %	61821.9
3	80	28.7%	0.0003 %	96272.6
4	30	13.73%	0.0003 %	46930.9
4	55	12.99%	0.0003 %	44398.3
4	80	3.07%	0.0003 %	10237.5
5	30	6.14%	0.0003 %	20993.8
5	55	8.78%	0.0003 %	30010.8
5	80	0.16%	0.0003 %	534

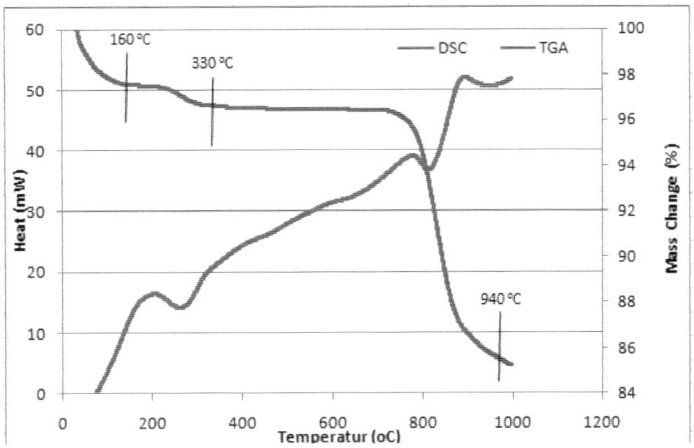

Figure 3. TGA result for Na_2CO_3 and nickel limonite mixture.

Table 3. Nickel and iron recovery from extraction with alkali roasting pre-treatment.

pH	Temperature (°C)	Nickel recovery	Iron recovery	Selectivity
2	80	100%	100%	1
3	30	20.42%	11.1%	1.8
3	55	3.95%	0.0158%	250.2
3	80	5.71%	0.0130%	438.7

The data distribution of iron recovery, nickel recovery, and selectivity levels can be seen in Table 3. Nickel recovery profile for roasting and alkali roasting pre-treatment is similar. However, the recovery obtained in alkali roasting pre-treatment is smaller than roasting pre-treatment (20.42% compared to 28.7%). When the extraction is done in pH 2, nickel recovery reach 100%, but the iron that dissolves with nickel is also quite high, reaching 100%. Therefore, this condition cannot be used further due to significant decreased in nickel selectivity compared to extraction at pH 3.

3.3. Comparison of the roasting and alkali roasting pre-treatment

Figure 4 shows the nickel recovery with and without pre-treatment at pH 3 and 80°C. Nickel recovery increases with time of extraction.

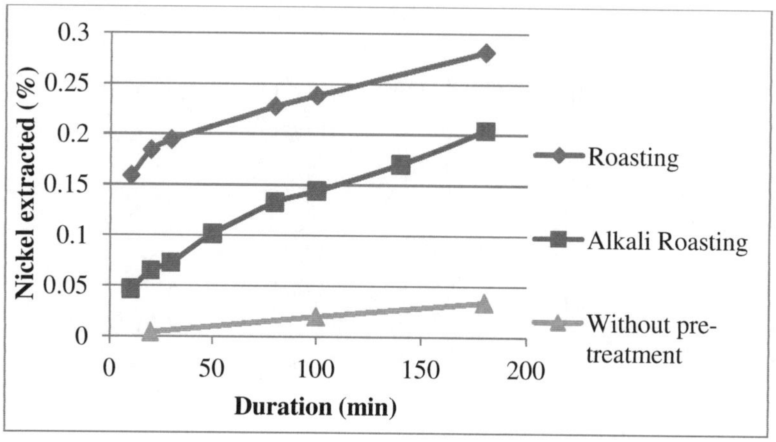

Figure 4. Result of nickel recovery from nickel limonite extraction with and without pre-treatment.

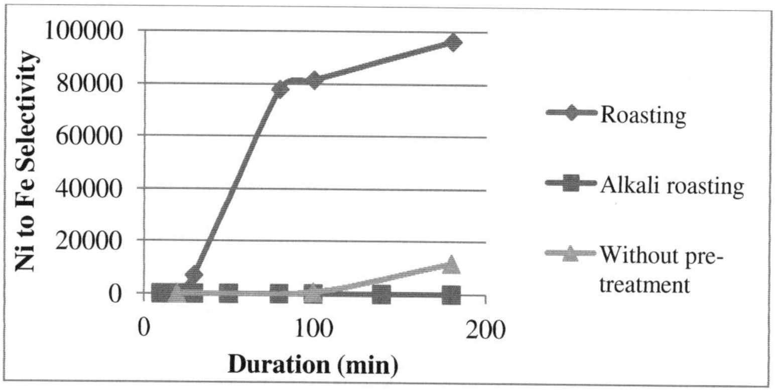

Figure 5. Nickel to iron selectivity with and without pre-treatment.

In addition, the curve also shows that pre-treatment is proven to increase nickel gain from 3.5% to 20.4% for alkali-roasting pre-treatment, and 28.2% for pre-treatment roasting. Figure 5 shows the selectivity of nickel to iron with and without pre-treatments. In general, selectivity for limonite nickel ore with pre-treatments increased with increasing extraction duration while selectivity for alkali-roasting pre-treatment tended to be constant. Roasting pre-treatment has the highest selectivity level of about 96272.6. Therefore, it can be concluded that the most suitable method of pre-treatment for limonite nickel extraction using SO_2-air is roasting at 650°C.

4. Conclusions

In this paper, the effect of pre-treatments to leaching of limonite ore using dissolved SO2-air was carried out. The roasting pre-treatment has shown better leaching recovery compared to the Na_2CO_3 roasting. The highest nickel recovery was obtained at pH condition of 2 at 80 °C using roasting pre-treatment, with nickel recovery of 61.39%. However, the selectivity of nickel over iron is 17.5, implying that many iron still dissolves in the leaching solution. The best process condition with highest nickel selectivity towards iron is at pH 3 and temperature of 80 °C, with nickel recovery of 28.7% and selectivity of 96000. Alkali roasting pre-treatment has shown nickel recovery to 100% at pH 2 and temperature of 80° C process condition, but with trade-off of the dissolution of all iron to the leaching solution.

Acknowledgment

The authors would like to thank the Ministry of Research, Technology, and Higher Education of the Republic of Indonesia to fund this research through Program Insentif Riset Nasional (National Research Incentive Program) 2017.

References

[1] Grimanelis D, Neou-Syngouna P and Vazarlis 1992 *Hydrometall.* **31** 139-146
[2] Naik P K, Sukla L B and Das S C 2000 *Hydrometall.* **54** 217-228
[3] Gbor P K, Ahmed I B, and Jia C Q 2000 *Hydrometall.* **57** 13–22
[4] Wulandari W, Soerawidjaja T H, Joshua S and Isradi H R 2017 *AIP Conference Proc.* **1805** 0700004
[5] Li J H, Li X H, Hu Q Y, Wang Z X, Zhou Y Y, Zheng J C, Liu W R and Li, L 2009 *Hydrometall.* **99** 84-88
[6] Astuti W, Hirajima T, Sasaki K, Okibe N 2016 *Miner. Eng.* 85 1-16
[7] Swamy Y V, Kar B B, and Mohanty J K 2003 *Hydrometall.* **69** 89–98

Leaching behavior of lanthanum, nickel and iron from spent catalyst using inorganic acids

W Astuti[1], N M Prilitasari[1], Y Iskandar[2], D Bratakusuma[2] and H T B M Petrus[3]

[1]Research Division for Mineral Technology, Indonesian Institute of Sciences (LIPI)
Jl. Ir. Sutami Km. 15, Tanjung Bintang, Lampung Selatan, Indonesia
[2]PT. Pertamina Tbk, UP Balongan, West Java, Indonesia
[3]Department of Chemical Engineering, Gadjah Mada University
Jl. Grafika No. 2, Yogyakarta, Indonesia

Email : widi004@lipi.go.id

Abstract.Highly technological applications of rare earth metals (REs) and scarcity of supply have become an incentive torecover the REs from various resources, which include high grade and low grade ores, as well as recycledwaste materials.Spent hydrocracking catalyst contain lanthanum and a variety of valuable metals such as nickel and iron. This study investigated the recovery of lanthanum, nickel and iron from spent hydrocracking catalyst by leaching using various inorganic acid (sulfuric acid, hydrochloric acid, and nitric acid). The effect of acid concentration, type of acid and leaching temperature was conducted to study the leaching behavior of each valuable metal from spent-catalyst. It has been shown that it is possible to recover more than 90% of lanthanum, however the leaching efficiency of nickel and iron in this process was very low. It can be concluded that the leaching process is selective for lanthanum recovery from hydrocracking spent-catalyst.

1. Introduction

The growing demand for metals particularly rare earth metals (REs)around the world havealigned the current research on metal extraction mostly onsecondary metal resources together with newest technologiespromoting primary metal resources. The huge tonnage ofindustrial wastes generated having high metal values haveprompted the industries to look forward for recycling options,where primary mineral resources are limited or exhausted. Allthe recycling methodologies have to be eco-friendly as well ascost-effective to be scaled up for a full scale operation forrecovery of valuable metal. Among all the secondary metalresources, spent hydrocracking catalysts generated from oilrefineries is well known to be rich in valuable metals including rare earth metals like La, Al, V,Mo, Co, Ni and Feetc, which are in the form of metal, metaloxides and metal sulfides. In an oil refinery the solid catalysts areused to treat the crude oil to achieve higher desired fuelproducts[1]. With the extensive use of thesolid catalyst during the processing of crude oil, this catalyst'slose their potential for its further use resulting as a wasteproduct referred as spent catalyst [2].

Spent catalysts can be classified as hazardous materials[3]. Due to their toxic nature, the disposal of spent catalysts canpollute the environment since heavy metals are leached out. Landfill of the spentcatalyst is not an environment friendly or eco-friendly process which could be leached out creating harmful effects on the lifeprocess existing on the earth crust [1].To avoid pollution in landdisposal as well as to minimize landfill space, the spent catalysts are subjected tometal extraction by various solubilization processes and reused in variety ofapplications [1, 4, 5]. Therefore prior to landfillits worthy to recover the metal values present in the spentcatalyst, for which extensive research works has been carriedout to find out a suitable method to extract the metal values. Pyro-metallurgical together with hydrometallurgicaltechniques has been well applied and investigated in past torecover metal values from the spent catalyst[2].

In this study the hydrometallurgical method was investigated to extract valuable metals especially lanthanum, nickel and iron from the spent hydrocracking catalyst originated from PT. Pertamina Tbk that is public oil and gas company in Indonesia. Chemical leaching tests were performedusing solutions of sulphuric acid, nitric acid, as well as hydrochloric acid in some leaching parameters to study the leaching efficiency and leaching behavior of lanthanum, nickel and iron.

2. Materials and Methods

2.1. Material

Spent hydrocracking catalyst was collected from oil and gas company in Indonesia (PT. Pertamina Tbk). The sample was mineralogically and chemically characterized by X-ray *fluorescence*(XRF) andX-raydiffraction (XRD). Table 1 shows the chemical composition of spent catalyst. It can be seen from Table 1 that spent catalyst contains significant value of lanthanum and nickel content. The XRD pattern of spent catalyst that was presented in Fig. 1 shows that spent catalyst contains nickel alumosilicate (zeolite) and nickel hydrogen tecto-alumisilicate (ZSM-5) in its mineral fraction.

Table 1. Chemical composition of spent catalyst by X-ray Fluoroscence (XRF).

Element	La	Fe	Ni	Si	Al	Ti
Wt %	9.73	1.77	8.22	17.11	15.72	1.11

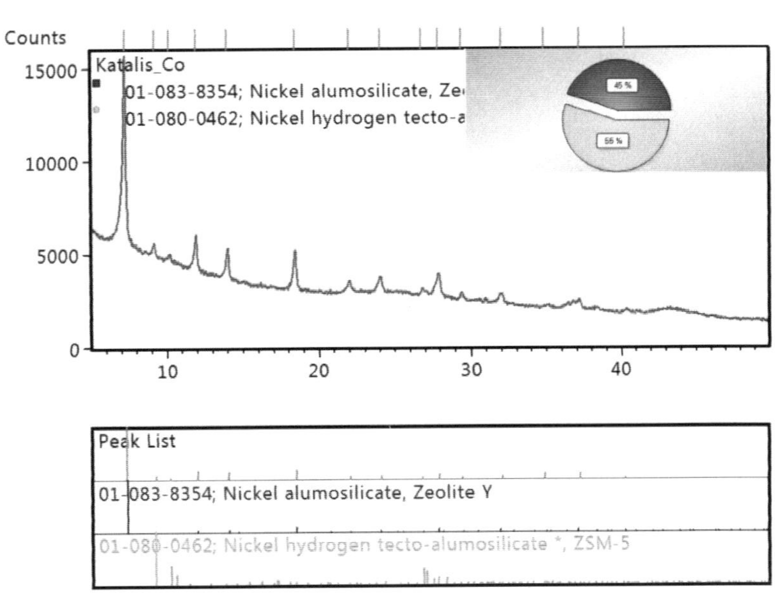

Figure 1. X-ray diffraction (XRD) pattern of raw spent-catalyst.

2.2. Leaching of the metals

Considering the high metal yields and the economic prospects for selective recovery of valuable metals, sulfuric acid, nitric acid, and hydrochloric acid were considered as a lixiviant. Leaching study

IOP Conf. Series: Materials Science and Engineering **285** (2017) 012007 doi:10.1088/1757-899X/285/1/012007

was carried out in a glass reactor over a hot plate. A known amount of the spent catalyst was transferred to the leach solution containing the desired concentration of the acids maintained at a present temperature. The slurry was stirred with a magnetic needle at 500 rpm for the entire duration of the leaching. In order to examine the progress of leaching, samples were withdrawn at different time intervals and were analyzed for the metals of interest. At the end of the leaching, the slurry was filtered over the Whatman filter paper (42 No), the residue was washed with distilled water and dried overnight at 353 K in an electric oven. The leaching efficiency was calculated by analyzing the metals in the filtrate after necesssary dilution and acidification. Satisfactory mass balance was obtained for each set of experiment. All the solutions generated in the leaching were analyzed by AAS (Atomic Absorption Spectrometer). Residue samples were characterized by XRFand XRD.

3. Result and Discussion

3.1. Effect of initial acid concentration

Concentration of leaching reagents is one of significant parameters in leaching process. The effect of initial acid concentration of three leaching reagents (nitric acid, sulfuric acid, and hydrochloric acid) on the leaching efficiencies of nickel and iron was investigated in the leaching conditions of leaching temperature (30 °C), leaching period (8 hours), stiring speed (500 rpm), pulp density (10% w/v), and particle size (< 75 μm).The results, shown in Fig.2 and Fig. 3, indicate that acid concentrationdid not have appreciable effect on the recovery of iron and nickel. This result can be explained more using XRD analysis of leaching residue that will be demonstrated in further section.

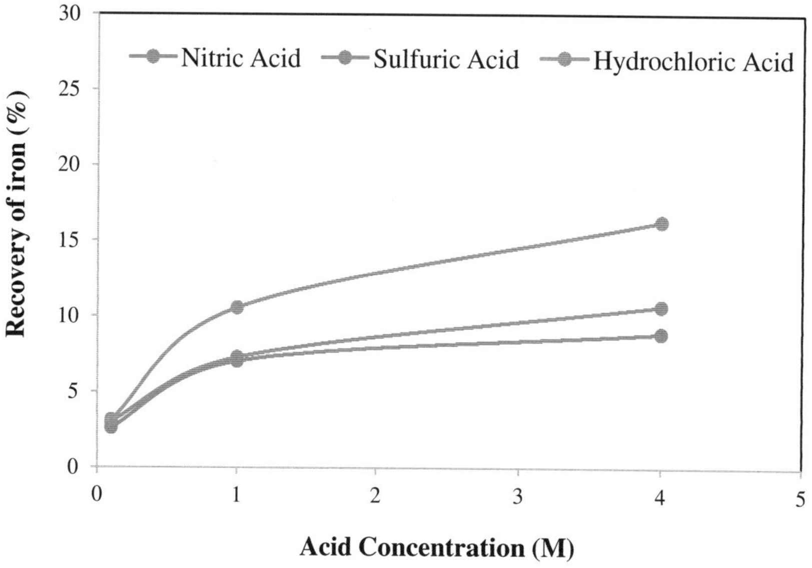

Figure 2. Effect of initial acid concentration on the recovery of iron using various acid types (leaching temperature: 30°C, leaching period: 8 hours).

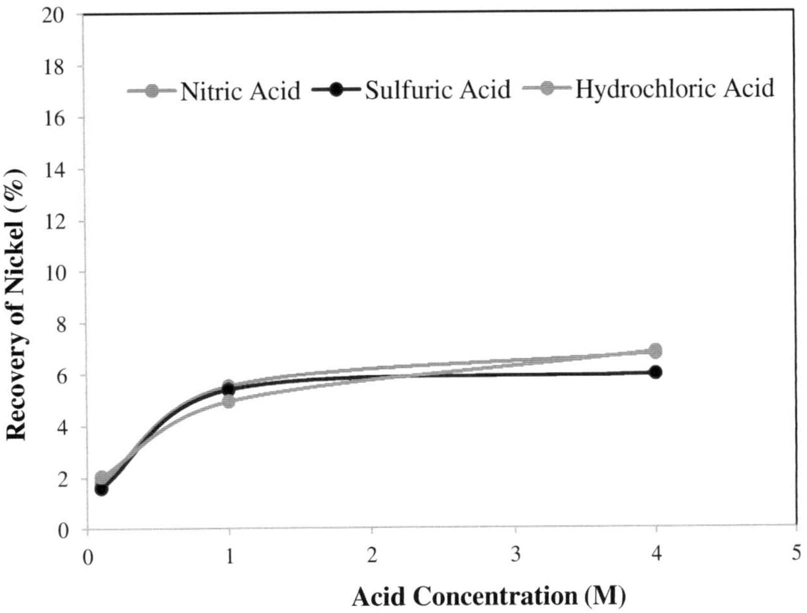

Figure 3. Effect of initial acid concentration on the recovery of nickel using various acid types (leaching temperature: 30°C, leaching period: 8 hours).

3.2. Effect of leaching temperature

Leaching temperature was also the significant parameter on the leaching eficiency. The effect of leaching temperature on the recovery of nickel and iron from spent hydrocracking catalyst was conducted using three leaching reagents (nitric acid, sulfuric acid, and hydrochloric acid) with 4 M of acid concentration, 8 hours of leaching period, 500 rpm of stiring speed, 10% w/v of pulp density, and < 75 μm of particle size.Fig. 4 and 5 show the effect of leaching temperature on the iron and nickel recovery from spent hydrocracking catalyst. It can be seen from those figures that temperature influenced significantly on the leaching efficiency of iron and nickel. The iron and nickel recoveries increasealong with the increase of leaching temperature. The maximum iron and nickel recovery resulted from this experiment were around 33% and 19% respectively at 90°C.

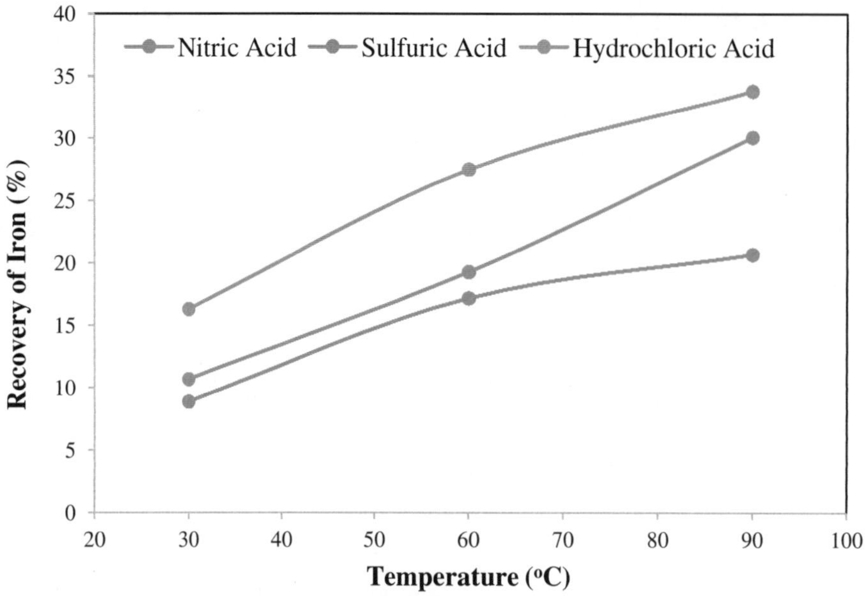

Figure 4. Effect of leaching temperature on the recovery of iron using various inorganic acids (initial acid concentration: 4 M, leaching period: 8 hours).

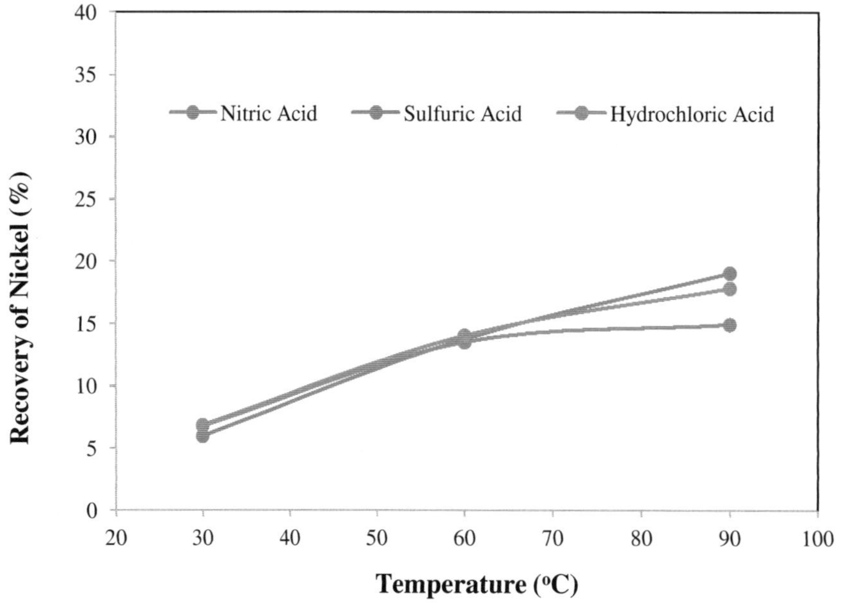

Figure5. Effect of leaching temperature on the recovery of nickel using various inorganic acids (initial acid concentration: 4 M, leaching period: 8 hours).

3.3. Effect of leaching reagents

Leaching reagents also have influence on the leaching efficiency. In this study, three leaching reagents (nitric acid, sulfuric acid, and hydrochloric acid) were compared to get the best leaching reagent that produce the highest leaching recovery of metals. The leaching tests were carried out using 4 M of acid concentration, 8 hours of leaching period, 90 °C of leaching temperature, 10% w/v of pulp density,

<75 μm of particle size, and 500 rpm of stiring speed. The recovery of lanthanum was also analyzed using XRF analysis from solid residue of leaching. Fig. 6 depicted the effect of leaching reagents on the recovery of lanthanum, iron, dan nickel. From Fig. 6, it can be concluded that all leaching reagents used have similar effect on the recovery of lanthanum, iron, and nickel.

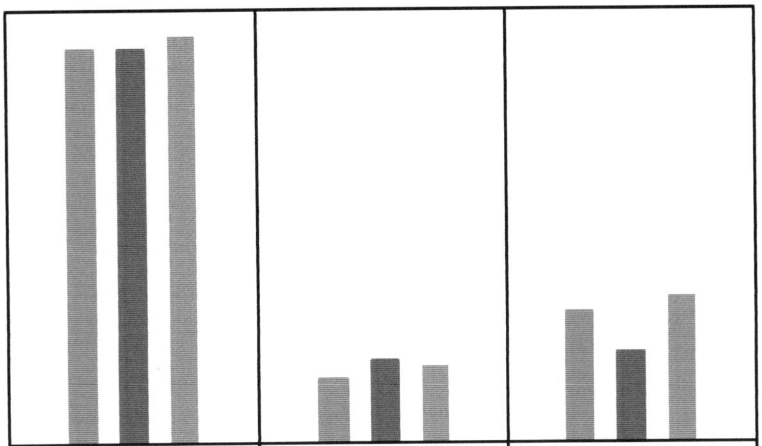

	Lanthanum Recovery	Nickel Recovery	Iron Recovery
Nitric Acid	90.9836	14.9521	30.0913
Sulfuric Acid	90.9836	19.0664	20.6986
Hydrochloric Acid	94.2623	17.8515	33.7974

Figure 6. Effect of leaching reagents on the recovery of lanthanum, nickel and iron (initial acid concentration: 4 M, leaching period: 8 hours, leaching temperature: 90°C).

3.4. Leaching behavior of metals

Leaching behavior of metals was also investigated from the XRD pattern of solid residue after leaching.Figure 7 showed the XRD pattern of solid residue after leaching using sulfuric acid, hydrochloric acid, and nitric acid with 4 M of acid concentration, 8 hours of leaching period, 90 °C of leaching temperature, 10% w/v of pulp density, <75 μm of particle size, and 500 rpm of stiring speed. From this figure, it can be seen that XRD pattern of solid residue is as similar as the XRD pattern of raw material (Fig. 1). This result showed that nickel content in raw material was not leached effectively. It was also fitted with the result from the leaching recovery of nickel. This phenomena occured is because of the inertness of nickel form in the raw material as nickel aluminosilicate.

IOP Conf. Series: Materials Science and Engineering **285** (2017) 012007 doi:10.1088/1757-899X/285/1/012007

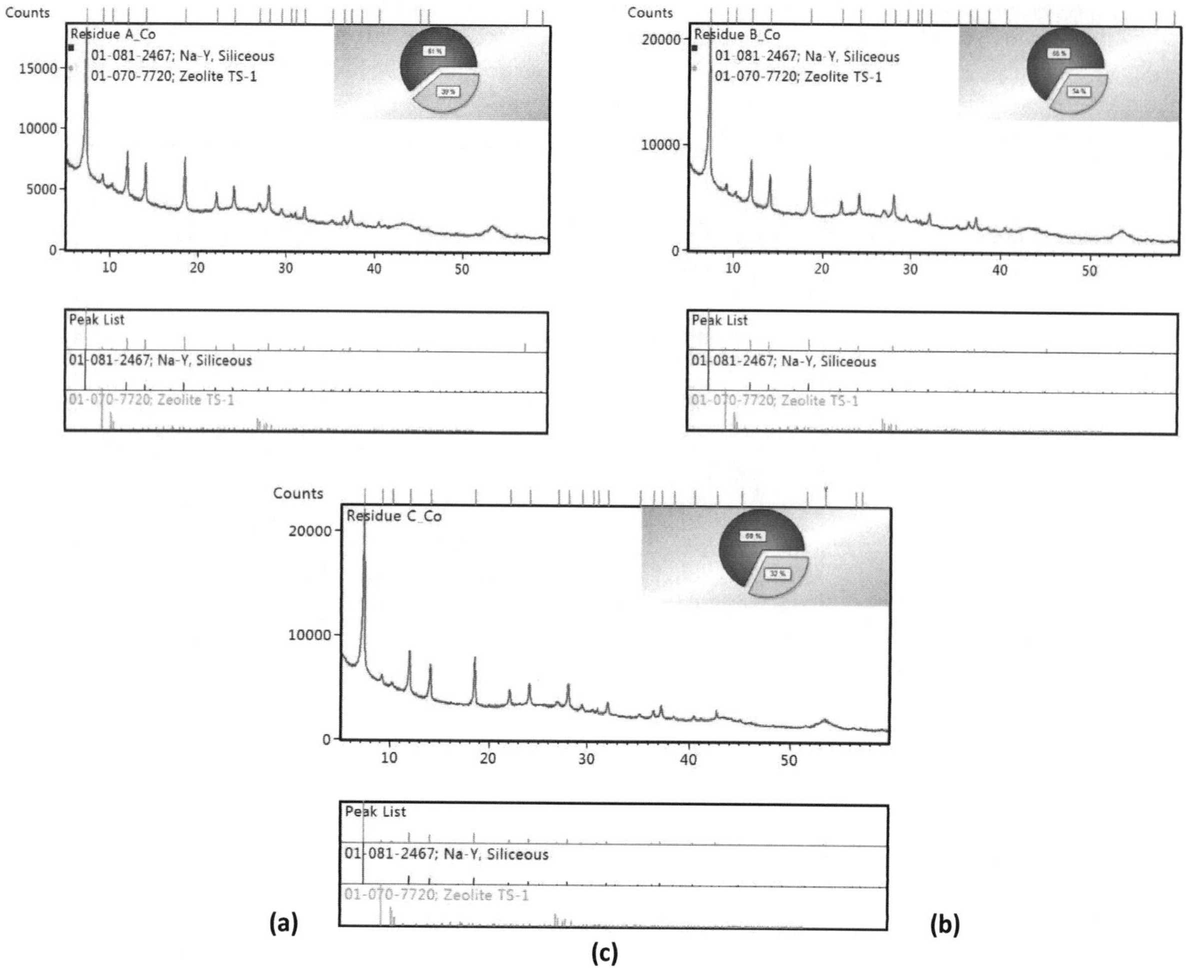

Figure7. X-ray diffraction pattern of solid residue after leaching from (a) Sulfuric acid reagent; (b) Hydrochloric acid reagent; (c) Nitric acid reagent.

4. Conclusion

It is possible to recover more than 90% of lanthanum, however the leaching efficiency of nickel and iron in this process was very low. The leaching process is selective for lanthanum recovery from hydrocracking spent-catalyst.

References

[1] Marafi M and Stanislaus A2003*J. Hazard. Mater***101** 123 – 32
[2] Marafi M and Stanislaus A 2008*Conserv. Recycling.***52**859 – 73
[3] Loehr RC, Rogers LA and Erickson DC 1993*Water Res.***27** 1127 – 38
[4] Furimsky E 1996*Catalysis Today***30**223 – 86
[5] Angelidis TN, Tourasanidis E, Marinou E, Stalidis GA 1995*Resources,Conservation and Recycling* **13** 269 – 82

IOP Conf. Series: Materials Science and Engineering **285** (2017) 012008 doi:10.1088/1757-899X/285/1/012008

Membrane separation for non-aqueous solution

S Widodo[1,2], Khoiruddin[2], D Ariono[2], Subagjo[2], I G Wenten[2]

[1]PPPTMGB LEMIGAS, Balitbang KESDM, Jl. Ciledug Raya, Kav. 109, Cipulir, Kebayoran Lama, Jakarta, 12230, Indonesia.
[2]Department of Chemical Engineering, Institut Teknologi Bandung, Jl. Ganesha 10, Bandung 40132, Indonesia

Email: igw@che.itb.ac.id

Abstract. Membrane technology has been widely used in a number of applications competing with conventional technologies in various ways. Despite the enormous applications, they are mainly used for the aqueous system. The use of membrane-based processes in a non-aqueous system is an emerging area. This is because developed membranes are still limited in separations involving aqueous solution which show several drawbacks when implemented in a non-aqueous system. The purpose of this paper is to provide a review of the current application of membrane processes in non-aqueous solutions, such as mineral oil treatment, vegetable oil processing, and organic solvent recovery. Developments of advanced membrane materials for the non-aqueous solutions such as super-hydrophobic and organic solvent resistant membranes are reviewed. In addition, challenges and future outlook of membrane separation for the non-aqueous solution are discussed.

1. Introduction

The rapid growth of membrane applications in water treatment [1-8], wastewater treatment [9-12], gas separation [13, 14], food processing and biotechnology [15-20], medical and chemical processing [11, 21-26], energy conversion [27], etc., is due to their attractive features. The membrane processes require a relatively lower energy consumption, are typically lower operating cost, have a higher selectivity, and can be operated at moderate conditions which make membrane an interesting alternative to conventional technologies in various processes [28-32]. Most of the commercial membranes are applied in the aqueous solution processing [1, 2, 30, 33]. Membrane is also potential to be applied in non-aqueous system, such as polystyrene treatment [34], aromatic solvents [35], production of an alkali-alkoxide from alcohol and salt [36], micro emulsion processing for drug delivery [37], purification of fuel [38, 39], purification of CdSe-CdS-ZnS, organic solvent regeneration and valuable component [40-43], ethanol-alkanes solutions [44-47], used oil processing [48-52], and vegetable oils processing [53-69]. However, the commercial applications are still limited. Most of the membranes were developed for aqueous system. For non-aqueous system they show several drawbacks such as low stability and low flux. Therefore, the development of membrane materials is still an emerging subject for a non-aqueous solution [52].

The purpose of this paper is to provide a comprehensive review of the membrane technology for non-aqueous solution processing. Recent developments of membrane application in non-aqueous system including mineral oil treatment, vegetable oil processing, and separation of solvent from the organic mixture are reviewed. The developments of new membrane material, such as super-hydrophobic and organic solvent resistant membranes are overviewed. Finally, the challenges and future outlook of membrane separation for the non-aqueous solution are discussed.

2. Application of membrane in non-aqueous solution

In the following sections, membrane processes for the non-aqueous solutions will be presented and discussed including the prospect and challenges of application in non-aqueous solutions.

2.1 Vegetable oil processing

In vegetable oil processing, a membrane could be applied in the stages of refining (Table 1) namely degumming, de-acidification, de-colorization, solvent recovery, metal removal, deodorizing, hydrolysis of fats and oils, synthesis of structured lipids, the concentration of certain compounds, and the separation of emulsions [70]. Several steps could be simplified into a single process using a membrane. For example, removal of undesirable compounds from the vegetable oil uses a single step without the addition of filter aids [71]. The removal of the damaging substance from vegetable oil caused an insignificant change of oil properties, minimized effluents, reduced the energy consumption, and preserved important nutritional components [70].

Table 1. Membrane processes in vegetable oil processing

Processes	Membrane processes	Remark	Ref.
Degumming	MF	Reduction of phosphorus, FFA and water	[66, 68]
	UF (ceramics, PVDF, PES, PSF); MF	Separate oil from the oil-hexane mixture. Permeate contents: hexane, triglycerides, FFAs and other small molecules; almost all the phospholipids are retained.	[58, 59, 62, 63, 65, 67, 69, 72-75]
	UF ceramics	Degumming & production of soy lecithin	[60]
De-acidification	UF, polymer (PAN, PES, PS), ceramics zirconia	Addition of an alkali followed by membrane filtration or by following an indirect route of selective solvent extraction of FFA followed by membrane separation.	[59, 61, 63, 76, 77]
De-colorization	Polymeric composite NTGS- 2100 & NTGS-1100 (active layer) PI, PSF (support layers) Nitto Denko Corp. Jpn.	Removal of chlorophyll up to 96% and 72% from the undiluted and oil-hexane (50wt%) solutions,	[78]
	MF, NF, RO	A 'membrane bleaching' process could save cost	[59, 65]
Solvent recovery	RO, NF, UF	pre-concentrate the miscella before evaporation; reduce the energy consumption	[59, 65]
	NF, polymeric (NF-99-HF, Alfa Laval)	Separation of FFA retention, 70%	[56, 79]
	NF, PVDF as a support & PDMS/CA as coating layer	Removal of hexane and FFA from crude soybean oil–hexane mixture. The PVDF-12% siloxane composite NF membrane reached the best results	[80]

The applications of membrane for degumming of the oil-solvent mixture were reported, i.e. nonporous polymeric membrane [81], inorganic solvent NF [82], ultrafiltration [72-75, 83, 84], and microfiltration [85]. Degumming of crude vegetable oil can be conducted directly using a membrane without the addition of solvent [70]. De Moura et al. [74] used UF for degumming of soybean oil, and the characteristics of the oil, i.e. phospholipids content, viscosity, color, FFAs and tocopherols were analyzed. The study found that the higher oil viscosity resulted in a lower flux. The corn oil processing in alumina multichannel ceramic MF showed that phosphorus rejection was significantly affected by trans-membrane pressure (TMP) [73]. The increasing TMP increase the phosphorus rejection, significantly reduce color and wax, and also remove 93.5% of phospholipids. The minor component e.g. tocopherols and tocotrienols were successfully preserved [74]. More than 99.7% of the phospholipids were rejected by a ceramic membrane [72], while almost complete rejection of oil was obtained using a cellulosic UF membrane in a dead-end filtration [86].

Fatty acid removal is an important step in vegetable oil processing, but the application of membrane still remains a great challenge. Bhosle et al. [79] reported that in the undiluted system, the solubility of triacylglycerols in oleic-acid determines the selectivity. The NF membrane exhibited good selectivity for oleic-acid and excellent solvent stability for acetone. However, the significant improvements are still required. De-acidification using NF composite membrane showed that PVDF materials were physically-resistant to solvent as well as a higher flux and oil recovery [80]. Removal of acid by forming a dispersion of saponified fatty acids/water/isopropanol in oil through a PAN UF membrane was investigated [77]. The study concluded that the flux reached the highest value of 95 $L.m^{-2}.h^{-1}.bar^{-1}$ at the fatty acid/water/isopropanol ratio of 1:6.5:3 (v/v). Purwasamita et al. [28] used hollow fiber poly-sulfone UF membrane contactor to reduce the acidity of crude palm oil (CPO). They found that membrane contactor was highly potential for de-acidification of CPO because of the higher surface contact provided by the membrane. Fractionation of fatty acid using UF membrane could decrease the acid chain while it increased the mass transfer coefficient and distribution coefficient [76].

Decolorization of vegetable oil using membrane processes has been conducted by several works. Reddy et al. [78] used a batch membrane cell using two polymeric composite membranes (NTGS-2100 and NTGS-1100) and one polyethylene MF (PE-30) to remove chlorophyll from sunflower oil and crude soybean oil. They found that the removal of chlorophyll from the undiluted and oil-hexane (50%-w) solutions using NTGS-2100 membrane achieved 96% and 72%, respectively. NTGS-1100 membrane exhibited higher flux rather than NTGS-2100 but resulted in a lower rejection [78]. Subramanian et al. [87] reported that the color compounds and oxidation products could be effectively reduced by using hydrophobic nonporous membranes while the beneficial compounds could be retained.

In the vegetable oil processing, the recovery of solvent from oil extraction (micelle) is the most critical steps regarding the economic, environmental, and safety reasons [70]. The controlling parameters of the permeation rate through membranes were solvent solubility parameter, dielectric constant (polarity), surface tension, and solvent viscosity [88]. The membranes processes like UF and RO have been used to recover the solvent from the micelle [89]. The important characteristics of membrane for solvent recovery are their solvents-resistance and the rejection performance [90]. The flux of permeate depends on the applied TMP, temperature, feed concentration, and cross-flow velocity. Higher pressures, temperatures, and cross-flow velocities yielded better permeate flux.

2.2 Waste oil processing.

Several types of membrane materials have been used to remove contaminants and impurities in waste oil, e.g. poly-ethersulphone, poly-vinylidenefluoride, poly-acrylonitrile [91], polyimide [92], and ceramic membranes [31, 32, 49-51]. Inorganic membranes have a higher rejection of contaminants and a better ash removal [93]. The properties of waste oil significantly affect the filtration performance [94], e.g. a higher flux of permeate obtained at a lower oil viscosity [95]. The usage of pressurized supercritical CO_2 as viscosity reduction will improve the permeability of membrane [32, 96, 97]. The performance of membrane process is also influenced by the filtration mode. For example, the dead-end filtration mode causes a cake formation onto the membrane surface, drastically flux reduction, and a low separation performance. The larger pore size results in a higher flux but has a lower rejection. Therefore, the flux to quality ratio was used as the indicator of filtration performance [49]. The light hydrocarbon in waste oil could be recovered using polyimide membrane from the lube filtrates at purity better than 99%. Combining a membrane system with a conventional process increases the energy efficiency and the capacity of solvent recovery [52, 98]. In used frying oil treatments, the removal of degradation materials using membrane processes will improve the life of usage [99, 100]. Lai and Smith [101] used ceramic MF membrane to remove asphaltene from heavy oil. They found that the pore-restriction mechanism initiates fouling, following by gel layer formation on the membrane surface. Consequently, the pore size has no significant impact on flux.

2.3 Other chemical processing

Membrane have been used in chemical processing like crude biodiesel treatment for catalyst removal [102] and glycerol reduction [38]. The ceramic MF produced a high-quality biodiesel by removing the residual catalyst. The retention of the catalyst was 93.64% and potassium was reduced from 8.3 mg/L to 0.3 mg/L [102]. Torres et al. [38] evaluated two UF membranes (PVDF and PSF) to reduce glycerol content from crude biodiesel. This study revealed that the membrane material is an important factor determining the separation performance. The PVDF membrane exhibited a higher stability and lower tendency of fouling than PSF membrane. Application of UF in the clarification of glycerin-rich solutions in the oleo-chemical industry demonstrated a declining flux because of fouling onto the membrane surfaces [83]. Similar to aqueous solutions, membrane rejection increased with applied pressure [83]. In solvent exchange, Livingston et al. [82] reported that NF exhibited a more effective process compared to distillation in swapping a high boiling point solvent to a lower one. By means of NF, the two distinct types of homogeneous catalysts, namely phase transfer catalysts and organometallic catalysts, can be separated from the main reaction products. Tsuru et al. [41] used porous silica-zirconia NF membrane for non-aqueous solution of ethylene glycol, diethylene glycol, triethylene glycol, and polyethylene glycol (PEG) in ethanol and methanol as solvent. They also concluded that the rejection strongly depended on the type of solvent used [43].

3. The development of membrane materials for non-aqueous solution

During the filtration of non-aqueous solution, the solvent significantly influences the retention of solute in the organic solvents. Yang et al. [43] reported that the lower retention value might be because of another transport mechanism in an organic solvent, not due to membrane deterioration. The flow of organic solvents through ceramic-oxide membranes was influenced by interfacial interactions [40]. The flux was affected by surface tension and viscosity of the solvent, furthermore the flux of solvents in the mixture was found to be highly non-linear [103]. In the case of oil filtration, flux is the main challenge due to the high viscosity of the oil. Improving membrane-oil interaction by modification or fabrication of hydrophobic or even super-hydrophobic membrane may be used to solve the problems. In the following sections, preparation of special membrane for non-aqueous systems including solvent-resistant and super-hydrophobic membrane will be presented.

3.1 Solvent-resistant membrane

Membranes can be classified into six categories base on their resistance to the solvent and the stability under extreme operation, i.e. based on phosphazene, imide, methacrylate, inorganic, liquid crystalline and cross-linked polymers [33]. The presence of certain groups like imide in the backbone required for the solvent-resistant properties of polymers [52]. A polyimide membrane manufactured from a co-polymer of diaminophenylindane with benzophenone tetra-carboxylic di-anhydride has an excellent chemical resistance and economically viable flux and rejection characteristics for the separation of light hydrocarbon solvents from lube oil filtrates. Ochoa et al. [75] used three polymeric materials (PVDF, PES, and PSF) and concluded that PVDF has the higher stability to hexane among others. A significant decrease of flux at the beginning of filtration was due to the concentration polarization and internal fouling. Fabrication of silica-zirconia porous membranes by a sol-gel process were conducted to produce organic-inorganic hybrid membranes in the nanometer scale [41, 46]. Alicieo et al. [84] used ceramic tubular membrane (0.01 μm) and a polysulfone hollow fiber membrane (100kDa) and found that the polysulfone membrane gives a higher flux than the ceramic membrane. The clarification of glycerin-rich solutions in the oleo-chemical industry using PVDF and PES membrane tend to be seriously fouled and PES exhibited a higher fouling than PVDF [83].

Tsuru et al. [47] modified the membrane with trimethylchlorosilane (TMS) to increase the performance as showed in a relatively constant permeability. The filtration of ethanol solutions with alkanes (hexane, decane, tetradecane) and alcohols (hexanol, octanol, decanol) as solutes showed that the rejection of alcohols decreased while alkanes rejection were nearly unchanged regardless of the temperatures. Cliff [39] used composite membranes consisted of a PDMS (polydimethylsiloxane) active layer bonded to a commercially available support layer of PAN. Silva [104] used integrally asymmetric polyimide and polyetherimide materials to develop organic solvent NF (OSN) membranes. The degradation of polyimide membrane in polystyrene-ethyl acetate solution shows that flow-induced deformation of the polystyrene chains highly affects the membrane retention [34]. Cliff [39] reported that membrane composition and the extent of polymer swelling were found to be the two key factors which influenced the solvent flux and the solute rejection. Polymer swelling was the most affected parameter when using a solvent with a solubility parameter close to the polymers. Therefore, development of membrane materials that suitable for those applications is required.

3.2 Super-hydrophobic membrane

The performances of membranes filtration are varied as a function of the feed, the contaminations, the temperature, working pressure, and other factors [31]. The previous studies reported that the polymer swelling was the most affected when using solvents with a solubility parameter close to the polymers [39], the formation of micro-cracks at the bend result in the loss of functionality [34], and several membranes tend to be seriously fouled [83]. The separation performance could be improved by developing the membrane materials for non-aqueous system, such as PES, sulfonated poly-sulfone (SPS), PAN, PP, PE, PVDF or PTFE [105, 106].

Generally, the hydrophobic membranes have higher tendency to be fouled in aqueous systems and tend to be wetted in non-aqueous applications. The increasing of hydrophobicity of the membranes materials improves permeability and increases the wetting resistance. In the non-aqueous processing such as vegetable oil filtration, solvent recovery, and waste oil treatments, increasing the hydrophobicity of the membrane is one of promising strategy to improve membrane performance. A super-hydrophobic membrane has a value of water contact angle higher than 150°, while contact angle hysteresis and sliding angle are smaller than 10° [107-109]. The development of super-hydrophobic membrane is an attractive strategy for the non-aqueous solution processing. The super-hydrophobic membrane has anti-wetting properties which resist the aqueous solution to flow through the membrane pores. The super-hydrophobic membranes also have antifouling properties which are result in a high permeability and high separation ability to the non-aqueous solution. In some cases, super-hydrophobic membranes also exhibit super-oleophilic characteristic, performing a high efficiency in non-aqueous separation. These advantages will spread out the applications of super-hydrophobic membrane especially for the non-aqueous processing [110, 111].

4. Summary and Future Outlook

Even though membrane processes have been used in almost industrial sectors, most of the commercial membranes are applied in the aqueous solution treatment. Meanwhile, in the non-aqueous solution, their applications are still limited. For non-aqueous processing, the stable material is needed, because membrane deterioration may occur during the operation which leads to a high operating cost. In addition, in case of oil filtration, membrane-oil interaction can be improved by fabricating a more hydrophobic or oleophilic membrane. Numerous studies on preparation of solvent resistant and super-hydrophobic membranes have been reported. Solvent resistant membranes have been prepared generally based on phosphazene, imide, methacrylate, inorganic, liquid crystalline, and crosslinked polymers as the backbone to support the solvent-resistant properties of polymers. Meanwhile, membranes with high hydrophobicity or super-hydrophobic are generally produced from polymers which have the low energy surfaces such as PP, PE, PVDF, or PTFE. In addition, the super-hydrophobic membrane can be obtained by modifying membrane surface roughness and coating hydrophobic material on the membrane surface. With those advanced membranes, membrane applications can be spread out for non-aqueous processing in the future.

References

[1] Wardani A K, Hakim A N, Khoiruddin, Wenten I G 2017 *Water Science and Technology*
[2] Wenten I G, Khoiruddin 2016 *Desalination* **391** 112-25
[3] Aryanti P, Joscarita S R, Wardani A K, Subagjo S, Ariono D, Wenten I G 2016 *Journal of Engineering and Technological Sciences* **48** 135-49
[4] Aryanti P T P, Yustiana R, Purnama R E D, Wenten I G 2015 *Water Treatment* **6** 379-92
[5] Camacho L M, Dumée L, Zhang J, Li J-d, Duke M, Gomez J Gray S 2013 *Water* **5** 94-196
[6] Wenten I G, Khoiruddin, Arfianto F, Zudiharto 2013 *Desalination* **314** 109-14
[7] Arar Ö, Yüksel Ü, Kabay N, Yüksel M 2013 *Desalination* **310** 25-33
[8] Wood J, Gifford J, Arba J, Shaw M 2010 *Desalination* **250** 973-6
[9] Ariono D, Purwasasmita M, Wenten I G 2016 *Journal of Engineering and Technological Sciences* **48** 367-87
[10] Adham S, Hussain A, Matar J M, Dores R, Janson A 2013 *Desalination* **314** 101-8
[11] Wenten I G, Julian H, Panjaitan N T 2012 *Desalination* **306** 29-34
[12] Dermentzis K 2010 *Journal of Hazardous Materials* **173** 647-52
[13] Himma N F, Wenten I G 2017 *AIP Conference Proceedings* **1840** 090005
[14] Yeon S-H, Lee K-S, Sea B, Park Y-I, Lee K-H 2005 *Journal of Membrane Science* **257** 156-60
[15] Wenten I G, Dharmawijaya P T, Aryanti P T P, Mukti R R, Khoiruddin 2017 *RSC Advances* **7** 29520-39
[16] Purwasasmita M, Kurnia D, Mandias F C, Khoiruddin, Wenten I G 2015 F*ood and Bioproducts Processing* **94** 180-6
[17] Khoiruddin, Widiasa I N, Wenten I G 2014 *Journal of Food Engineering* **133** 40-5
[18] Widiasa I N, Sutrisna P D, Wenten I G 2004 *Separation and Purification Technology* **39** 89-97
[19] Souchon I, Athès V, Pierre F-X, Marin M 2004 *Desalination* **163** 39-46
[20] Wenten I G, Widiasa I N 2002 *Desalination* **149** 279-85
[21] Ariono D, Khoiruddin, Subagjo, Wenten I G 2017 *Materials Research Express* **4** 024006
[22] Khoiruddin, Wenten I G 2016 *Journal of Engineering and Technological Sciences* **48** 1-11
[23] Wenten I, Khoiruddin 2016 *Journal of Engineering Science and Technology* **11** 916-34
[24] Khoiruddin, Hakim A N, Wenten I G 2014 *Membrane Water Treatment* **5** 87-108
[25] Xie Z, Duong T, Hoang M, Nguyen C, Bolto B 2009 *Water research* **43** 1693-9
[26] Ding Z, Liu L, Li Z, Ma R, Yang Z 2006 *Journal of membrane Science* **286** 93-103
[27] Khoiruddin, Aryanti P T P, Hakim A N, Wenten I G 2017 *AIP Conference Proceedings* **1840** 090006
[28] Purwasasmita M, Nabu E B P, Khoiruddin, Wenten I G 2015 *Journal of Engineering and Technological Sciences* **47** 426-46
[29] Purwasasmita M, Juwono P B, Karlina A M, Khoiruddin K, Wenten I G 2014 *Reaktor* **14** 284-90
[30] Yang X, Wang R, Fane A G, Tang C Y, Wenten I G 2013 *Desalination and Water Treatment* **51** 3604-27
[31] Mynin V, Smirnova E, Katsereva O, Komyagin E, Terpugov G, Smirnov V 2004 *Chemistry and technology of fuels and oils* **40** 345-50
[32] Gourgouillon D, Schrive L, Sarrade S, Rios G M 2000 *Environmental science & technology* **34** 3469-73
[33] Razdan U, Joshi S, Shah V 2003 *Current Science* 761-71
[34] Beerlage M, Heijnen M, Mulder M, Smolders C, Strathmann H 1996 *Journal of membrane science* **113** 259-73
[35] Tarleton E, Robinson J, Millington C, Nijmeijer A 2005 *Journal of membrane science* **252** 123-31
[36] Sridhar S 1996 *Journal of membrane science* **113** 73-9
[37] Jadhav C, Shinde S, Kate V, Payghan S 2014 *Asian Journal of Biomedical and pharmaceutical sciences* **4**

[38] Torres J J, Rodriguez N E, Arana J T, Ochoa N A, Marchese J, Pagliero C 2017 *Journal of Cleaner Production* **141** 641-7
[39] Cliff K T 2011 A *study of membrane swelling and transport mechanisms in solvent resistant nanofiltration.* © Kevin Terry Cliff)
[40] Guizard C, Ayral A, Julbe A 2002 *Desalination* **147** 275-80
[41] Tsuru T, Sudoh T, Yoshioka T, Asaeda M 2001 *Journal of Membrane Science* **185** 253-61
[42] Bhanushali D, Kloos S, Kurth C, Bhattacharyya D 2001 *Journal of Membrane Science* **189** 1-21
[43] Yang X, Livingston A, Dos Santos L F 2001 *Journal of Membrane Science* **190** 45-55
[44] Li X, Chen B, Cai W, Wang T, Wu Z, Li J 2017 *RSC Advances* **7** 11381-8
[45] Tres M V, Racoski J C, Di Luccio M, Oliveira J V, Treichel H, de Oliveira D, Mazutti M A 2014 *Food research international* **63** 33-41
[46] Tsuru T, Kondo H, Yoshioka T, Asaeda M 2004 *AIChE Journal* **50** 1080-7
[47] Tsuru T, Miyawaki M, Kondo H, Yoshioka T, Asaeda M 2003 *Separation and Purification Technology* **32** 105-9
[48] Cao Y, Yan F, Li J, Liang X, He B 2009 *Desalination and Water Treatment* **11** 73-80
[49] Psoch C, Wendler B, Goers B, Wozny G, Ruschel B 2004 *Journal of Membrane Science* **245** 113-21
[50] Ciora R, Liu P K 2003 *Separation Journal* **15** 51-60
[51] Wenten I G 2002 *J Sci Technol Membrane Sci Technol* **24** 1010-24
[52] White L S, Nitsch A R 2000 *Journal of Membrane Science* **179** 267-74
[53] Werth K, Kaupenjohann P, Knierbein M, Skiborowski M 2017 *Journal of Membrane Science* **528** 369-80
[54] Li X, Cai W, Wang T, Wu Z, Wang J, He X, Li J 2017 *Separation and Purification Technology* **181** 223-9
[55] Kong W, Miao Q, Qin P, Baeyens J, Tan T 2017 *Frontiers of Chemical Science and Engineering* **11** 166-76
[56] Fornasero M, Marenchino R, Pagliero C 2013 *Advances in Materials Science and Engineering* **2013**
[57] Tres M, Nobrega R, Carvalho R, Oliveira J, Luccio M D 2012 *Brazilian Journal of Chemical Engineering* **29** 577-84
[58] de Souza Araki M, de Morais Coutinho C, Gonçalves L A, Viotto L A 2010 *Separation and Purification Technology* **71** 13-21
[59] Ladhe A R, Kumar N K 2010 *Membrane Technology* 63-78
[60] Basso R C, Gonçalves L A G, Grimaldi R, Viotto L A 2009 *Journal of Membrane Science* **330** 127-34
[61] Lai L L, Soheili K C, Artz W E 2008 Journal of the American Oil Chemists' Society **85** 189-96
[62] de Carvalho C C, de Souza M P, da Silva T D, Gonçalves L A G, Viotto L A 2006 *Desalination* **200** 543-5
[63] Manjula S, Subramanian R 2006 *Critical Reviews in Food Science and Nutrition* **46** 569-92
[64] Kong Y, Shi D, Yu H, Wang Y, Yang J, Zhang Y 2006 *Desalination* **191** 254-61
[65] Cheryan M 2005 Technology **2005** 5-7
[66] Hafidi A, Pioch D, Ajana H 2005 *Innovative Food Science & Emerging Technologies* **6** 203-12
[67] Kim I-C, Kim J-H, Lee K-H, Tak T-M 2002 *Journal of Membrane Science* **205** 113-23
[68] Pioch D, Larguèze C, Graille J, Ajana H, Rouviere J 1998 *Industrial Crops and Products* **7** 83-9
[69] Lin L, Rhee K, Koseoglu S 1997 *Journal of Membrane Science* **134** 101-8
[70] de Morais Coutinho C, Chiu M C, Basso R C, Ribeiro A P B, Gonçalves L A G, Viotto L A 2009 *Food Research International* **42** 536-50
[71] Bottino A, Capannelli G, Comite A, Ferrari F, Marotta F, Mattei A, Turchini A 2004 *Journal of Food Engineering* **65** 303-9
[72] Ribeiro A P B, Bei N, Gonçalves L A G, Petrus J C C, Viotto L A 2008 *Journal of Food Engineering* **87** 514-21

[73] De Souza M P, Petrus J C C, Gonçalves L A G, Viotto L A 2008 *Journal of Food Engineering* **86** 557-64
[74] de Moura J M, Gonçalves L A, Petrus J C C, Viotto L A 2005 *Journal of Food Engineering* **70** 473-8
[75] Ochoa N, Pagliero C, Marchese J, Mattea M 2001 *Separation and Purification Technology* **22** 417-22
[76] Keurentjes J T, Sluijs J T, Franssen R J H, Van't Riet K 1992 *Industrial & Engineering Chemistry Research* **31** 581-7
[77] Keurentjes J, Doornbusch G, Riet K V 1991 *Separation Science and Technology* **26** 409-23
[78] Reddy K K, Subramanian R, Kawakatsu T, Nakajima M 2001 *European Food Research and Technology* **213** 212-8
[79] Bhosle B M, Subramanian R, Ebert K 2005 *European Journal of Lipid Science and Technology* **107** 746-53
[80] Firman L R, Ochoa N A, Marchese J, Pagliero C L 2013 *Journal of Membrane Science* **431** 187-96
[81] Arora S, Manjula S, Krishna A G, Subramanian R 2006 *Desalination* **191** 454-66
[82] Livingston A, Peeva L, Han S, Nair D, Luthra S S, White L S, Freitas Dos Santos L M 2003 *Annals of the New York Academy of Sciences* **984** 123-41
[83] Amin I N H M, Mohammad A W, Markom M, Peng L C 2010 *Journal of Food Engineering* **101** 264-72
[84] Alicieo T, Mendes E, Pereira N, Lima O M 2002 *Desalination* **148** 99-102
[85] Koris A, Vatai G 2002 *Desalination* **148** 149-53
[86] Ebert K, Petrus Cuperus F 1999 *Membrane Technology* **1999** 5-8
[87] Subramanian R, Nakajima M, Raghavarao K, Kimura T 2004 *Journal of the American Oil Chemists' Society* **81** 313
[88] Darvishmanesh S, Degrève J, Van der Bruggen B 2009 *Chemical Engineering Science* **64** 3914-27
[89] Pagliero C, Ochoa N, Martino P, Marchese J 2011 *Journal of the American Oil Chemists' Society* **88** 1813-9
[90] Köseoglu S, Engelgau D 1990 *Journal of the American Oil Chemists' Society* **67** 239-49
[91] Abro R, Chen X, Harijan K, Dhakan Z A, Ammar M 2013 *ISRN Chemical Engineering* **2013**
[92] Cao H, Gu W, Fu J, Liu Y, Chen S 2017 *Applied Surface Science*
[93] Siswanti 2010 *Eksergi* **10** 27-31
[94] Mohammadi T, Safavi M A 2009 *Desalination* **249** 83-9
[95] Dominguez-Rosado E, Pichtel J 2003 *Proceedings of the Indiana Academy of Science* **112** 109–16
[96] Yoo J R, Sung J H, Park K H 2005 *Transactions of the Korean Nuclear Society*
[97] Sarrade S, Schrive L, Gourgouillon D, Rios G 2001 *Separation and Purification Technology* **25** 315-21
[98] Gould R M, White L S, Wildemuth C R 2001 *Environmental Progress & Sustainable Energy* **20** 12-6
[99] Miyagi A, Nakajima M, Nabetani H, Subramanian R 2001 *European Journal of Lipid Science and Technology* **103** 208-15
[100] Subramanian R, Nandini K, Sheila P, Gopalakrishna A, Raghavarao K, Nakajima M, Kimura T, Maekawa T 2000 *Journal of the American Oil Chemists' Society* **77** 323
[101] Lai W-C, Smith K 2001 Fuel **80** 1121-30
[102] Atadashi I, Aroua M K, Aziz A A, Sulaiman N 2014 *Alexandria Engineering Journal* **53** 705-15
[103] Machado D o R, Hasson D, Semiat R 1999 *Journal of Membrane Science* **163** 93-102
[104] Silva M S F 2007 *Polyimide and polyetherimide organic solvent nanofiltration membranes.* FCT-UNL)

[105] Albrecht W, Hilke R, Kneifel K, Weigel T, Peinemann K-V 2005 *Journal of Membrane Science* **263** 66-76
[106] Nishikawa N, Ishibashi M, Ohta H, Akutsu N, Matsumoto H, Kamata T, Kitamura H 1995 *Energy Conversion and Management* **36** 415-8
[107] Himma N F, Wardani A K, Wenten I G 2017 *Materials Research Express* **4** 054001
[108] Celia E, Darmanin T, de Givenchy E T, Amigoni S, Guittard F 2013 *Journal of Colloid and Interface Science* **402** 1-18
[109] Roach P, Shirtcliffe N J, Newton M I 2008 *Soft matter* **4** 224-40
[110] Himma N F, Wardani A K, Wenten I G 2017 *Polymer - Plastics Technology and Engineering* **56** 184-94
[111] Himma N F, Anisah S, Prasetya N, Wenten I G 2016 *Journal of Polymer Engineering* **36** 329-62

IOP Conf. Series: Materials Science and Engineering **285** (2017) 012009 doi:10.1088/1757-899X/285/1/012009

Monosodium glutamate for simple photometric iron analysis

E Prasetyo

Research Unit for Mineral Technology, Indonesian Institute of Sciences
Jl. Ir. Sutami Km.15 Tanjung Bintang, Bandar Lampung 35361, Indonesia

Email: erik005@lipi.go.id

Abstract. Simple photometric method for iron analysis using monosodium glutamate (MSG) was proposed. The method could be used as an alternative method, which was technically simple, economic, quantitative, readily available, scientifically sound and environmental friendly. Rapid reaction of iron (III) with glutamate in sodium chloride-hydrochloric acid buffer (pH 2) to form red-brown complex was served as a basis in the photometric determination, which obeyed the range of iron (III) concentration 1.6 – 80 μg/ml. This method could be applied to determine iron concentration in soil with satisfactory results (accuracy and precision) compared to other photometric and atomic absorption spectrometry results.

1. Introduction

Iron (Fe) is one of the elements most frequently analyzed in laboratory for many different purposes or studies. For geological and mining studies, Fe analysis is executed using a variety of methods depending on the sensitivity or range of Fe content in the sample, precision and accuracy required, which is considering also the cost and time consumed in the analysis [1]. In geological studies, the Fe content in samples typically would be in the range from ppm to percent, although the most interest would be put in the range per mile and up [2, 3]. This is the case if the geological studies were conducted for prospecting and mining purposes.

In laboratory, the samples obtained from prospecting and mining (quality control) usually would be analyzed using atomic absorption spectrometer (AAS) or standard volumetric method, while samples obtained from geological studies would be analyzed using AAS or induced coupled plasma mass spectrometer (ICP-MS). AAS and ICP are generally preferred due to its sensitivity and simplicity, but notoriously sumptuous since they involved high capital cost. Other method used is photometry. This method has advantages in terms of versatility and sensitivity and even the sensitivity could compete with the sensitivity of AAS method. In geochemical and environmental studies, this method is still widely applicable to determine the trace element content in samples. The method has advantages since it involves low capital cost and could be executed directly in the field by a relatively inexperienced operator [4]. These advantages are considered significant due to the increasing interest in geochemical and environmental studies especially in poor and developing countries where the exploration program to find new mineral resources and the environmental problem are still of major concern. On the other hand, the increasing interest has to be balanced with the development of simple, quantitative, readily available, economic and scientifically sound analytical method [5].

The analysis cost in photometry directly depends on the special chemical reagents used (chromogenic reagent). For iron analysis, some of chromogenic reagents include thiocyanate [6, 7],

1,10-phenanthroline [8, 9] and bathophenanthroline [10]. Iron analysis using these chromogenic reagents has disadvantages due to its complicated procedure and very time consuming. Furthermore, in economic terms, these chromogenic reagents are expensive and not readily available. Besides, the application of these reagents could pose problems in disposal due to their toxic and non-biodegradable nature [11, 12].

One of chemical compound which could be proposed to address the problem of complicated procedure, economic factor, availability and disposal problem is monosodium glutamate (MSG). MSG is sodium salt of glutamic acid and easily obtained in market as flavour enhancer with relatively high purity (min. 99%). Glutamate could react with iron (III) in aqueous solution to form red-brown complex (Figure 1). This reaction further could be exploited as a basis in determining iron content in sample. Therefore in this paper the possibility of MSG application in photometric determination of iron (III) content would be discussed including the optimum condition, accuracy and reproducibility.

Figure 1. Standard solution used for calibration in photometric determination of iron (III) from left to right: 20, 40, 60, 80, 100, 160 and 200 μg/ml iron (III)

2. Experimental

2.1Reagent and apparatus
The chromogenic reagent employed in this experiment is MSG (min. 99%), Ajinomoto brand, produced by PT Ajinomoto Indonesia. Powdered iron, 1,10-phenanthroline monohydrate, sodium chloride, nitric acid, sulfuric acid and hydrochloric acid were provided by Merck Chemical Indonesia all in analytical grade. The apparatus consists of Spectroquant NOVA 60 spectrophotometer, including 1 cm quartz cuvette, and pH 0-14 universal indicator paper, all provided by Merck Chemical. Atomic absorption spectrometry to determine the iron content in soil samples was carried out in Laboratory of Analytical Chemistry, Department of Chemistry, Institute of Technology Bandung, Indonesia using Shimadzu AA-630 atomic absorption spectrophotometer.

2.2 General procedures
Standard solution of iron (III) 1000 μg/ml was prepared by dissolving 1 g of dried iron powder in 5 ml of concentrated nitric acid and heated to nearly dryness. The result was then transferred to 1 liter volumetric flask and diluted to mark with deionised water. Working solutions (standard solution for calibration and sample solution) were made by transferred accordingly the standard solution of iron (III) or sample solution into 25 ml volumetric flask and buffered to pH 2 by adding 2 ml of NaCl/HCl buffer. MSG 20% was then added as much as 2 ml into the flask and finally the mixture was dilute to mark with deionised water. The iron (III) content of the working solution is determined based on

absorbance measured by spectrophotometer at wavelength 340 nm and 1 cm cuvette. Besides, the iron (III) content could be also determined visually by matching the colour of sample solution with the calibrated standard solution (figure 1).

2.3 Soil samples

The sample used in this study was soil samples obtained from coffee plantation Ranau area, West Lampung, Indonesia. The soil sample was dominated by silicate minerals (clay) and contained minor oxides. For analysis the soil samples were comminuted to less than 80 mesh particle size and as much as 0.2 gram would be destructed using mixture of concentrated nitric acid, hydrochloric acid and sulphuric acid and heated, leaving only silica and silicate. The destruction results were filtered using Whatman 41 filter paper and the filtrate were transferred to 50 ml volumetric flask and diluted to mark with deionised water. Further, as much as 2 ml of this diluted filtrate would be used in each analysis.

3. Results and discussions

In order to determine the optimum wavelength where the absorbance was maximum a working solution (standard solution for calibration) was prepared. The solution used contained 500 µg/ml iron (III) ion and 1.6 % glutamate concentration in final solution which was buffered with NaCl/HCl to pH 2. Absorbance measurement was performed in wavelength range 340 – 810 nm using spectrophotometer. The correlation between wavelength and absorbance (absorption spectra) of iron (III)-glutamate complex is shown in figure 2 which maximum absorbance occurred at 340 nm.

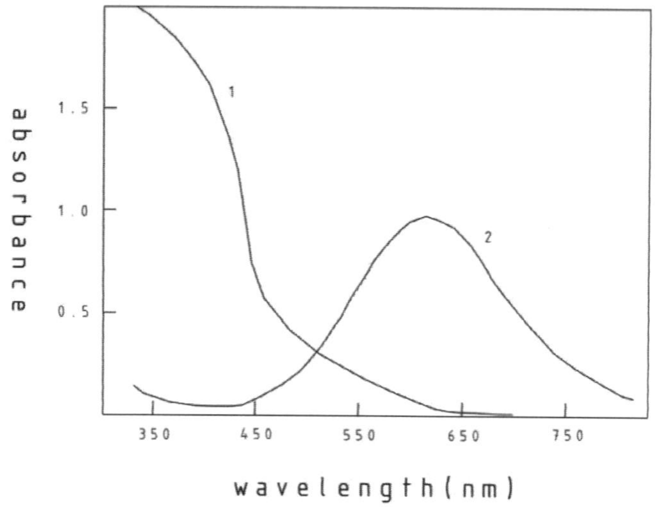

Figure 2. Absorption spectra of (1) iron (III)-glutamate complex, iron (III) concentration 500 µg/ml, pH 2, MSG 1.6% (2) copper-glutamate complex, copper concentration 200 µg/ml, pH 10, MSG 1.6%

Acidity was significantly influenced the complex formation and its stability. Qualitatively it appeared that iron (III) ion would be precipitated as hydroxides if pH was greater than 3, marked by turbidity of working solution. At pH less than 2.5 and in extreme acid condition the complex would not be formed and colour intensity faded due to protonation of glutamate. In this case, pH 2 was chosen to avoid the precipitation of iron (III) ion and to cope with this problem, excess of glutamate was added to guarantee the maximum colour intensity.

To study the effects of concentrations of glutamate, a series of working solution was made which iron (III) concentration remained constant (200 µg/ml) at pH 2 using NaCl/HCl buffer. The glutamate/iron (III) molar concentration in working solution was varied between 1 and 13 (figure 3). The graph shows that absorbance is positively correlated with the molar ratio. This condition applied until the molar ratio 1L while at the higher ratio the absorbance remained constant. This indicates that the addition of more MSG do not affect the absorbance significantly and also indicates the molar ratio of iron (III)-glutamate complex in working solution to be 1:11.

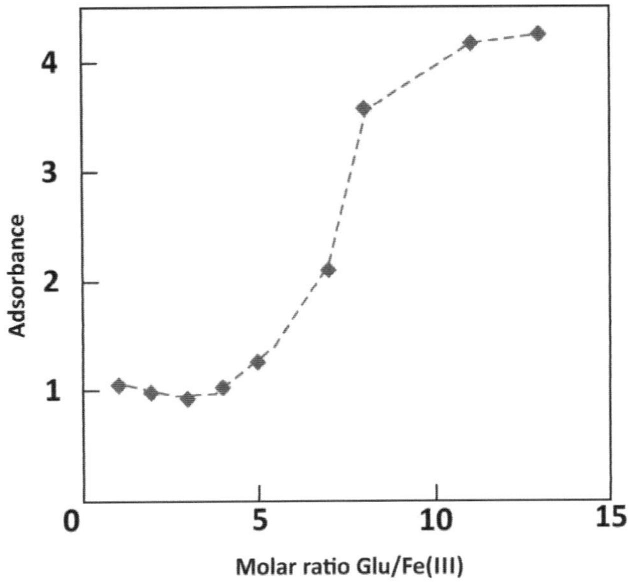

Figure 3. Relation between absorbance and molar ratio of glutamate/iron (III)

Calibration curve representing the association between absorbance and iron (III) concentration was required for iron (III) determination in sample solution. The curve was constructed using series of standard solution (working solution) with different iron (III) concentration (Figure 4). The curve was linear and correlation between absorbance and iron (III) concentration could be formulated with linear regression equation A = 0.034c (µg/ml) – 0.031 with correlation coefficient r^2 = 0.999, where A and c stood for absorbance and iron (III) concentration respectively. The equation was conformed if iron (III) concentration fell on the range of 1.6 – 80 µg/ml. Forhigher concentration the correlation of absorbance – iron (III) concentration lost its linearity. Theoretically, if detection limit could be calculated using the value of molar absorption and lowest measured absorbance (≈ 0.020) then the concentration limit would be 1.47 µg/ml. The numerical value of sensitivity of proposed method, as molar absorption coefficient (ε) and specific absorbance (a) were calculated to be $1.4 \cdot 10^3$ $l.mol^{-1}.cm^{-1}$ and 0.03 $ml.\mu g^{-1}.cm^{-1}$ respectively.

Working solution could be directly measured its absorbance immediately at room temperature. Based on observation, a working solution was stable for at least 2 days, which was shown by insignificance change in absorbance (decreasing less than 3% from initial absorbance). For higher concentration range, linearity could be obtained using wavelength analysis 410 nm, within the range 10 – 200 µg/ml (r^2 = 0.994, concentration limit 2.43 µg/ml, ε = 424 $l.mol^{-1}.cm^{-1}$).

Observation on the possibility of interference by various ions as source of matrix error in this determination was still qualitative. So far copper (II) or cupric ion caused interference by forming blue coloured complex with glutamate. Absorption spectra of the complexes of iron (III)-glutamate is shown in Figure 2. However, the interference of cupric ion could be minimized since based on the Figure 2, the absorbance of cupric-glutamate complex in wavelength 340 nm was negligible. The complex cupric-glutamate has a maximum absorbance at wavelengtharound600nm. Besides, the cupric-glutamate complex formation could be prevented in acidic solution (pH < 7). Other ions including alkaline metals K, Na, Ca and Mg, and transition metals Mn (II), Fe (II), Cr, Zn, Co and Ni so far did not form coloured complex with glutamate, therefore, it is qualitatively indicated that these ions would not interfere in the iron (III) determination using MSG.

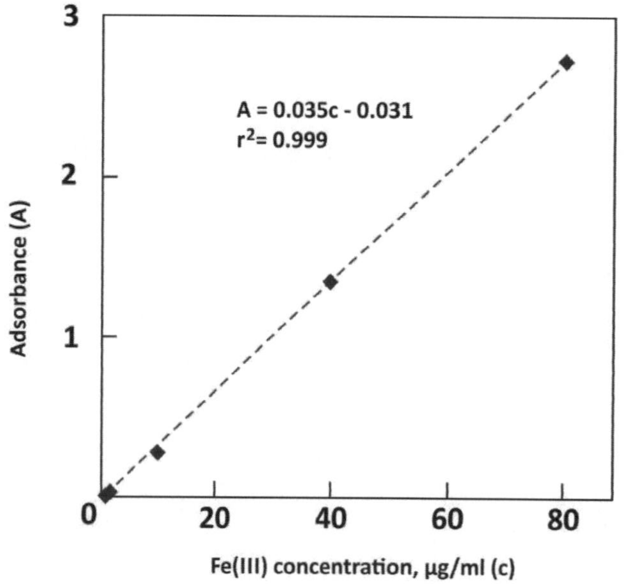

Figure 4. Calibration curve iron (III) concentration (c) – absorbance (A)

This proposed method was successfully applied in analysis of soil sample (section 2). The filtrate as much as 2 ml as result from destruction process was transferred to 25 ml volumetric flask and buffered to pH 2 by adding 2 ml NaCl/HCl buffer. MSG 20% solution was added as much as 2 ml into the flask and finally the mixture was dilute to mark with deionised water and homogenized. The absorbance could be directly measured at wavelength 340 nm using 1 cm cuvette. As comparisons, the same sample would be analyzed using standard photometric which 1, 10-phenanthroline was applied as chromogenic reagent. The procedure of the standard photometric determination referred to Marczenko and Balcerzak [13]. To determine the accuracy of both photometric determinations, iron (III) concentration in samples was also determined using atomic absorption spectrometer (AAS) which results were summarized in table 1.

Tabel 1. Determination of iron (III) in soil sample. Photometric and AAS results in µg/ml, recovery in % (ratio of photometric/AAS results). MSG and phen are proposed monosodium glutamate and standard 1,10 phenanthroline methods respectively

No	Sample	Photometric		AAS	Accuracy	
		MSG	phen		MSG	phen
1	ER/011/SL	6.85	6.93	6.71	102.2	103.4
2	ER/015/SL	5.85	6.03	6.21	94.3	97.1
3	ER/022/SL	7.09	6.96	7.85	90.3	88.7
4	ER/040/SL	5.62	5.52	6.11	91.9	90.3
5	ER/054/SL	6.56	6.76	6.32	103.7	106.8

Based on table 1, the result of proposed photometric method using MSG as chromogenic reagent of 5 samples analyzed had accuracy between 90.32 % - 103.72 % compared to AAS results which were set as true value. These results even better than standard photometric method incorporated 1,10 phenanthroline which accuracy varied between 88.70 % - 106.83 % compared to AAS results. To assess the reproducibility (precision), five samples was analyzed twice using proposed method (table 2). Low RSD value (ratio between standard deviation and mean value), which is less than 5% signifies the high precision in analysis using proposed method.

Table 2. Reproducibility of proposed MSG photometric method to determine iron (III) concentration after two measurements (MSG 1 and MSG 2 in μg/ml)

No	Sample	MSG1	MSG2	RSD (%)
1	ER/028/SL	4.74	4.68	0.89
2	ER/038/SL	4.59	4.71	1.80
3	ER/051/SL	7.15	7.27	1.16
4	ER/053/SL	6.06	6.18	1.35
5	FK/068/SL	13.50	13.38	0.62

4. Conclusions

Monosodium glutamate could be proposed as alternative chromogenic reagent in photometric determination of iron (III). Preconcentration and dilution could be carried out accordingly to meet the concentration range of 1.6 – 80 μg/ml with limit detection of 1.47 μg/ml. The proposed photometric method incorporating monosodium glutamate could be used as an alternative in determining iron (III) content in sample which is technically simple, quantitative, readily available, economic, scientifically sound and environmental friendly. This method is applicable to analyze iron content in soil sample with high accuracy and precision.

References

[1] Moon C J, Whateley M K G and Evans A M 2006 *Introduction to Mineral Exploration* 2nd ed. Blackwell Publs
[2] Dunn C E 2007 *Biogeochemistry in Mineral Exploration* Elsevier Amsterdam
[3] Laznicka P 2006 *Giant metallic deposit: Future Sources of Industrial Metals* Springer Verlag Berlin.
[4] Jungreis E 1997 *Spot test analysis: clinical, environmental, forensic, and geochemical applications* J. Wiley
[5] Lottermoser B G, Schnug E and Haneklaus S 2011 *Science of the Total Environment* **409** 3512 – 19
[6] Barreto I S, Lima M B, Andrade S I E, Araújo M C U and Almeida L F 2012 *Food Anal. Methods* **5** 1040 – 5
[7] Niedzielski P, Zielinska-Dawidziak M, Kozak L, Kowalewski P, Szlachetka B, Zalicka S and Wachowiak W 2014 *Food Anal. Methods* **7** 2023 – 32
[8] Braunschweig J, Bosch J, Heiste K, Kuebeck C and Meckenstock R U 2012 *Journal of Microbiological Methods* **89(1)** 41 – 8
[9] Serra-Mora P, Moliner-Martínez Y, Herráez-Hernández R, Verdú-Andrés J and Campíns-Falcó P 2016 *Food Analytical Methods* **9(5)** 1150 – 4
[10] Abdel Ghafar H H, Mohamed G G, Abd El-Salam M and El-Sayed Abd. A Y 2015 *Desalination and Water Treatment* **55(1)** 96 – 106

[11] www.sigma-aldrich.com, 01 November 2011
[12] Gos S and Andreas R 2011 *The relevance of alternative lixiviants with regard to technical aspects, work safety and environmental safety* Degussa AG, Rodenbacher Chaussee 4, 63457 Hanau, Germany, prepared by CyPlus® (www.cyplus.com) accessed 10 August 2011
[13] Marczenko Z and Balcerzak M 2000 *Separation, Preconcentration and Spectrophotometry in Inorganic Analysis* Elsevier Amsterdam

IOP Conf. Series: Materials Science and Engineering **285** (2017) 012010 doi:10.1088/1757-899X/285/1/012010

Flue gas carbon capture using hollow fiber membrane diffuser-separator

D Ariono[1], A S Chandranegara[1], S Widodo[1,2], Khoiruddin[1], I G Wenten[1]

[1]Department of Chemical Engineering, Institut Teknologi Bandung, Jl. Ganesha 10, Bandung 40132, Indonesia
[2]PPPTMGB LEMIGAS, Balitbang KESDM, Jl. Ciledug Raya, Kav. 109, Cipulir, Kebayoran Lama, Jakarta, 12230, Indonesia

Email: igw@che.itb.ac.id

Abstract. In this work, CO_2 removal from flue gas using membrane diffuser–separator was investigated. Hollow fiber polypropylene membrane was used as the diffuser while pure water was used as the absorbent. Separation performance of the membrane diffuser–separator as a function of CO_2 concentration (6-28%-vol.) and flow rate (gas: 0.8-1.55 $L.min^{-1}$ and liquid: 0.2-0.7 $L.min^{-1}$) was investigated and optimized. It was found that CO_2 removal was significantly affected by CO_2 concentration in the feed gas. On the other hand, CO_2 flux was more influenced by flow rates of liquid and gas rather than concentration. The optimized CO_2 removal (64%) and flux (1 x 10^{-4} $mol.m^{-2}.s^{-1}$) were obtained at the highest gas flow rate (1.55 $L.min^{-1}$), the lowest liquid flow rate (0.2 $L.min^{-1}$), and 6.2%-vol. of CO_2 concentration. Outlet gas of the membrane diffuser system tends to carry some water vapor, which is affected by gas and liquid flow rate. Meanwhile, in the steady-state operation of the separator, the gas bubbles generated by the membrane diffuser take a long time to be completely degassed from the liquid phase, thus a portion of gas stream was exiting separator through liquid outlet.

1. Introduction

Carbon Dioxide (CO_2) is a major source of global warming phenomenon. One attempt to control the emission of greenhouse gas is by removing CO_2 from flue gas. On the other hand, the separation of CO_2 from the gas stream is an important step in some industrial processes and is needed for technical, economic, and environmental reasons. In the liquefied natural gas (LNG) industry, CO_2 must be separated to prevent several problems such as a freezing problem in low-temperature chillers, corrosion in the process equipment, reduction of the caloric value of gas flow, and a decrease of pipeline capacity [1].

Generally, CO_2 removal is performed by solvent absorption, solid adsorption, and membrane-based separation. Solvent-based separation is currently the well-established technology. It has high selectivity and capacity as well as simplicity in solvent regeneration [2,3]. However, solvent absorption which traditionally uses packed columns, has some disadvantages such as high regeneration energy as well as difficulty in obtaining an accurate estimate of the gas–liquid mass transfer area and a limited range of gas and liquid flow rates due to operational problems including flooding, loading, channeling, and foaming [4,5]. Membrane technology has been considered as the promising alternative for conventional technologies in various processes including separation, purification, chemical reaction, and contacting processes [6–17]. Membrane technology offers interesting features such as lower energy consumption, relatively low operating expenditure, smaller footprint, easy to scale-up, easy to operate, and more process intensive [18–21]. Recently, using membrane as a contactor of two streams has been widely studied [22–24]. The gas-liquid contacting process using hollow fiber membranes is a hybrid process that combines the advantages of conventional reactive absorption technology with membrane contactor processes. This process is expected to be used as an alternative technology that has the potential to overcome operational problems encountered in conventional absorption processes.

Although much research has been done regarding the separation of CO_2 using hollow fiber membrane contactor, there is still little commercial plant found with such technology. This is associated with several drawbacks that hinder their commercialization [25–29]. According to Luis et al. [30], there are four main factors that need to be considered to make the contactor membrane a competitive technology against conventional contactor: solving membrane wetting phenomenon, fabricating a high durability and low cost membrane, selecting an appropriate and environmentally friendly solvent or absorbent, and investigating the influence of other component in the gas stream.

In this study, a new configuration of a membrane-based contacting process is proposed. A unique configuration, a series membrane diffuser–separator is used for CO_2 removal from flue gas. Hollow fiber membrane serves as a micro bubble generator resulting in a two-phase fluid. In this configuration, the mass transfer will occur on the bubble surface. This is different from the membrane contactor which keeps the inter phase area remained in the membrane pore. By filling the pore with gas it is expected that the membrane wetting does not occur and the CO_2 separation performance is expected to be stable in a long-term operation.

2. Materials and Method

CO_2 and N_2 gasses were obtained from PT BOC. Polypropylene membrane was supplied by GDP Filter Indonesia (average pore size: 0.04-0.05 μm). For all of the experiments, pure water was used as the solvent (absorbent).

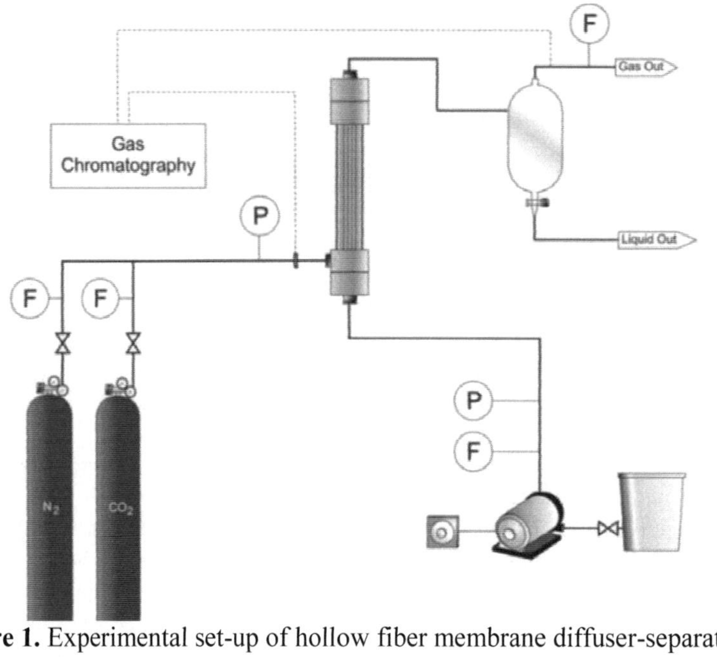

Figure 1. Experimental set-up of hollow fiber membrane diffuser-separator

The equipment used in this research was a set of tools that can be divided into three main tools, namely the absorbent tank, diaphragm pump, membrane module, separator, and gas chromatography (GC Shimadzu). The CO_2/N_2 gas mixture was passed to the membrane module via a dead-end shell, in which the lumen was continuously absorbed by water. The gas stream was dispersed in a micro bubble in the water stream, where the mass transfer occurred on the bubble surface. This two-phase flow was then separated through the gravity separator. To obtain the gas-liquid separation data, the gravity separator was operated under both closed and steady valve conditions with a 14 cm liquid height. The absorption experiments were conducted at room temperature and atmospheric pressure. Gas in and out of the system was analyzed using GC after 10 minutes of operation. The equipment scheme used in the study is shown in Figure 1. Analysis of reactant and product samples was carried out with Gas Chromatograph GC-Shimadzu with Porapax column equipped with TCD system. Argon was used as carrier gas.

3. Results and Discussion

3.1. Effect of operating parameters on CO_2 removal

Data on the effect of operating variables on CO_2 separation were obtained from the analysis of inlet gas and outlet gas in a closed separator system. In the absence of a liquid outflow, it was assumed that all the gas exit through the gas outlet and the absorbed CO_2 still remained in the liquid phase. Modeling using regression for the effect of CO_2 concentration, gas flow rate, and liquid flow rate on CO2 separation resulted in the following equation:

$$Y_1 = 1.51757 - 0.05841\,(A) - 0.77880\,(B) - 0.63684\,(C) + 0.00047\,(A^2) + 0.31381\,(B^2) + 0.80335\,(C^2) + 0.01091\,(AB) + 0.03337\,(AC) - 0.42762\,(BC) \quad (1)$$

where Y_1 is CO_2 separation (0-1), A is CO_2 concentration (%Volume), B is gas flow rate (L.min^{-1}.), and C is liquid flow rate (L.min^{-1})

The validity of the equation was evaluated by comparing actual data distribution to the model shown in Figure 2. With R2 = 89.81%, the equation was able to represent the actual data distribution of the experiment. Looking at the normal probability plot curve in Figure 2, the actual distribution of the data is close to the normal distribution line which indicates that the model is satisfied. Plots of residual versus fitted value does not show the tendency of a particular pattern.

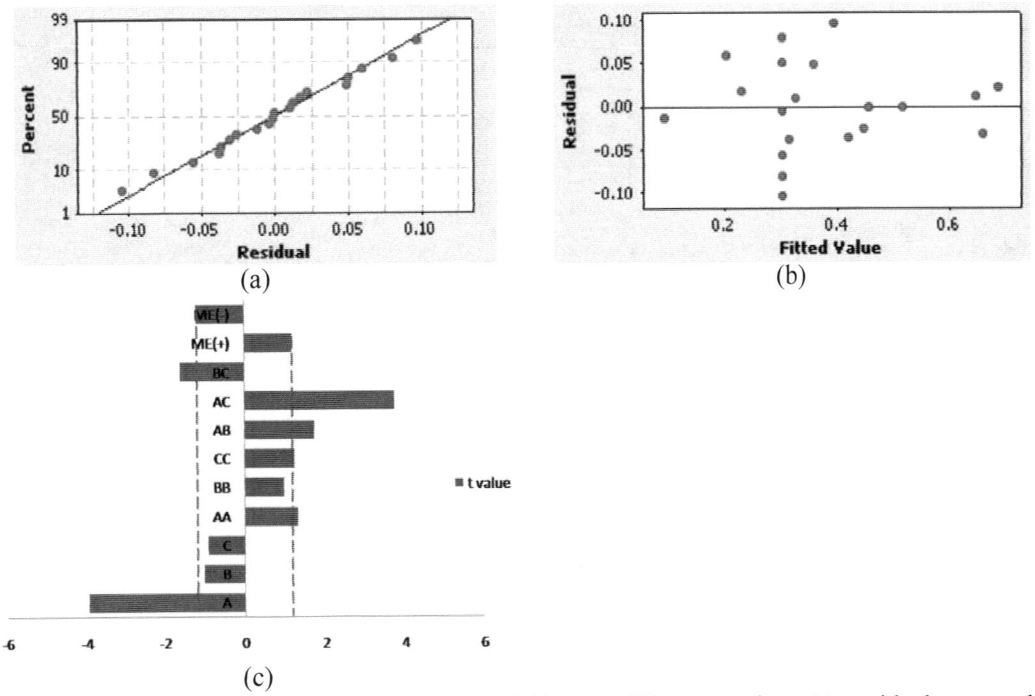

Figure 2. Analysis the effect of operating variables on CO_2 separation. (a) residual vs. predicted response, (b) normal probability plot of residual, and (c) Pareto diagram

A Pareto diagram is constructed based on the model to determine the effect of experimental variables on the CO_2 separation (Figure 2). The diagram shows that the variable A (CO_2 concentration in the feed) has the highest effect on the separation of CO_2. Variable B (gas flow rate) and variable C (liquid flow rate) have no significant effect on CO_2 separation result. This shows that within the range of operating condition used in this study, the increase in CO_2 concentration in the feed from 6-28%-vol will decrease CO_2 separation significantly, while the increase of gas flow rate in the range of 0.82-1.55 $L.min^{-1}$ and the increase of liquid flow rate in the range 0.2-0.72 $L.min^{-1}$ only able to slightly increase CO_2 separation. In other words, a higher CO_2 concentration requires larger flow rate to achieve the desired separation value. The interaction of CO_2 concentration feed and the liquid flow rate (AC) shows a significant value in the positive direction. It means the larger liquid flow rate tends to bind the higher CO_2 from the gas which results in an increased CO_2 separation. The behavior of CO_2 separation due to the change of operating variables is shown in **Figure 3**.

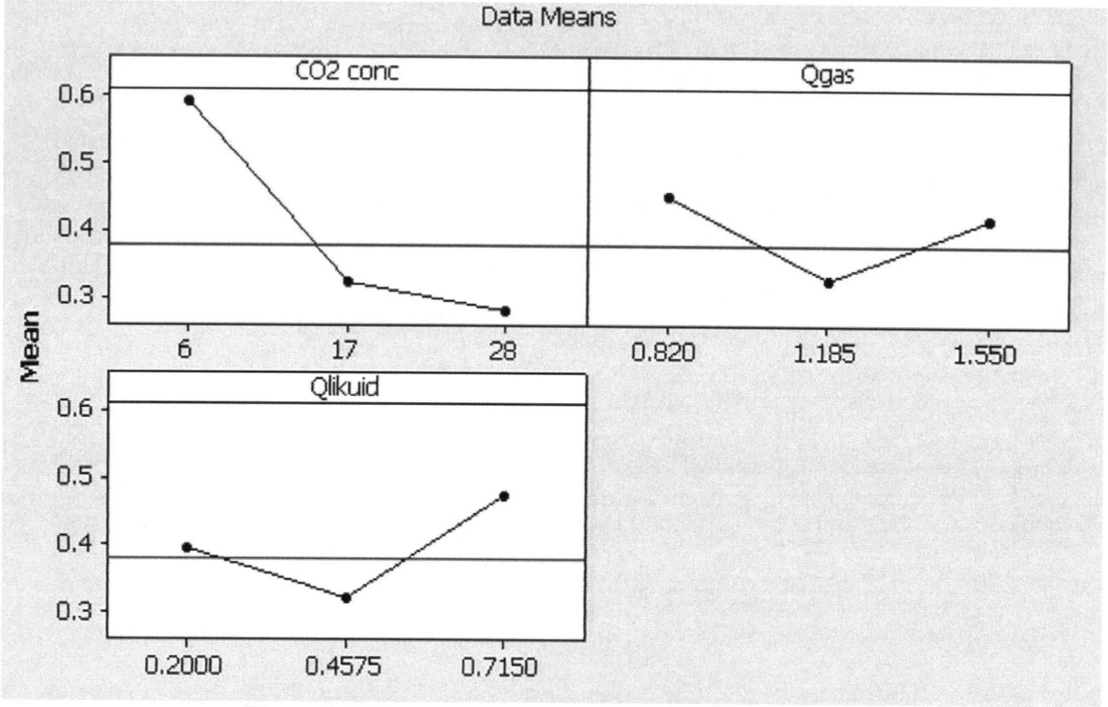

Figure 3. Effect of operating variables on CO_2 separation

Figure 4 shows that the highest CO_2 separation will be achieved under larger liquid flow rate and smaller gas flow rates. At low CO_2 concentration (6%-vol) the increase of gas flow rate will decrease CO_2 separation, especially at a higher liquid flow rate. While at higher feed CO_2 concentration, the CO2 separation will decrease significantly, but with the same gas flow rate and liquid flow rate profile. At a low feed concentration conditions, CO_2 is well absorbed and has reached equilibrium. It appears that the increase in CO_2 concentration in the feed greatly affects the separation of CO_2. Water as absorbent has a finite physical absorption capacity expressed by the Henry constant of 0.8317. At low concentrations of feed CO_2, water is able to absorb CO_2 until it reaches equilibrium. At high gas flow rates, the formation of the gas bubble will increase so the gas-liquid contact area will decrease. The increase in liquid flow rate causes greater turbulence in the lumen side of the membrane resulting in lower gas-liquid contact time. At high gas and liquid flow rates, separation becomes inefficient due to the shorter contact time.

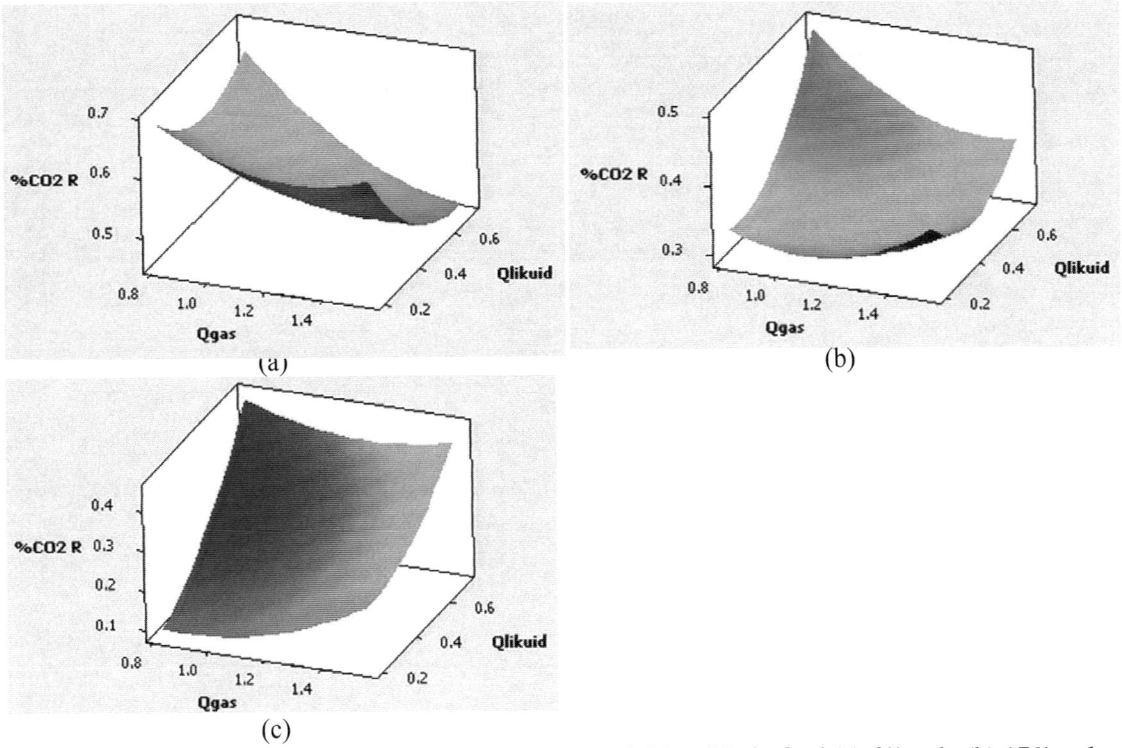

Figure 4. Influence of operation variable on separation of CO_2. CO_2 in feed (a) 6%-vol., (b) 17%-vol., and (c) 28%-vol

3.2. CO_2 flux

The effect of variables on CO_2 flux is shown in
Figure 5. It is showed that the flux of CO_2 is more influenced by the gas and liquid flow rates than the CO_2 concentration in the feed. The increase in gas flow rate initially causes the flux to decrease due to the formation of larger air bubbles in the liquid which reduces gas-liquid contacting area. However, at a further increase of gas flow rate, there is a significant increase in CO_2 flux. This indicates that at high gas flow rates, a greater gas intensity is more dominant than the reduction of contact area due to the magnification of gas bubble size. In other words, a high gas flow rate creates a large gas-liquid contact area with a larger number of bubbles of large size. While at a lower gas flow rate, small diameter bubbles are produced with fewer amounts. This is similar to the effect of higher liquid flow rate. At a larger liquid flow rate, the decrease of CO_2 flux is due to the shorter contact time which associated with a higher superficial velocity in the lumen side of the membrane. Meanwhile, the increase in CO_2 concentration in the feed tends to increase the flux value due to the increasing amount of transferred CO_2. However, the flux increase is not significant.

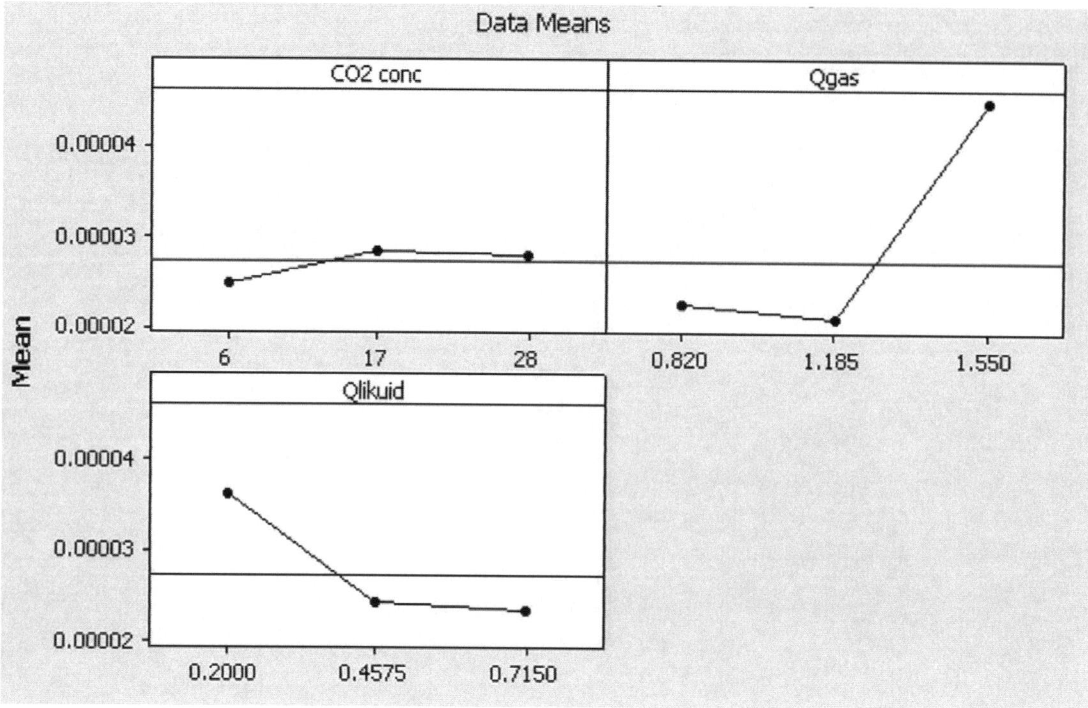

Figure 5. Effect of operating variable on CO_2 flux

3.3. Gas-liquid separation

The observation of the separator performance (gas-liquid separation) was evaluated from two parameters, namely how much water is incorporated into the carried-over/re-entrainment phases and how many gas bubbles are carried into the liquid phase outlets. The data retrieval for the water which is incorporated in the gas phase was carried out under closed separator conditions. The effect of the gas flow rate and the liquid flow rate on the gas difference is given in
Figure *6*.

The positive differential values of flow rate indicate that the feed gas flow rate is higher than the output gas. This is due to the presence of CO_2 in the liquid phase so that the flow rate of the outlet gas is decreased. The negative value of gas flow rate difference indicates that the feed gas is smaller than residue gas. The increasing gas flow rate at the outlet is caused by carried water vapor after contact between gas and liquid in the separator. The positive gas difference value indicates that the reduction of gas volume due to the transfer of CO_2 to the liquid phase is more dominant than the amount of water vapor that moves to the gas stream. Meanwhile, the negative gas difference value shows that the water vapor carried to the gas phase is more dominant than the transfer of CO_2 to the liquid phase.

The increase of the gas flow rate causes the larger bubbles in liquids and decrease the contact area between gas-liquid. Thus the displacement of water into the vapor phase is lower and the transfer of CO_2 to the liquid phase is more dominant than the movement of water to the vapor phase. Conversely, an increase in the liquid flow rate will increase the turbulence in the lumen causing the bubble gas to be more easily fragmented and dispersed in the liquid, so the contact area is increased. This increase in contact area causes the water vapor carried in the gas phase to be more dominant than the transfer of CO_2 to the liquid phase. Meanwhile, the increase of CO_2 concentration in the feed did not significantly affect the gas flow rate difference. The amount of gas loss in the liquid flow is seen from the decrease in the gas flow rate as the separator is operated steadily against the gas flow rate (the separator is closed). The effect of operating variables on gas loss is depicted in

Figure 7. The graph shows that the CO_2 concentration in the feed almost has no effect on the amount of gas involved in the liquid phase. The most influential variable is the liquid flow rate. The increase of the liquid flow rate in the fixed volume of the separator will reduce the residence time in the separator. The lower residence time causes gas bubbles have not been fully degassed. They remain in the liquid phase and are carried in the liquid stream.

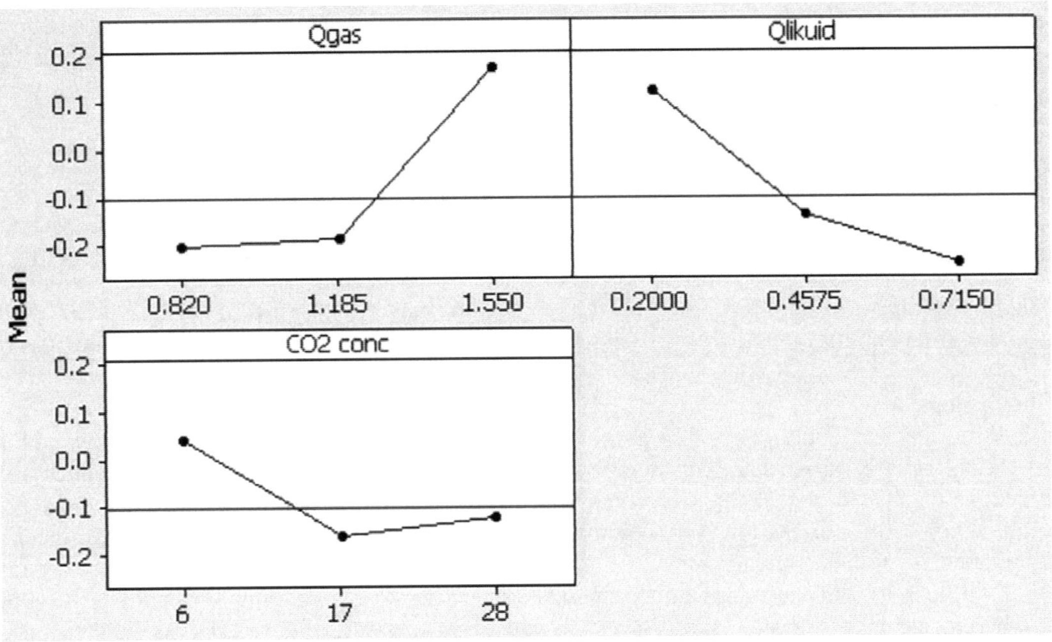

Figure 6. Effect of operating variable on the difference of gas flow rate

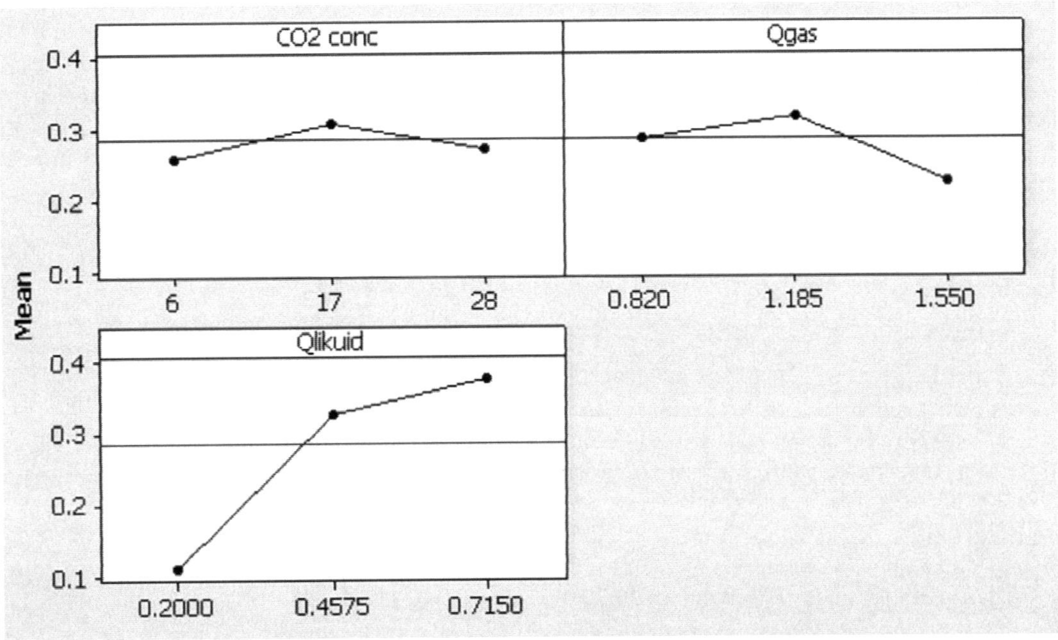

Figure 7. Effect of operating variable on gas loss in the liquid phase

The residence time needed for the gas bubble to be released from the liquid is expressed in Table 1. It shows that the residence time of 0.57-3.4 minutes in this experimental range is not sufficient for degassing of less than 5 microns bubbles. It takes 5.13 minutes for all 5-micron gases to be discharged from the liquid. This data also shows that the bubbles diameter generated by the membrane diffuser are distributed at <10 microns. This is also similar to the effect of the gas flow rate on the amount of gas loss. Initially, the increase in gas flow rate causes higher gas loss since the gas bubbles are dispersed in the liquid. However, the increase in gas flow rate further causes the larger bubbles which make it easier to separate and require only a short degassing time. This causes the loss of gas carried in the liquid phase decreases significantly.

Table 1. Degassing time of gas bubbles in the liquid

dp (micron)	1	5	10	12	15	20	25	50
Uc (gas velocity) (m/s)	1.8E-07	4.6E-04	0.013	0.033	0.097	0.40	1.17	34.6
Degassing time (s)	794.7	307.8	10.4	4.3	1.4	0.35	0.12	0.004
Degassing time (minute)	13.2	5.13	0.17	0.07	0.024	0.006	0.002	6.74E-05

4. Conclusions

The hollow fiber gas-liquid contacting process, a hybrid process combining the advantages of both conventional reactive absorption and membrane contactor processes, is considered as an alternative technology which can potentially overcome the drawbacks of conventional processes. However, membrane contactor still has not been commercialized for CO_2 removal due to wetting problem. A distinct concept of using membrane technology as contacting device is introduced to overcome wetting problem. In series membrane diffuser-separator is used to generate micro-bubble in the lumen side wherein the mass transfer takes places instead of maintaining interfacial area in the pore of membrane. Separation performance of the membrane diffuser-separator as function of CO_2 concentration (6-28 %-vol), gas flow rate (0.82-1.55 $L.min^{-1}$), and liquid flow rate (0.2-0.72 $L.min^{-1}$) was investigated and optimized. It was found that CO_2 removal was significantly affected by CO_2 concentration in feed gas. On the other hand, CO_2 flux was more influenced by flow rates of liquid and gas than CO_2 concentration. The optimized CO_2 removal (64%) and flux (1×10^{-4} $mol.m^{-2}.s^{-1}$) were obtained at the highest gas flow rate (1.55 L/min), the lowest liquid flow rate (0.2 L/min), and 6.2%-vol. of CO_2. Outlet gas of the membrane diffuser system tend to carry several water vapor which is affected by gas and liquid flow rate. Meanwhile in the steady-state operation of the separator, the gas bubbles generated by the membrane diffuser take a long time to be completely degassed from the liquid phase, thus a number of gas was exiting separator through the outlet of liquid stream.

References

[1] Boucif N, Corriou J P, Roizard D, Favre E 2012 *AIChE J.* **58** 2843-55
[2] Wang M, Lawal A, Stephenson P, Sidders J, Ramshaw C 2011 *Chem. Eng. Res. Des.* **89** 1609-24
[3] Aaron D, Tsouris C 2005 *Sep. Sci. Technol.* **40** 321-48
[4] Mansourizadeh A, Ismail A F 2009 *J. Hazard. Mater.* **171** 38-53
[5] Ahmad A L, Sunarti A R, Lee K T, Fernando W J N 2010 *Int. J. Greenh. Gas Control* **4** 495-8
[6] Drioli E, Curcio E 2007 *J. Chem. Technol. Biotechnol.* **82** 223-7
[7] Wenten I G, Dharmawijaya P T, Aryanti P T P, Mukti R R, Khoiruddin 2017 *RSC Adv.* **7** 29520-39
[8] Ariono D, Purwasasmita M, Wenten I G 2016 *J. Eng. Technol. Sci.* **48** 367–87
[9] Khoiruddin, Widiasa I N, Wenten I G 2014 *J. Food Eng.* **133** 40–45
[10] Strathmann H, Grabowski A, Eigenberger G 2013 *Ind. Eng. Chem. Res.* **52** 10364–79
[11] Rathore A S, Shirke A 2011 *Prep. Biochem. Biotechnol.* **41** 398–421

[12] Wenten I G, Widiasa I N 2002 *Desalination* **149** 279–85
[13] Ariono D, Khoiruddin, Subagjo, Wenten I G 2017 *Mater. Res. Express* **4** 24006
[14] Wenten I G, Khoiruddin 2016 *J. Eng. Sci. Technol.* **11** 916–34
[15] Khoiruddin, Aryanti P T P, Hakim A N, Wenten I G 2017 *AIP Conf. Proc.* **1840** 90006
[16] Purwasasmita M, Kurnia D, Mandias F C, Khoiruddin, Wenten I G 2015 *Food Bioprod. Process.* **94** 180–6
[17] Khoiruddin K, Hakim A N, Wenten I G 2014 *Membr. Water Treat.* **5** 87–108
[18] Ma L, Dong X, Chen M, Zhu L, Wang C, Yang F, Dong Y 2017 *Membr.* **7**
[19] Yang X, Wang R, Fane A G, Tang C Y, Wenten I G 2013 *Desalin. Water Treat.* **51** 3604–27
[20] Wardani A K, Hakim A N, Khoiruddin, Wenten I G 2017 *Water Sci. Technol.* **75** 2891–9
[21] Wenten I G, Khoiruddin, Arfianto F, Zudiharto 2013 *Desalination* **314** 109–14
[22] Himma N F, Anisah S, Prasetya N, Wenten I G 2016 *J. Polym. Eng.* **36** 329–62
[23] Purwasasmita M, Nabu E B P, Khoiruddin, Wenten I G 2015 *J. Eng. Technol. Sci.* **47** 426–46
[24] Wenten I G, Julian H, Panjaitan N T 2012 *Desalination* **306** 29–34
[25] Lv Y, Xu C, Yan G, Guo D, Xiao Q 2013 *Adv. Mater.* Res. **616–618** 1541–5
[26] Li J-L, Chen B-H 2005 *Sep. Purif. Technol.* **41** 109–22
[27] Zhang Z, Wu X, Wang L, Zhao B, Li J, Zhang H 2017 *RSC Adv.* **7** 13451–7
[28] Lv Y, Yu X, Tu S-T, Yan J, Dahlquist E 2010 *J. Memb. Sci.* **362** 444–52
[29] Bottino A, Comite A, Costa C, Di Felice R, Varosio E 2015 *Sep. Sci. Technol.* **50** 1860–9
[30] Luis P, Van der Bruggen B, Van Gerven T 2011 *J. Chem. Technol. Biotechnol.* **86** 769–75

Humic acid provenance influence to the adsorption capacity in uranium and thorium removal

E Prasetyo

Research Unit for Mineral Technology, Indonesian Institute of Sciences
Jl. Ir. Sutami km. 15 Tanjung Bintang, Lampung 35361, Indonesia

Email: erik005@lipi.go.id

Abstract. It is common knowledge that humic acid is organic compound without certain chemical composition since it is derived from different organic materials. Further this raises question whether the different humic acid sample used could lead to different adsorbent properties e.g. adsorption capacity. To address the problem, this paper is aimed to clarify the relation between the provenances of humic acid and synthesized adsorbent properties especially adsorption capacities by quantitative and qualitative functional groups determination including discussion on their effect to the metal ion adsorption mechanism using three humic acid samples. Two commercial samples were derived from recent compost while the other extracted from tertiary carbonaceous mudstone strata.

1. Introduction

Environmental pollution caused by the release of heavy metal or radioactive elements needs serious attention, especially in developing countries. To resolve the problems, the removal of these pollutants from the wastewater is stipulated as mandatory. Several technologies have been proposed for pollutant removal including electrochemical treatment, chemical separation, liquid-liquid separation, intracellular sequestration e.g. phytoextraction and solid phase extraction. Among these techniques, solid liquid separation or adsorption is considered as the most promising due to its advantages including versatility, cost effective [1] and ability to manage the low concentration pollutant [2].

Numerous adsorbent had been proposed to remove heavy metals and radioactive elements, but unfortunately most of these adsorbents are not available for wide scale operation due to complicated and high cost synthesis procedure. This problem inspired the researchers to synthesize low cost adsorbents using widely and inexpensive materials, including humic acid. Humic acid is major organic component in soil, coal and peat, also could be found in aquatic environment e.g. lakes and sea water [3]. Humic substance could be classified based on solubility into fulvic acid (FA, soluble fraction), humin (HUM, insoluble fraction) and humic acid (HA, soluble only in basic solution). According to standard protocol, HA and FA are extracted from soil using strong basic solution (sodium or potassium hydroxide), and HA further is separated from FA through precipitation by changing the pH of solution to acidic. HA has ability to bind metal ions due to the presence of functional group such as carboxyl and hydroxyl, which act as ligand [4].

Several HA based adsorbent had been reported [1, 5-9]. In these reports, only single HA sample was used and this evokes the question whether the different HA sample used could lead to different adsorbent properties e.g. adsorption capacity. This concern rose due to common knowledge that HA is organic compound without certain chemical composition since it is derived from different organic materials. To put this indecision into rest, this paper is aimed to clarify the relation between the provenances of HA and synthesized adsorbent properties especially adsorption capacities by quantitative and qualitative functional groups determination including discussion on their effect to the metal ion adsorption mechanism.

2. Material and Method

2.1. Humic acid samples

Three HA sample used in this research, 2 of them are commercial HA obtained from Acros Organic, USA (HA-AC) and Wako Chemical, Japan (HA-WK) and the other was extracted from carbonaceous mudstone of tertiary coal strata, Mesuji, Lampung, Indonesia (HA-MJ). Before used further, these HA samples were purified according to the protocol by Koopal et al. [10]. In brief, 10 g HA was dissolve in 1 L sodium hydroxide 0.1 M and stirred for 12 h. Mineral and undissolved materials were separated from supernatant by centrifugation. HA in supernatant phase was precipitated by changing the pH to 2 using hydrochloric acid and finally the HA was recovered by centrifugation and dried at 50°C.

2.2. Functional group determination

HA functional groups were determined by titration method according to Khil'ko et al. [11]. Total acidity (exchangeable proton) was quantified using barium method. Weighed HA sample was introduced into flask containing 50 ml barium hydroxide 0.1 N, stirred for 18 h and settled for 4 days. Five ml supernatant solution was taken and diluted to 100 ml using boiled distillate water. This mixture was titrated by 0.1 N hydrochloric standard solutions by using phenolphthalein indicator to detect the end point. Blank titration was also carried out using distillate water instead of HA sample. Total acid acidity was calculated using following equation (1).

$$[H^+] = \frac{(V_0 - V_{HCl})M_{HCl}V_{Ba}}{mV_a} \quad (1)$$

Where $[H^+]$, V_0, V_{HCl} , M_{HCl} , V_{Ba}, m, V_a are total acidity (meqiv/g), volume of hydrochloric acid used for blank solution titration (ml), volume of hydrochloric acid used for HA sample solution titration, hydrochloric acid concentration (meqiv/ml), initial volume of barium hydroxide (ml), HA in aliquot portion (g) and aliquot volume used for each titration (ml), respectively.

In carboxylic determination, weighed HA sample was mixed with 10 ml sodium acetate 0.1 M and 40 ml boiled distillate water. The mixture was homogenized at 100°C for 4o min. The precipitate formed was filtered and washed thoroughly using 100 ml boiled distillate water and liberated acetic acid was titrated using potassium hydroxide 0.05 N, using phenolphthalein indicator to detect end point. Carboxylic content was calculated using following equation.

$$[COOH] = \frac{V_{KOH}M_{KOH}V_1}{mV_2} \quad (2)$$

Where [COOH], V_{KOH}, M_{KOH}, m, V_1 and V_2 are carboxylic content (mequiv/g), volume of potassium hydroxide used for titration (ml), potassium hydroxide concentration (mequiv/ml), HA in each aliquot portion (g), total volume (152 ml) and aliquot volume for each titration (ml), respectively. The hydroxyl content [OH] of HA sample was obtained by subtraction of total acidity with carboxylic content (3).

$$[OH] = [H^+] - [COOH] \quad (3)$$

2.3. Adsorption capacity determination

Adsorption capacity of each HA sample was carried out using batch method. Before used further in the experiments, HA needs to be stabilized by attaching it to the silica gel particle through silylation process according to Prado et al. [12] as shown in Figure 1. This step was necessary considering HA solubility in aqueous phase. The adsorbent synthesized using HA-AC, HA-WK and HA-MJ are referred as SiHA-AC, SiHA-WK and SiHA-MJ, respectively.

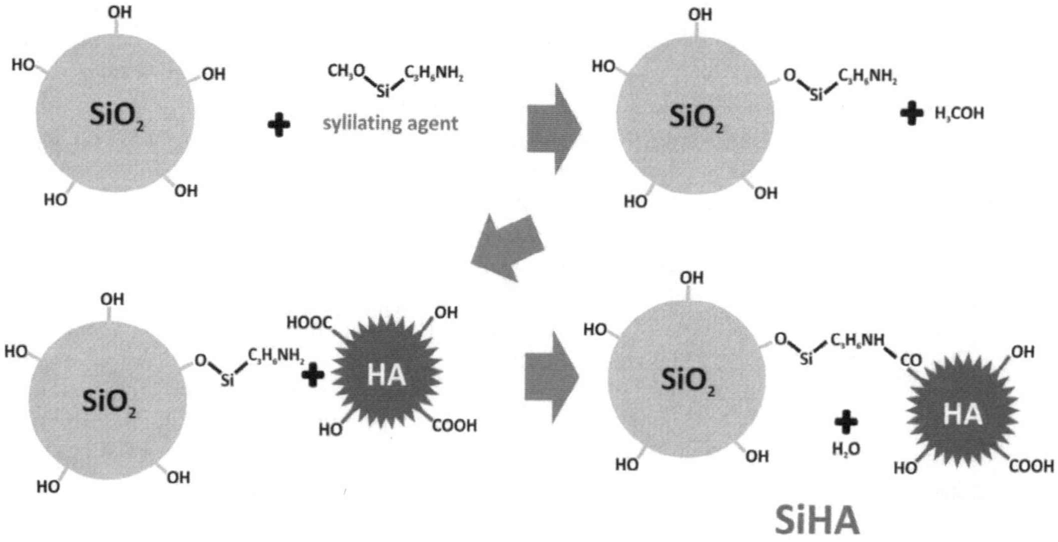

Figure 1. Silylation process to attach HA on silica gel surface (modified from Prado et al. [12])

Batch experiment to determine adsorption capacity was done by equilibrating 10 mg of SiHA adsorbent with 5 ml uranyl (UO_2^{2+}) or thorium (Th^{4+}) 250 mg/L (pH 3). The mixture was equilibrated for 24 h and supernatant solution was separated by centrifugation. Optimum parameters such as pH, solid-liquid ratio, uranyl or thorium concentration and equilibrium time for adsorption studies were set according to the previous research [9]. Uranyl and thorium concentration in liquid phase before and after equilibration were measured spectrophotometrically using Arsenazo III as chromogenic reagent using Jasco V-530 UV/Vis spectrophotometer. Adsorption capacity was calculated using following equation (4).

$$q = (C_0 - C_E)\,{}^{V}\!/_{m} \quad (4)$$

where q, Co, CE, V and m are adsorption capacity (mmol/g), uranyl or thorium initial concentration (mmol/L), uranyl or thorium equilibrium concentration (mmol/ml), volume of liquid phase (ml) and adsorbent added (g), respectively.

2.4. Humic acid and adsorbent characterization

HA and Si-HA characterization were carried out using FT-IR in order to determine the functional group present in the surface of HA and SiHA, also to clarify the differences among three HA samples and SiHA adsorbents. Other characterization is SEM to reveal the morphology of HA and adsorbent particle, also to confirm the silica gel surface modification by humic acid. FT-IR data was obtained using Jasco FT/IR 4100 spectrometer by the KBr method with a wavelength range of 400 to 4000 cm^{-1}, while surface morphology was captured using Hitachi S-2400 scanning electron microscope operated at 15 kV after gold coating in Graduate School of Environmental Science, Hokkaido University.

IOP Conf. Series: Materials Science and Engineering **285** (2017) 012011 doi:10.1088/1757-899X/285/1/012011

3. Results and discussions

3.1. Humic acid samples and their purification

HA as previously explained is derived from organic material (especially plant material) degradation [13]. Generally, commercial HA is extracted from compost (recent age) i.e. HA-AC and HA-WK, while HA-MJ was extracted from organic matter of carbonaceous mudstone with tertiary age (2.58 – 65 million years ago). The organic matter had been through biodegradation process by microorganism in early phase and sedimentation or diagenesis (temperature and pressure) in advance phase. Since derived from compost, organic matter as precursor only experienced biodegradation while HA-MJ, besides decomposition by microorganism, the organic matter had been subject to high temperature and pressure. Advance phase would affect HA structure through cracking process of organic compound with high molecular mass [14] into product with lower mass, and it is assumed that HA-MJ would have lower molecular mass and more functional group due to cracking process (Figure 2).

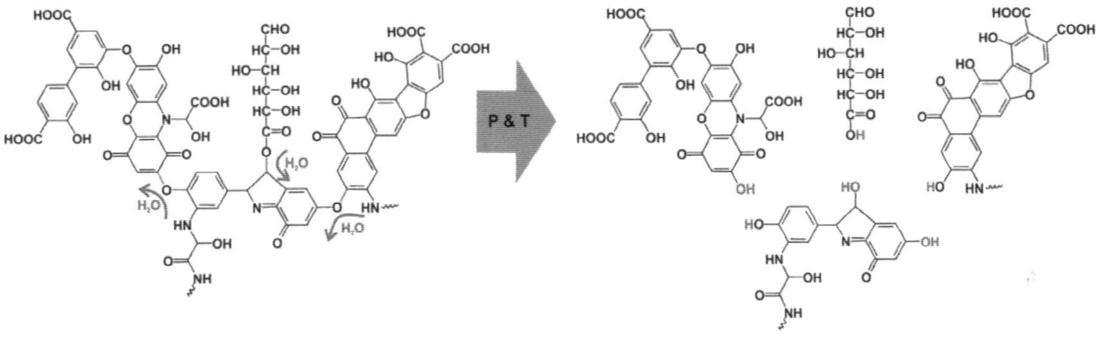

Figure 2. Cracking of humic substance through increasing pressure (P) and temperature (T)

3.2. Functional group determination

Functional group determination (total acid and carboxylic) was carried out while hydroxyl was obtained from the subtraction of total acidic by carboxylic fraction. The results are shown in Table 1.

Table 1. The functional group (total acidic, carboxylic and hydroxyl) present in each HA sample and HA-silica gel adsorbent

Sample	Total acidity (meq/g)	Carboxylic (meq/g)	Hydroxyl (meq/g)
HA-AC	5.35±0.16	3.54±0.05	1.81
HA-WK	4.47±0.00	2.37±0.06	2.10
HA-MJ	4.73±0.00	4.48±0.11	0.24
SiHA-AC	4.59±0.00	0.28±0.00	4.31
SiHA-WK	3.86±0.06	0.44±0.03	3.42
SiHA-MJ	4.23±0.00	0.28±0.05	3.95

Table 1 shows significant difference in case of total acidity (total exchangeable proton) for two commercial HA samples while HA-MJ value of total acidity close to the value of HA-WK. Total acidity for each HA sample mostly contributed by carboxylic functional group. In this respect there is contrast between commercial humic acid (HA-AC and HA-WK), which carboxylic group contributed 66% and 53% respectively to the total acidity compared to the value of HA-MJ. In the tertiary humic acid sample, 95% of total acidity comprise of carboxylic acid. This is probably caused by biodegradation and advance process (diagenesis), which leads the ester or fatty acid of organic material as HA precursor were transformed (cracked) into carboxylic acid (Figure 2).

After modification of HA by silica gel to form SiHA adsorbent, the total acidity decrease around 15% from the original HA sample values. As predicted, carboxylic content in each adsorbent significantly decrease (90 % of initial value), since this group reacted with amine group during silylation process (Figure 1) to form peptide bonding which attached the humic acid to the silica gel surface. Interestingly, the hydroxyl group was increasing after HA modification. This is due to the presence of hydroxyl of silanol group on the surface of silica gel (Figure 1).

3.3. Adsorption capacity determination

Adsorption capacity of HA modified silica gel to remove uranyl and thorium from aqueous solution using batch method had been carried out and the results are shown in Table 2.

Table 2. Adsorption capacity of HA modified silica gel (SiHA). Carb and hyd mean carboxylic and hydroxyl capacity for each ion, respectively

Adsorbent	UO_2^{2+} (mmol/g)	Carb. UO_2^{2+} (mmol/g)	Hyd. UO_2^{2+} (mmol/g)	Th^{4+} (mmol/g)	Carb. Th^{4+} (mmol/g)	Hyd. Th^{4+} (mmol/g)
SiHA-AC	0.10 ±0.01	0.14	2.16	0.14± 0.00	0.07	1.08
SiHA-WK	0.07 ±0.00	0.22	1.71	0.13 ±0.00	0.11	0.86
SiHA-MJ	0.11 ±0.02	0.24	1.98	0.18 ±0.18	0.07	0.96

Based on Table 2, all HA modified silica gel adsorbents had more affinities toward thorium than to uranyl ions, which are shown by the capacity values. For each ion species, HA-MJ modified silica gel (SiHA-MJ) possesses highest capacity compared to the both commercial HA modified adsorbent. Interestingly, statistic analysis using t-test two-sample assuming unequal variance revealed insignificant difference among three HA acid sample used for uranyl and thorium removal in terms of adsorption capacity. Stochiometrically, by using functional group data in Table 1 and adsorption capacity for uranyl and thorium in Table 2, the maximum theoretic capacity of each adsorbent could be calculated. To bind a uranyl ion, the adsorbent has to provide two carboxylic or two hydroxyl groups and in the case of thorium, four carboxyl or hydroxyl are required to bind a thorium ion (Figure 3).

Based on dissociation constant data, carboxyl would release proton at pH 4.28 [15], while hydroxyl which consists of phenolic or silanol would be dissociated at higher pH at basic condition between 10 and11, hence in this experiment condition (pH 3), it could be said that only carboxyl group plays important role in binding mechanism. Based on carboxyl capacity for uranyl ion in Table 2, the carboxyl capacity is higher compared to uranyl ion adsorbed. This indicated that not all carboxyl groups on the adsorbent could bind uranyl ion due to incomplete dissociation at pH 3. In contrast with thorium, carboxyl capacity is lower compared to the adsorption capacity, which inferred that all carboxylic had been bound to thorium ion, and fraction of thorium ion was bound electrostatic ally to the hydroxyl group. According to these results, it could be hypothesized that two interaction mechanisms occurred: ion exchange and electrostatic force. Interaction to bind uranyl ion was exclusively contributed by ion exchange, while in the case of thorium, the interaction was dominated by ion exchange and to lesser extent contributed by electrostatic force. The ability of thorium ion to form electrostatic binding is partly caused by ionic radii and charge factors [5, 8].

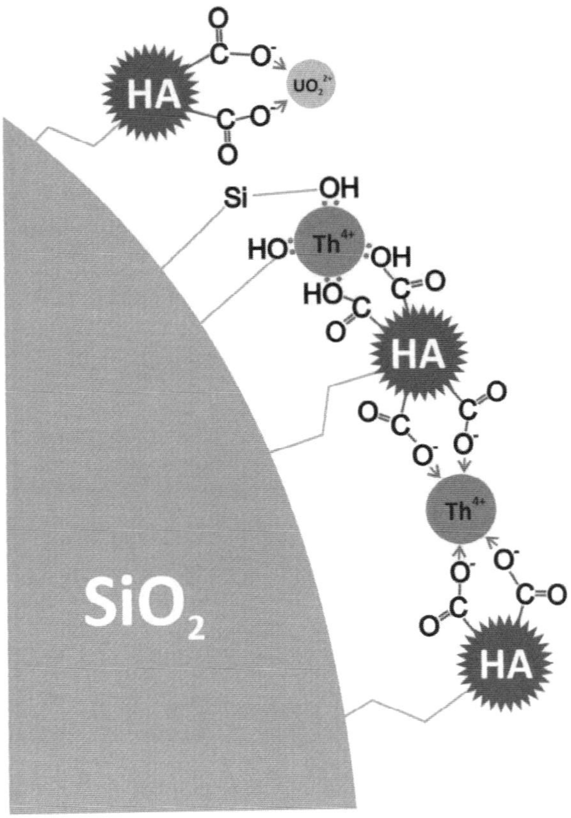

Figure 3. Possible adsorption mechanism, bonding between ligand (functional group) and target ion (uranyl and thorium), double red dot and red arrow signify the electrostatic interaction and ion exchange respectively

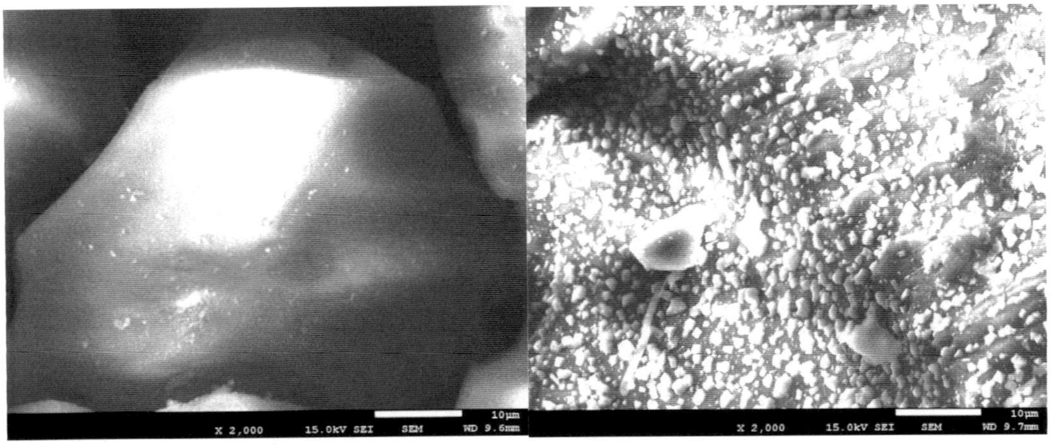

Figure 4. SEM image of unmodified silica gel (top) and humic acid modified silica gel (SiHA, bottom)

3.4. Humic acid and adsorbent characterization
Characterization using FT-IR and SEM was done to confirm the adsorbent formation by comparing the raw materials (silica gel and HA) with the product (SiHA adsorbent). The results of SEM characterization are shown in Figure 4.

Figure 4 show the difference between silica gel and SiHA in terms of surface morphology. Silica gel shows smooth surface morphology and the modification is marked by surface encrustation. FT-IR was done to prove the presence of certain functional group and to confirm the surface modification. The results in Figure 5 show that silica gel (SiO_2) reveals the peak at 796 cm^{-1} (tetrahedral silica matrix), adsorption at 1080 cm^{-1} (Si-O bonding) and 954 cm^{-1} (silanol functional group). HA modified silica gel (SiHA) spectra confirm that SiHA is hybrid material between HA and silica gel through the presence of carboxyl group (3500-3800 cm^{-1}) of HA, while silica gel spectra are still retained at 796 and 1080 cm^{-1}. Two commercial HA samples based on Figure 5 show no significant deference in terms of FT-IR spectra.

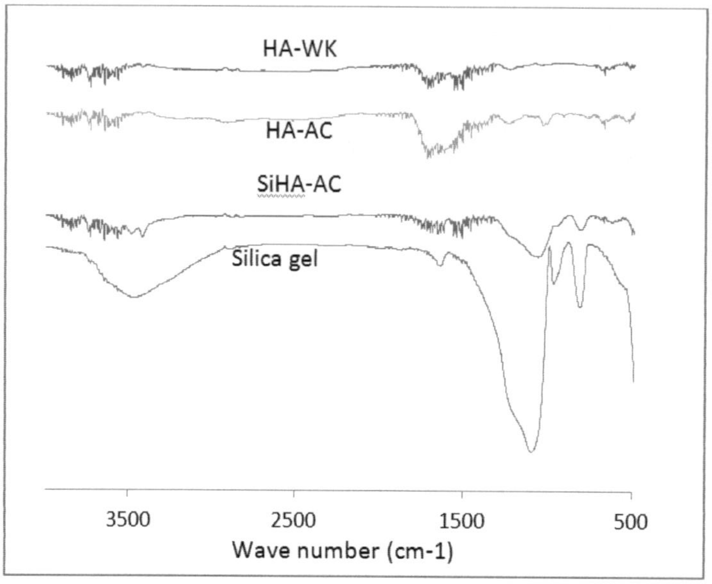

Figure 5. FT-IR spectra of Silica gel, SiHA-AC, HA-AC and HA-WK

4. Conclusions
Three commercial humic acids had been purified and used in adsorbent synthesis. The adsorbents were used in uranyl and thorium removal. Based on titration analysis, total acidity for each HA sample mostly contributed by carboxylic functional group. In this respect there is contrast between commercial humic acid (HA-AC and HA-WK), compared to the value of HA-MJ. After modification of HA by silica gel to form SiHA adsorbent, the total acidity decrease and as predicted, carboxylic content in each adsorbent significantly decrease by 90 % of initial value, while hydroxyl group was increasing due to the presence of hydroxyl of silanol group on the surface of silica gel. All SiHA adsorbents had more affinities toward thorium than to uranyl ions. For each ion species, HA-MJ modified silica gel (SiHA-MJ) possesses highest capacity compared to the other adsorbents. However, statistic analysis using t-test two-sample assuming unequal variance revealed insignificant difference among three HA acid sample used for uranyl and thorium removal in terms of adsorption capacity. Stochiometrically, by using functional group data, it could be hypothesized that two interaction mechanisms occurred. Interaction to bind uranyl ion was exclusively contributed by ion exchange, while in the case of thorium, the interaction was dominated by ion exchange and to lesser extent

contributed by electrostatic force. Characterization using FT-IR and SEM confirmed the adsorbent formation, while FT-IR reveals no significant deference between HA samples.

Acknowledgement
The research is supported by MEXT scholarship of Japan for foreign students and under supervision of Kazuhiro Toyoda of Graduate School of Environmental Science, Hokkaido University. All instrumentation and characterization were performed in Graduate School of Environmental Science, Hokkaido University.

References

[1] Anirudhan T S, Bringle C D, Rijith S 2010 *J. Environ. Radioactiv.* **101** 267-76
[2] Gasser M S 2007 Sep. Purif. Technol. **53** 81-8
[3] Stevenson F J 1994 *Humus Chemistry: Genesis, Composition, Reactions* New York: John Wiley & Sons
[4] Davies G, Fataftah A, Cherkasskiy A, Ghabbour E A, Radwan A, Jansen S A, Kolla S, Paciolla M D, Buermann W, Balasubramanian M, Budnick J, Xing B 1997 *J. Chem. Soc. Dalton Trans.* 4047-60
[5] Erdogan S, Merdivan M, Hamamci C, Akba O, Baysal A 2004 *Anal. Lett.* **37** 2565-75
[6] Prado A G S, Miranda B S, Dias J A 2004 *Colloids and Surf A: Physicochem. Eng. Asp.* **242** 137-43
[7] Seki H, Suzuki A 1995 *J. Colloid Interf. Sci.* **171** 490-4
[8] Khalili F, Al-Banna G 2015 *J. Environ. Radioact.* **146** 16-26
[9] Prasetyo E, Toyoda K 2016 *J. Radioanal. Nucl. Chem.* **310** 69-80
[10] Koopal L K, Yang Y, Minnaard A J, Theunissen P L M, Riemsdijk W H V 1998 *Colloids and Surf. A: Physicochem. and Eng. Asp.* **141** 385-95
[11] Khil'ko S L, Kovtun A I, Rybachenko V I *2011 Solid Fuel Chemistry* **45** 337-48
[12] Prado A G S, Miranda B S, Dias J A 2004 *Colloids and Surf. A: Physicochem. and Eng. Asp.* **242** 137-43
[13] Ponomarenko E V, Anderson D W 2001 *Canadian Journal of Soil Science* **81** 285-97
[14] Diessel C F K 1992 *Coal-Bearing Depositional Systems* Berlin: Springer
[15] Andjelkovic T, Perovic J, Purenovic M, Blagojevic S, Nikolic R, Andjelkovic D, Bojic A 2006 *Eclet. Quim.* **31** 39-46

IOP Conf. Series: Materials Science and Engineering **285** (2017) 012012 doi:10.1088/1757-899X/285/1/012012

Ceramic membrane ozonator for soluble organics removal from produced water

U W R Siagian[1], A S Dwipramana[2], S B Perwira[2], Khoiruddin[2], I G Wenten[2]

[1]Department of Petroleum Engineering, Institut Teknologi Bandung, Jl. Ganesha 10, Bandung 40132, Indonesia
[2]Department of Chemical Engineering, Institut Teknologi Bandung, Jl. Ganesha 10, Bandung 40132, Indonesia

Email: igw@che.itb.ac.id

Abstract. In this work, the performance of ozonation for degradation of soluble organic compounds in produced water was investigated. Tubular ceramic membrane diffuser (with and without a static mixer in the lumen side) was used to facilitate contact between ozone and produced water. The ozonation was conducted at ozone flow rate of 8 L.min^{-1}, ozone concentration of 0.4 ppm, original pH of the solution, and pressure of 1.2 bar, while the flow rates of the produced water were varied (192, 378 and 830 mL.min^{-1}). It was found that the reduction of benzene, toluene, ethylbenzene, and xylene were 85%, 99%, 85%, and 95%, respectively. A lower liquid flow rate in a laminar state showed a better component reduction due to the longer contacting time between the liquid and the gas phase. The introduction of the static mixer in the lumen side of the membrane as a turbulence promoter provided a positive effect on the performance of the membrane diffuser. The twisted static mixer exhibited the better removal rate than the spiral static mixer.

1. Introduction

Produced water is the largest waste stream co-produced during oil extraction in oil production operations [1,2]. The properties of produced water depend on geological formation, geographical location of the reservoir, and residual production chemicals [3]. The major compounds of produced water include dissolved and dispersed oil compounds, dissolved formation minerals, production solids, dissolved gases, and production chemical compounds [4]. The most soluble organic compounds in produced water are BTEX (benzene, toluene, ethylbenzene, and xylenes) and phenols [5]. Organic acids such as formic acid and propionic acid, aliphatic hydrocarbons, phenols, carboxylic acid, and low molecular weight aromatic compounds are also included as soluble compounds [1,6]. Dissolved organics, especially BTEX, are necessary to be removed when the option of produced water management is discharge, reuse, or consume in beneficial use [7]. BTEX is considered a serious problem since it has some acute and long term toxic effects [8]. Thus, numerous studies and methods have been explored in attempts to remove soluble organic compounds from produced water [9–12].

Ozone as a strong oxidant is easily dissolved in water and is able to dissociate soluble toxic organic compounds into harmless compounds. The aqueous reaction between ozone and substances to be treated is known to be a complex process, including the direct reaction of molecular ozone and indirect reaction in which OH radicals formed as an oxidizing agent during ozone auto-decomposition [13]. Because of strong oxidation properties, ozone has become a superior agent for disinfection, detoxification, deodorization, and organic removal of water and wastewater [14–16]. As a strong oxidant, ozone can be used to treat a dissolved organic with low concentration (e.g. BTEX in produced water) effectively which can hardly be achieved by other methods. Ozone ability to oxidize organics matter makes it compatible to treat produced water, which contains many organic compounds.

IOP Conf. Series: Materials Science and Engineering **285** (2017) 012012 doi:10.1088/1757-899X/285/1/012012

It has been found that the main limitation in ozonation system is the low mass transfer rate of ozone from the gas to the liquid phase [17,18]. The mass transfer efficiency depends on the mixing characteristics of the gas-liquid contactor used, the kinetics of ozone reaction in the water, and the number and size of bubbles produced [17–19]. The efficacy can be increased by creating a higher surface-area-to-volume ratio for the contact of ozone with the solution through the generation of smaller bubbles [20]. A smaller bubble size results in a higher surface-to-volume ratio. In addition, smaller bubbles have higher residence times in contactor, leading to higher gas volume fractions [21]. The floating of bubbles and the possible additional mechanical agitation of the liquid create shear forces at the boundary layer which improve the local mass transfer and also keep the bubbles from coalescing [22].

Membrane technology has been considered as the promising alternative for conventional technologies in various processes including separation, purification, chemical reaction, and contacting processes [23–34]. Membrane technology offers interesting features such as lower energy consumption, relatively low operating expenditure, smaller footprint, easy to scale-up, and easy to operate [35–38]. Recently, the use of a membrane as a contacting device between two process streams has been widely studied [39–41]. Ozonation through membrane contactor has been used in several studies for organics treatment such as ozonolysis of hazardous organic compounds [42], decolorization of NOM and indigo [43], and degradation of organic pollutants in the wastewater [14]. The inorganic membranes are more suitable for ozone contactor rather than polymeric membranes because of its higher durability and oxidation resistance [44].

Although ozonation process has been applied in many wastewater treatments, the use of ozonation process to remove soluble organic compounds from produced water still has not been investigated in detail. The aim of this study is to investigate the ozonation of soluble organic compounds from produced water. A ceramic membrane module was used as a contactor (ozonator) to increase the ozone mass transfer rate during the ozonolysis of soluble organic compounds.

2. Materials and method

Chemicals used in this experiment involved benzene, toluene, ethylbenzene and xylene as a model solution. In addition, acetonitrile (CH_3CN) was used as solvent for UV spectrophotometer analysis. Tubular ceramic membrane was used as diffuser in ozonation process (ozonator). The specification of the membrane is shown in Table 1. The High Efficiency Ozonizer (O_3&W OWA-200; AC 110/220 V 50/60 Hz) was used to generate ozone.

Table 1. Membrane specification

Parameter	Description
Material	Ceramic membrane
Module type	Tubular
Module number	1
Membrane length (cm)	34
Number of tube	7
Tube diameter (mm)	4
MWCO (kDa)	100
Surface area (m^2)	0.1

The experimental set up used in this study is depicted in Figure 2. BTEX solution was introduced into the membrane lumen and air stream contains ozone was passed through the shell side in dead-end mode. Ozone gas passed through membrane pores from shell side toward membrane lumen side creating micro bubbles. The contact between ozone and BTEX solution occurs in the bubbles surface, where the micro bubble can be assumed to be completely soluble in the liquid state inside membrane lumen. The gas flow rate is set at 8 L.min-1, 0.4 ppm of ozone concentration, and 1.2 bar of pressure. The flow rates of produced water were varied at 192, 378, and 830 $mL.min^{-1}$ to determine the effect of

IOP Conf. Series: Materials Science and Engineering **285** (2017) 012012 doi:10.1088/1757-899X/285/1/012012

different liquid capacity on the removal of BTEX. The effluent of ceramic membrane reactor was treated into separator vessel to give an adequate degassing time for the gas to be separated from the liquid. Unreacted residue gas was discharged from the top of separator while the effluent liquid was recirculated into the membrane diffuser system. A static mixer was introduced into membrane lumen to provide better hydrodynamic and increase the performance of BTEX reduction. The effect of static mixer addition was determined by comparing the BTEX reduction for each configuration at constant flow rate of 378 ml min^{-1}. Two different shapes of the static mixer, namely spiral and twisted static mixer, were used in this study. The shapes of both static mixers are shown Figure 1.

BTEX concentration was analyzed with UV Spectrophotometer based on Thomas method (Thomas and Burgess, 2007). BTEX concentration has been analyzed using UV spectra with the wave length of 256, 262, and 264 nm respectively. The standard solution used in UV spectrophotometer was prepared from 5%-volume of CH3CN (Acetonitrile) solution. This standard solution was also used as a solvent for the sample with the ratio of solvent to sample was 1:1, in order to compare the BTEX concentration with the pure standard solution.

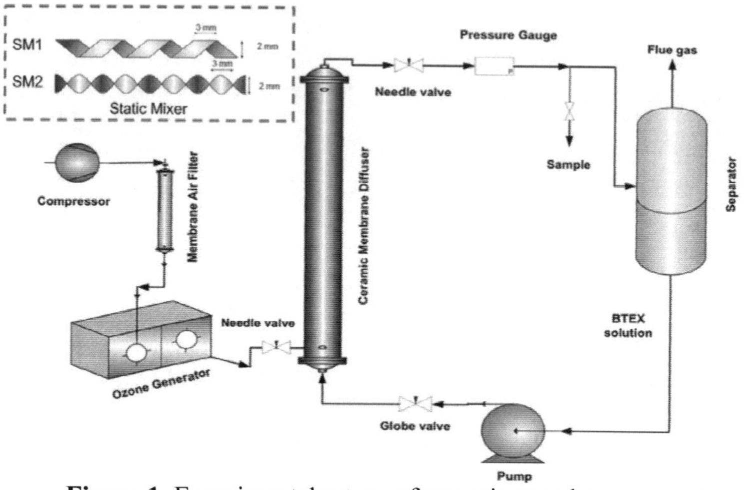

Figure 1. Experimental set-up of ceramic membrane ozonator

3. Results and discussion

3.1. Removal of BTEX

The contact of ozone and BTEX solution inside lumen membrane leads to the reaction of BTEX dissociation into the harmless compounds. The reduction of BTEX concentration within solution has been observed and the result is shown in Figure 2. The initial benzene concentration was 4095 ppm, which is reduced to the final concentration of 557 ppm after 250 minutes of ozonation. Total reduction of Benzene concentration reaches 86.4%, with the highest reduction rate occurred at 120 until 210 minutes. Toluene concentration decreased from initial concentration of 937 ppm to the final concentration of 6 ppm, with the total reduction of toluene reaches 99.4%. At the same time, ozonation process performance within membrane diffuser for ethylbenzene reduction reaches 85% for 94 ppm initial concentration. As for xylene concentration with the initial value of 4024 ppm, the concentration reduction after 210 minutes of ozonation process reaches 95%.

From Figure 2 it is observed that the concentration reduction of each component shows a similar trend. The difference in reduction rates is mainly due to the different initial concentration of the component in the solution. Higher initial concentration leads to the faster reduction of component following the reaction equilibrium rule. In addition, ethyl benzene is easier to destroy compared to the other organics. It is because ethyl benzene has ethyl groups attached to the benzene, which may be easier to cut by oxidant than the aromatic ring.

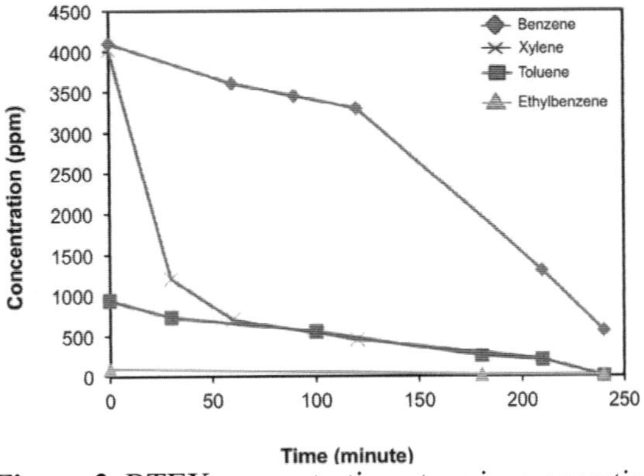

Figure 2. BTEX concentration at various operating time

3.2. The effect of liquid flow rate

In order to determine the effect of feed flow rate into the reduction performance of the BTEX component, three different flow rates were introduced into membrane diffuser, and the removal of toluene was observed for each flow rates. Toluene was chosen as the model for the following study because it has the best percent reduction. The flow rates of the liquid phase were 192 (Q_1), 378 (Q_2), and 830 ml/minute (Q_3). The effect of different feed flow rate on toluene reduction can be seen in Figure 3.

As shown in **Figure** *3*, higher feed flow rate leads to lower toluene reduction rate. It was found that the toluene reduction rate follows the following order: $Q_1 > Q_3 > Q_2$. It can be explained by the decrease of contacting time between the liquid and the gas phases as the flows increase. As a result of short contacting time in a higher flow rate, the longer time is needed for the reaction to reach its equilibrium. Thus, for a larger feed capacity (larger produced water stream) a higher membrane area will be required. However, flow rate Q_3 showed the better performance compared to Q_2 although Q_3 is higher. It could be associated with the increased Reynolds number at the higher flow rate. When the liquid stream on a turbulence regime, a better dispersion and mixing of gas in the liquid could be achieved. As the results, a better reaction of ozone and toluene was obtained.

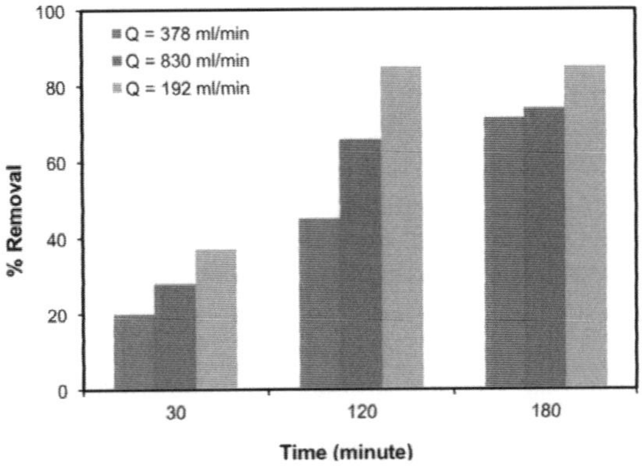

Figure 3. Removal of toluene at various liquid flow rates and operating time

3.3. The effect of static mixer

The main limitation in ozonation system is the low mass transfer rate of ozone from the gas phase to the liquid phase. In order to increase ozone mass transfer rate, higher turbulent flow is desired. The effect of static mixer presence on the lumen of ceramic membrane diffuser to the removal of toluene is shown in Figure 4. Figure 4 shows that the addition of static mixer into the lumen of ceramic membrane diffuser can increase toluene reduction rate. The liquid turbulent flow can increase the ratio of surface area to ozone volume within the solution.

Figure 4 also shows that SM2 (twisted static mixer) generate better toluene reduction performance compared with SM1 (spiral static mixer). The ability of twisted static mixer to increase the mass transfer rate of ozone from gas phase to liquid phase was better than spiral static mixer due to the higher turbulence induced by a spiral fluid channel.

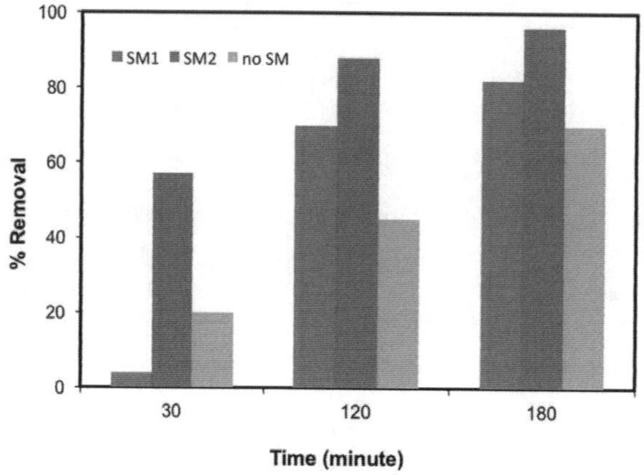

Figure 4. Toluene removal at various static mixer types and operating time

4. Conclusions

The performance of ozonation through ceramic membrane diffuser was investigated during the removal of BTEX from a synthetic produced water solution. The effect of ozonation time, liquid flow rate, and static mixer were investigated. The removal rate of benzene, toluene, ethyl benzene and xylene were 86%, 99%, 85%, and 95%, respectively (ozonation time: 250 min.; liquid flow rate: 378 mL.min^{-1}; without static mixer). A lower liquid flow rate in a laminar state showed a better component reduction due to the longer contacting time between the liquid and the gas phase. The introduction of static mixer in the lumen side of the membrane as the turbulence promoter provided a positive effect on the performance of the membrane diffuser. The twisted static mixer exhibited the better removal rate than the spiral static mixer.

References

[1] Fakhru'l-Razi A, Pendashteh A, Abdullah L C, Biak D R A, Madaeni S S, Abidin Z Z 2009 *J. Hazard. Mater.* **170** 530–51
[2] Silalahi S H D, Leiknes T 2009 *Desalination* **236** 160–9
[3] Bader M S H 2007 *Desalination* **208** 159–68
[4] BR H, SR D 1994 *Chem. Eng. Res. Des.* **72** 176–88
[5] Ekins P, Vanner R, Firebrace J 2007 *J. Clean. Prod.* **15** 1302–15
[6] Stephenson M T 1992 *A Survey of Produced Water Studies BT - Produced Water: Technological/Environmental Issues and Solutions* ed J P Ray and F R Engelhardt (Boston, MA: Springer US) pp 1–11
[7] Arthur J D, Langhus B G, Patel C 2005 *Technical summary of oil & gas produced water treatment technologies* (Tulsa, OK: All Consult. LLC)

[8] Mitra S, Roy P 2011 *Res. J. Environ. Sci.* **5** 394–8
[9] Tellez G T, Nirmalakhandan N, Gardea-Torresdey J L 2002 *Adv. Environ. Res.* **6** 455–70
[10] Fakhru'l-Razi A, Pendashteh A, Abidin Z Z, Abdullah L C, Biak D R A, Madaeni S S 2010 *Bioresour. Technol.* **101** 6942–9
[11] Liu N, Li L, McPherson B, Lee R 2008 *J. Memb. Sci.* **325** 357–61
[12] Cha Z, Lin C-F, Cheng C-J, Andy Hong P K 2010 *Chemosphere* **78** 583–90
[13] Staehelin J, Hoigne J 1985 *Environ. Sci. Technol.* **19** 1206–13
[14] Guha A K, Shanbhag P V, Sirkar K K, Vaccari D A, Trivedi D H 1995 *AIChE J.* **41** 1998–2012
[15] Selcuk H, Meric S 2006 *Glob. NEST J* **8** 95–102
[16] Magara Y, Itoh M, Morioka T 1995 *Prog. Nucl. Energy* **29** 175–82
[17] Gurol M D, Singer P C 1983 *Water Res.* **17** 1173–81
[18] Fuchun X, Cunli L 1990 *Sci. Eng.* **12** 269–79
[19] Hongde Z, W S D, J S S 1994 *J. Environ. Eng.* **120** 821–40
[20] Sharma M M 1993 *Chem. Eng. Res. Des.* **71** 595–610
[21] Klasson K T 2002 *Ozone treatment of soluble organics in produced water* (ORNL Oak Ridge National Laboratory (US))
[22] Picard C, Larbot A, Sarrazin J, Janknecht P, Wilderer P 2001 *Ann. Chim. Sci. des Matériaux* **26** 13–22
[23] Drioli E, Curcio E 2007 *J. Chem. Technol. Biotechnol.* **82** 223–7
[24] Wenten I G, Dharmawijaya P T, Aryanti P T P, Mukti R R, Khoiruddin 2017 *RSC Adv.* **7** 29520–39
[25] Ariono D, Purwasasmita M, Wenten I G 2016 *J. Eng. Technol. Sci.* **48** 367–87
[26] Khoiruddin, Widiasa I N, Wenten I G 2014 *J. Food Eng.* **133** 40–45
[27] Strathmann H, Grabowski A, Eigenberger G 2013 *Ind. Eng. Chem. Res.* **52** 10364–79
[28] Rathore A S, Shirke A 2011 *Prep. Biochem. Biotechnol.* **41** 398–421
[29] Wenten I G, Widiasa I N 2002 *Desalination* **149** 279–85
[30] Ariono D, Khoiruddin, Subagjo, Wenten I G 2017 *Mater. Res. Express* **4** 24006
[31] Wenten I G, Khoiruddin 2016 *J. Eng. Sci. Technol.* **11** 916–34
[32] Khoiruddin, Aryanti P T P, Hakim A N, Wenten I G 2017 *AIP Conf. Proc.* **1840** 90006
[33] Purwasasmita M, Kurnia D, Mandias F C, Khoiruddin, Wenten I G 2015 *Food Bioprod. Process.* **94** 180–6
[34] Khoiruddin K, Hakim A N, Wenten I G 2014 *Membr. Water Treat.* **5** 87–108
[35] Ma L, Dong X, Chen M, Zhu L, Wang C, Yang F, Dong Y 2017 *Membr.* **7**
[36] Yang X, Wang R, Fane A G, Tang C Y, Wenten I G 2013 *Desalin. Water Treat.* **51** 3604–27
[37] Wardani A K, Hakim A N, Khoiruddin, Wenten I G 2017 *Water Sci. Technol.* **75** 2891–9
[38] Wenten I G, Khoiruddin, Arfianto F, Zudiharto 2013 *Desalination* **314** 109–14
[39] Himma N F, Anisah S, Prasetya N, Wenten I G 2016 *J. Polym. Eng.* **36** 329–62
[40] Purwasasmita M, Nabu E B P, Khoiruddin, Wenten I G 2015 *J. Eng. Technol. Sci.* **47** 426–46
[41] Wenten I G, Julian H, Panjaitan N T 2012 *Desalination* **306** 29–34
[42] Shanbhag P V, Guha A K, Sirkar K K 1995 *J. Hazard. Mater.* **41** 95–104
[43] Phattaranawik J, Leiknes T, Pronk W 2005 *J. Memb. Sci.* **247** 153–67
[44] Janknecht P, Wilderer P A, Picard C, Larbot A, Sarrazin J 2000 *Sci. Eng.* **22** 379–92

Characterization and utilization potential of basalt rock from East-Lampung district

K Isnugroho, Y Hendronursito and D C Birawidha

Research Unit for Mineral Processing - Indonesian Institute of Sciences
Jl. Ir. Sutami Km. 15, Tanjung Bintang, Lampung Selatan, Indonesia

Email: kusn005@lipi.go.id

Abstract. The aim of this research was to study the petrography and chemical properties of basalt rock from East Lampung district, Lampung province. Petrography analysis was performed using a polarization microscope, and analysis of chemical composition using X-RF method. From the analysis of basalt rock samples, the mineral composition consists of pyroxene, plagioclase, olivine, and opaque minerals. Basic mass of basalt rock samples is, composed of plagioclase and pyroxene with subhedral-anhedral shape, forming intergranular texture, and uniform distribution. Mineral plagioclase is colorless and blade shape, transformed into opaque minerals with a size of <0.2 mm, whereas pyroxene present among the blades of plagioclase, with a greenish tint looked and a size of <0.006 mm. Mineral opaque has a rectangular shape to irregular, with a size of <0.16 mm. The chemical composition of basalt rock samples, consisting of 37.76-59.64 SiO_2; 10.10-20.93 Fe_2O_3; 11.77-14.32 Al_2O_3; 5.57-14.75 CaO; 5.37-9.15 MgO; 1.40-3.34 Na_2O. From the calculation, obtained the value of acidity ratio (Ma)= 3.81. With these values, indicate that the basalt rock from East Lampung district has the potential to be utilized as stone wool fiber.

1. Introduction

Basalt is a common variety of volcanic rocks. It is usually fine-grained due to its rapid crystallization as lava on the Earth's surface. Basalt is mainly used in the form of crushed rock in construction, as well as in industrial and highway engineering[1]. Gabbro-basalt rocks are raw material for production gravel, building stones, mineral fiber, and derivated heat-insulation materials, as well as the products of cast stone material[2], [3]. In the last decade, basalt has emerged as a contender in the fiber reinforcement of composite. These fibers perform more highly than another type of fiber such as glass and carbon fiber. Basalt combines three silicate mineral plagioclase, pyroxene, and olivine. Plagioclase describes a number of triclinic feldspars that consist of sodium and calcium silicates. Pyroxenes are a group of crystalline silicates that contain any two of three metallic oxides, magnesium, iron or calcium. Olivine is a silicate that combines magnesium and iron $(Mg,Fe)_2SiO_4$[4].

Technological processing of basalt into basalt fiber, basalt cast, and basalt ceramic have been developed in some Eastern European countries, China dan Germany. In Indonesia, basalt rock reserves reached > 1 billion tons. Its presence there in various regions in Indonesia start from Sumatera, Java, Kalimantan, Sulawesi and Papua. In the province of Lampung, basaltrock is distributed in sub districts ofMataram baru, Jabung, Bumi Agung, Marga Tiga, Sukadana and Labuhan Maringgai in East Lampung district. The total resources are greater than 10 million m^3. The potential of basalt rock in this location, it has not been used optimally. Results of surveys in mining locations provided descriptions that these minerals were mined with simple methods by people around the locations. There were some companies conducting mining processes, but these mining processes are not done continually. Heavy equipment and processing tools in mining locations were very limited. Even a crushing process was conducted traditionally by using man power. The end products are split rocks, which is used for building or civil construction, thus the added value is very low. The aim of this research was to characterization of basalt rock from East Lampung district, Lampung province. The results of this characterization will determine what products will be produced and what technology it will be used. This research is expected to give an important contribution to the added value of basalt rock.

2. Methodology

Basalt rock characterized using polarizing microscopy and X-Ray Fluorescence (XRF) method. Six rock samples have been collected from sub district of Mataram Baru, East Lampung district, Lampung province. Thin sections were prepared as follow: a slab of basalt with 30 mm x 20 mm x 10 mm thick was cut, polished using 200, 400, 800 and 1000 silicon carbide powder, in order to get a smooth surface easy to stack with glass slide (26 x 42 mm) using Canada balsam adherent. An automated thin section machine called multiplied Grinder housed was used for making the slide. Approximately the thickness result was 45 microns. Then using heavy duty polishing table and hand polishing with 1000 silicon carbide powder, the sample was ground to 30 microns[5].The next stage is the analysis using an optical polarizing microscope. Nikon optical microscope is used particularly to identify mineral constituents of the studied basaltic rocks and to determine the mineralogical properties and textures.

XRF was performed using an XRF BRUKER S2 Ranger. The process begins with grinding samples until it reaches a size of 100 mesh. Then the sample is formed using a hydraulic tablet press. Samples are heated at temperatures of 1000 ° C in the crucible. Furthermore, the sample is analyzed based on the identification and enumeration of characters of X-rays caused by the photoelectric effect. The photoelectric effect occurs because the electrons in the target atom in the sample exposed to high-energy rays (gamma radiation, X-rays). As a result of the sample exposed to X-ray radiation, the electron vacancy occurs, while the electrons originating from the skin surface samples will fall to fill the vacancy. XRF will use X-rays emitted by the material and arrested detector to be used to analyze the content of the element in the sample. After getting the chemical composition using XRF method, calculated method was performed to determined the modulus of acidity (Ma), where are according to Eqs.1,

$$M_a = \frac{mSiO_2+mAl_2O_2}{mCaO+mMgO} \quad (1)$$

Where m is the mass content of oxide (%) [3]. The modulus of acidity is the basis for determining whether the basalt rocks that can be processed into basalt fiber, basalt casting or basalt ceramic.

3. Results and discussion

A geological condition of East Lampung was a plateau with an altitude of 0 to 50 meters. In the eastern part was formed from sediment as regosol, in a swamp area as lacustrine sediment (hydromorf soil), and in the rivers as alluvium surface sediment (a mixture of clay and galuh sand). While in the river, there was contained quartz sand. At that altitude, there is an acid lava flow from volcanic rocks, namely Lampung tuffa which covers most of the land area of East Lampung with latosol and podsolic. There are also rock form that belong to basalt type (pleistocene). At an altitude of 50 to 500 meters, there is Lampung tuffa material composed of volcanic sediment (Pleistocene) which is located further to the west was getting higher. In the northern part of this region, there are Palembang bottom formation and top formation. While in the Palembang formation layer contained in Sukadana region was marked by the appearance of acidic deposition tuffa. The Palembang layer is the starter (plateau) which is accompanied by intrusidesit covered by alluvium deposition such as volcanic sand, silt, and clay derived from volcanic ash. Geological formations found in this region include the Alluvium (Qal), deposition swamp (Qs), quartz (Qak), Sukadana basalt (Qbs), Terbanggi formation (QPT), and clay formation (Qti) as seen in figure 1. Alluvium formation (Qal) containss rock which is consist of lump, gravel, and sand. Furthermore, it contains soil, silt and clay. Qs containssand, silt, mud, clay and plant residue. Qak was a rock in the form of fine quartz sand, Qbs was a hollow basalt rock, and QPT was a sand stone with clay stone inserts while Qti was consist ofa pumice stone, clay and sand stones[6].

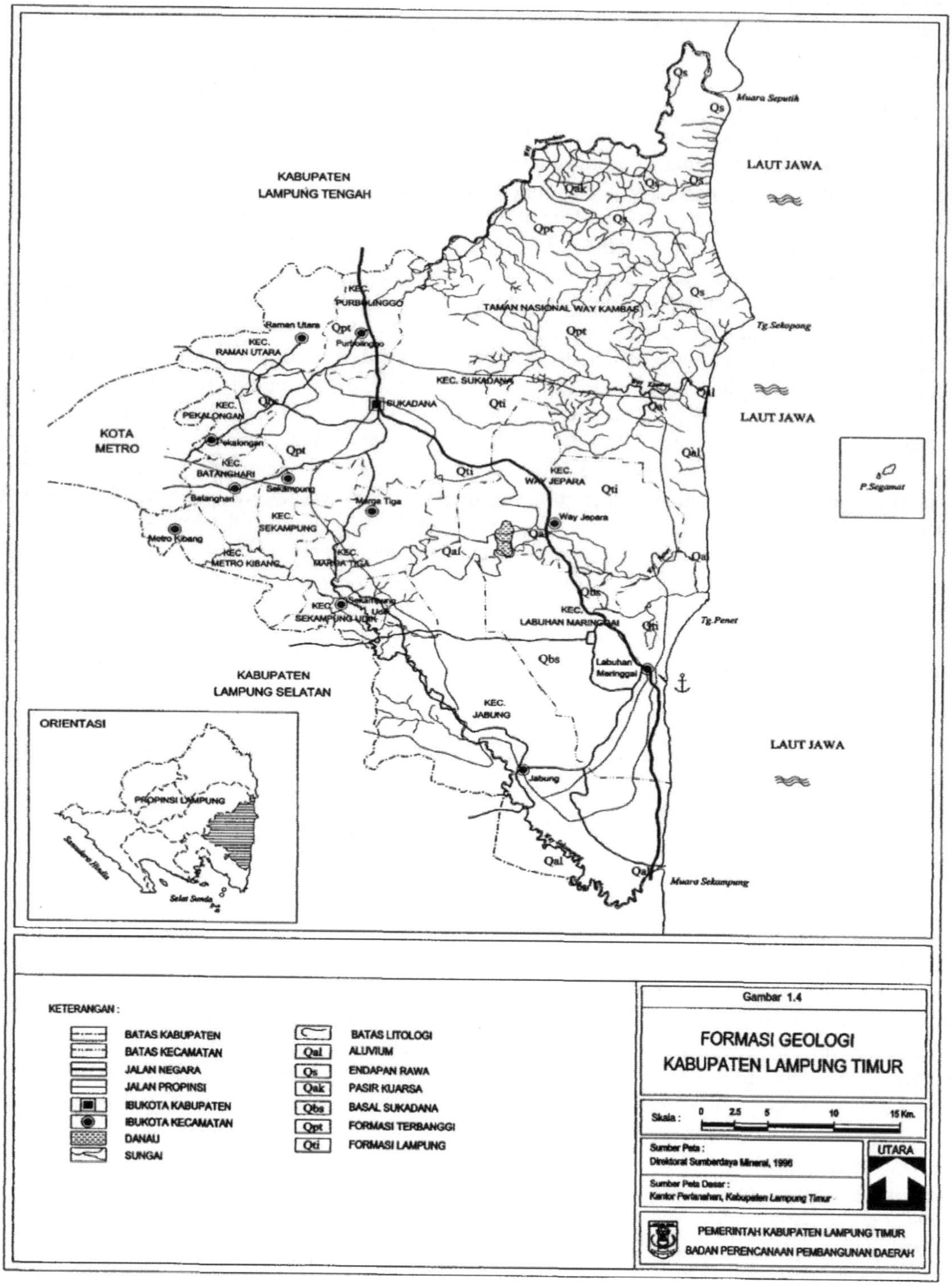

Figure 1. Geological formation map of East Lampung districts[6].

From the description of microscopic, thin section of rock had inter-granular texture holocrystaline, porphyritic, and consist of phenocrysts and groundmass. Phenocrysts composed of Olivine (Ol) with subhedral-anhedral form, with a size of 0.84 to 3.5 mm and an irregular fracture.Plagioclase was found and present as phenocrysts, with a maximum size of 0.8 mm. Groundmass composed by plagioclase (Pl) and pyroxene (Pi) has a form-anhedral subhedral and forming inter-granular texture with uniform distribution, as seen in figure 2. The appearance of plagioclase (Pl) is colorless, has a blade shape and transformed into opaque mineral, with a size of < 0.2 mm. Meanwhile, pyroxene (Pi) is present between the blades of plagioclase (Pl), greenish, moderate relief, with a size of < 0.006 mm. Isotropic opaque minerals in both the X-Nikol position and // - Nikol. Mineral opaque has a rectangular shape to irregular, with a size of < 0.16 mm. The main mineral consisting of plagioclase (Pl) 50%, olivine (Ol) 8%, pyroxene (Pi) 32%, mineral opaque 5%, and the common cavity 5%.

Figure 2. The appearance of petrography microstructure of basalt rock

The chemical compositions of the basalt rock from East Lampung district were analyzed using XRF method. The result of X-RF analysis was presented in table 1.

Table 1.X-RF analysis of basalt rock

Basalt	Mass Content
SiO_2	37,76-59,64
Al_2O_3	11,77-14,32
CaO	5,57-14,75
MgO	5,37-9,15
Fe_2O_3	10,1-20,93
K_2O	1,7-6,69
Na_2O	1,4-3,34
TiO_2	1,81-3,73

Based on the equation 1, the modulus of acidity of the raw materials from Mataram Baru sub district, East Lampung lies in the range (Ma) of 3.09 to 4.53 while the average value (Ma) is 3.81. The acidity modulus (Ma) of the basalt raw material is the main parameter to define the quality of the final fiber product. If the Ma <1.8, the fiber is considered to be a mineral wool for base materials are basic volcanic rock and cider. These fibers are brittle but have acceptable insulation properties. If the Ma > 1.8, the fiber is called rock wool and if its base materials are basalt then it is called basalt wool (basalt fiber, BF). The higher acidity module indicates that the basalt fibers are good quality with excellent properties, they have a high strength, corrosion resistance, temperature resistance, along with an extended operating temperature range[1].

4. Conclusion

Basalt rocks from Mataram Baru sub-district, Lampung Timur district, Lampung province, could have added value if it processed into products other than rubble stone and macadam. This rock has the potential to be utilized as stone wool fiber.

Acknowledgement

Authors acknowledge financial support and facilities from research unit of mineral processing, Indonesian Institue of Sciences and Darsono as basalt mine owner at Mataram Baru sub district, East Lampung for the assistance provided.

References

[1] Pisciotta A, Perevozchikov B V, Osovetsky B M, Menshikova E A and Kazymov K P 2014 *Nat. Resour. Res.*
[2] Drobot N F, Noskova O A, Steblevskii A V, Fomichev S V and Krenev K A 2013 *Theoretical Foundations of Chemical Engineering* **47**(4) 484 - 8
[3] Fomichev S V, Babievskaya I Z, Dergacheva N P, Noskova O A and Krenev V A 2010 *Inorganic Materials* **46(10)** 1121 - 5
[4] Perevozchikova B V, Pisciotta A, Osovetsky B M and Menshikov E A 2014 *Energy Procedia* **59** 309 - 14
[5] Odat S 2014 *Open J. Geol.* **4** 198–205
[6] BPBD 2000 *Rencana Tata Ruang Wilayah Kabupaten Lampung Timur* Bandung

IOP Conf. Series: Materials Science and Engineering **285** (2017) 012015 doi:10.1088/1757-899X/285/1/012015

The effect of entry into force of the mining law 2009 began January 2014 in the production of NPI (Nickel Pig Iron) in China

P Prasetyo

Research Center of Metallurgy and Material – LIPI
Puspiptek, Serpong, Indonesia

Email: pprasetiyo2002@yahoo.com

Abstract. China is the largest producer NPI (Nickel Pig Iron) obtained from the processing of the low grade laterite (oxide ore) with pyrometallurgical process. Furthermore NPI is used as substitute FeNi (Ferro Nickel) to make stainless steel (SS). The low grade laterite are imported from Philippines and Indonesia because China does not have the natural resources of laterite. China mainly imported limonite contains Ni < 1.5 % from Philippines and the low grade saprolite contains Ni > 1.5 % from Indonesia. The entry into force of the Mining Law 2009 began January 12, 2014 which prohibits the export of the raw materials of mineral, and obliged to process minerals in Indonesia. Applicability of the legislation automatically affects the production of NPI in China. The effects in the short term since the enactment of the Act, the production of NPI declined and imports FeNi increased in China in 2014/2015. So that in anticipation, the NPI producers in China are trying to relocate the NPI plant out of China especially to Indonesia. For the producers who do not relocate the NPI plant, they tried to get the low grade saprolite from the other country especially from Philippines.If the NPI factory from China is relocated to Indonesia, it seems that China is likely to relocate the NPI plant to Indonesia. So in anticipation of the government must make government regulations (GR/PP: Peraturan Pemerintah) which in principle do not harm Indonesia. The government regulation (GR/PP) is to be mutually beneficial to both parties or win-win solution. This reality will be studied in this paper.

1. Introduction

The Law Number 4 of 2009 about Mining and Coal give mandate dismiss export the raw materials of mineral, and mandatory for processing the raw materials of mineral within the country. Prior to The Law Number 4 of 2009 enacted began January 12, 2014, Indonesia has been export the large quantities of nickel oxide ore called laterite especially the low grade laterite to China. For Indonesia the low grade laterite consists of limonit and the low grade saprolite contains Ni < 1.8 %. While the high grade laterite is saprolite contains Ni $\geq$ 1.8%.

Laterite mining primarily come from Southeast Sulawesi especially Pomalaa and Halmahera in North Maluku especially Teluk Buli (Buli Bay). The laterite mining done to take limonite and saprolite for export and processed within the country. Limonite with certain requirements exported to Queensland Nickel in Australia. The high grade saprolite contains Ni $\geq$ 1.8% exported mainly to Japan, processed to produce FeNi in Pomalaa by PT Aneka Tambang, and to producing Ni matte in Sorowako by PT Vale Indonesia. Laterite contains Ni > 1.5% exported to China, and this laterite are the low grade saprolite. As for the Indonesian laterite export data to foreign countries from 2005 - 2013, can be seen in table 1 below.

China is the largest producer NPI (Nickel Pig Iron) which is derived from the processing of the low grade laterite imported from Philippines and Indonesia. China was forced to import the low-grade laterite because it has no the natural resources laterite. Limonite contains Ni < 1.5% from Philippines is processed in China using blast furnace (BF) to produce the low grade NPI or the medium grade NPI. While laterite from Indonesia contains Ni > 1.5% processed using the EAF (Electric Arc Furnace) or RKEF (Rotary Kiln Electric Furnace) to produce the high grade NPI. In 2013 China imported the low grade laterite ± 41 million tons from Indonesia and ± 29.6 million tons from Philippines. The low grade laterite is used to produce NPI contains ± 460 kt Ni which equivalent to about 70% production nickel (Ni) in China [2]. Furthermore NPI contains Ni used as substitute FeNi for making stainless steel (SS). As for the production of NPI in China from 2006 until 2013, can be seen in table 2 below.

Table 1. Export Laterite Indonesia 2005 – 2013 (million ton) [1]

Year	China (+Hk)	Australi a	Yunani	Japan	Ukraine	Etc	Total	% change p.a
2005	0,07	0,95	0,10	2,03	0,49	0,06	3,70	13,6 %
2006	0,66	0,71	0,15	2,07	0,73	0,08	4,39	18,6 %
2007	5,43	0,37	0,39	1,85	0,59	0,39	9,03	105,4 %
2008	6,59	0,51	0,35	1,83	1,01	0,31	10,59	17,3 %
2009	7,88	0,00	0,39	1,46	0,58	0,12	10,44	- 1,5 %
2010	14,35	0,84	0,24	1,45	0,63	0,21	17,73	69,9 %
2011	36,23	0,96	0,57	1,86	0,71	0,51	40,84	130,4 %
2012	43,54	1,45	0,64	1,55	1,20	0,06	48,45	18,6 %
2013	59,17	1,57	0,46	1,98	1,48	0,14	64,80	33,8 %
Eksport 2013	91,3 %	2,4 %	0,7 %	3,1 %	2,3 %	0,2 %	100 %	

Table 2. The production NPI in China 2006 - 2013 Based The Grade Nickel [3]

Year	Low	Medium	High		Total Ni (kilo ton)	Growth
			EAF	RKEF		
2006	9	17	3	-	29	-
2007	29	57	10	-	96	231 %
2008	10	39	36	-	85	11 %
2009	45	24	43	-	112	32 %
2010	38	62	108	-	208	86 %
2011	42	64	148	35	289	39 %
2012	49	64	148	35	289	28 %
2013 F	52	21	205	232	510	38 %

Outside China, PT Indoferro in Cilegon Banten Indonesia is the first factory of NPI commissioning April 2012 uses blast furnace (BF) to producing NPI. The production capacity of PT Indoferro ± 250,000 ton NPI p.a then the capacity increased to 500,000 tonnes NPI p.a in 2014 [4]. Ingots a weight 8.5 kg are the form of NPI product from PT Indoferro.The size of ingot 230 mm X 160 mm X 50 mm with a composition of Ni: 6.0 % min, Cr: 2.0 % min, C: 5.0 % max , S: max 3.0 %, Mn: 1.5 % max, P: 0.08 % max, S: 0.3 % max, Cu: 0.10 % max, Pb: 0.05% max, Fe rest (± 80.97%) [5].

The entry into force of the Mining Act 2009 on January 12, 2014 which prohibits export the raw materials of mineral, and obliged to process the raw materials of mineral in Indonesia. The enactment of laws will automatically affects the production of NPI and stainless steel (SS) in China. For example the effect in the short term from 2013 until 2015, the production of NPI decreased and FeNi imports increased. This is shown in figure 1 below [6].

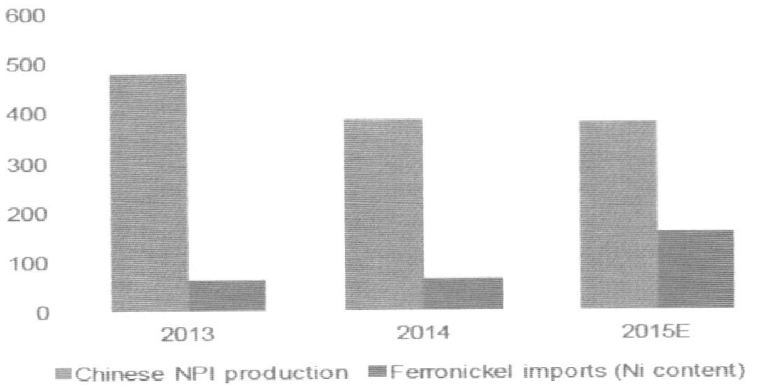

Figure 1. Decrease of NPI Production And Increased Imports FeNi (Ni kt)

For long-term anticipation the producers NPI in China are trying to relocate the NPI plant to outside China mainly in Indonesia. For the producers who do not relocate the NPI plant they tried to get the low grade saprolite from the other countries especially from Philippines.

It seems that NPI producers from China are more interested in relocating the NPI plant to Indonesia. This is proven in early 2014 the government has given permission to the investors especially from China to build smelters to process laterite. The smelters are mainly used to process the low grade laterite for producing NPI (Nickel Pig Iron). As for the definition the low grade laterite for Indonesia consists limonite and the low grade saprolite contain Ni < 1.8 %. It is estimated that there are 60 companies from China partnership with the local companies will build the smelter. Of the 60 companies estimated 13 companies started producing NPI (Nickel Pig Iron) by the end 2017. Of thirteen (13) companies turned out two (2) already in operation before the Mining Law 2009 was enacted January 2014, namely PT Indoferro has been in operation since April 2012 and CIMM since late 2013. As for the 13 companies producers NPI in Indonesia, can be seen in figure 2 below [7].

IOP Conf. Series: Materials Science and Engineering **285** (2017) 012015 doi:10.1088/1757-899X/285/1/012015

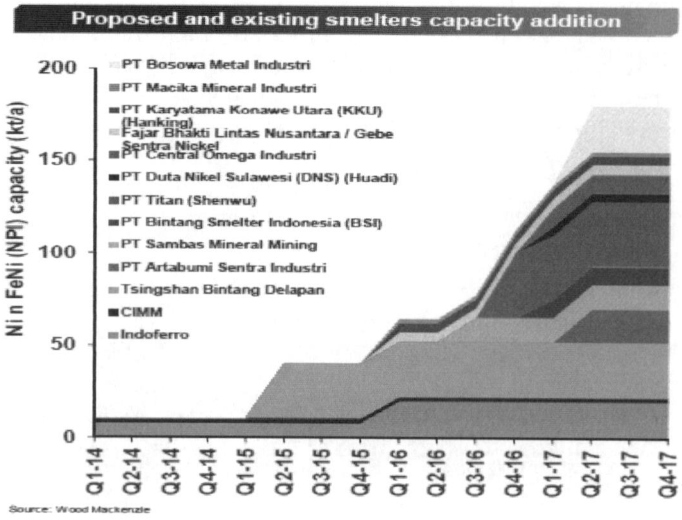

Source: Wood Mackenzie

Figure 2. The Companies Producer NPI in Indonesia

According to government writers should be smart to address the fact of relocation of NPI plant, namely by issuing government regulations (GR/PP : Peraturan Pemerintah) which in principle the GR/PP does not harm Indonesia. The regulation is to be mutually beneficial to both parties or win-win solution. Do not let the natural resources of laterit (nickel oxide) in Indonesia is spent by China, and the Indonesian nation is only a spectator in his own country. The reality will be assessed in the article with the above title.

2. Why China Producing NPI (Nickel Pig Iron)?

Starting in 2004 global nickel supply is limited while the price is high and fluctuate, and reached the highest price US $ 50 / kg in 2007. As for the fluctuations in nickel prices from 2000 - 2014 with the highest price in 2007 and laterite exports from Indonesia, can be seen in figure 3 below.

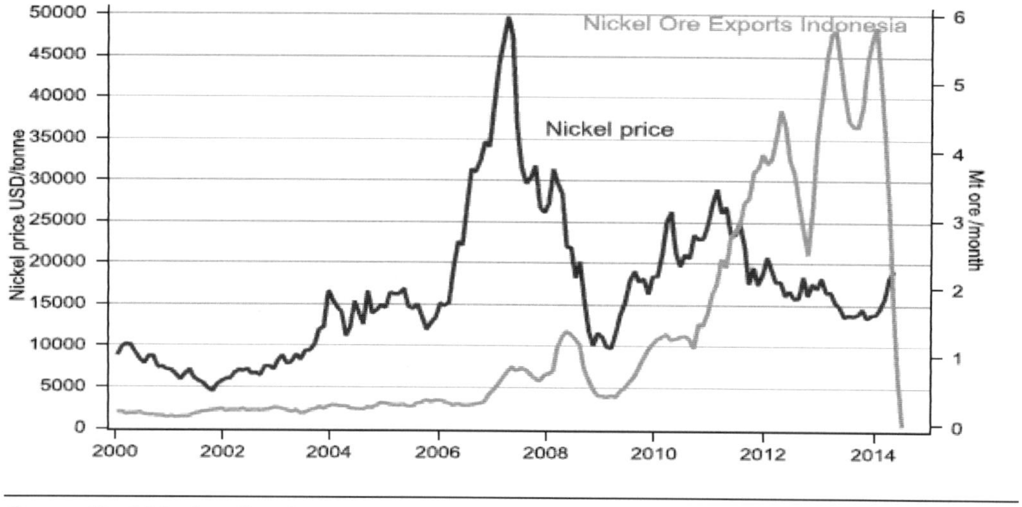

Sources: Handelsbanken, Bloomberg

Figure 3. Price of Nickel 2004-2014 And Export laterite From Indonesia [8]

IOP Conf. Series: Materials Science and Engineering **285** (2017) 012015 doi:10.1088/1757-899X/285/1/012015

At that time nickel very needed by China to make stainless steel (SS) to build infrastructure within the country in connection with the Beijing 2008 Olympic Games. Maybe this fact cause the Chinese government made policy all smelters in China must made NPI (Nickel Pig Iron) from the low grade laterite. As for the purpose of the policy is to get nickel (Ni) at price "relatively cheap" to make stainless steel (SS). Furthermore the whole product NPI contains 4-13% Ni bought by the government at the price of nickel (Ni) according to standard International price LME (London Metal Exchange). The price of nickel (Ni) "relatively cheap" is derived from the low grade laterite imported mainly from the Philippines and Indonesia. Because China does not have the natural resources of laterite. Furthermore the low grade laterite are enhanced add value to be NPI (Nickel Pig Iron) by means used the existing facilities for producing iron / steel in the country. These facilities are BF (Blast Furnace), EF (Electric Furnace), and RKEF (Rotary Kiln Electric Furnace). Furthermore product NPI (from BF, EF, and RKEF) are used replace of FeNi to make stainless steel (SS).

The Chinese government issued policy to make NPI (Nickel Pig Iron) because China including the country in the world with the high enough consumption for stainless steel (SS) per capita (kg / year). The high enough consumption for stainless steel (SS) per capita (kg / year) means the consumption of nickel (Ni) per capita (kg / year) also high. Because to make stainless steel (SS) always required nickel (Ni). As for the comparison of the production and the consumption of stainless steel (SS) per capita (kg / year) in China with the other countries especially the Western countries (the European Union), shown in figures 4 and 5 below.

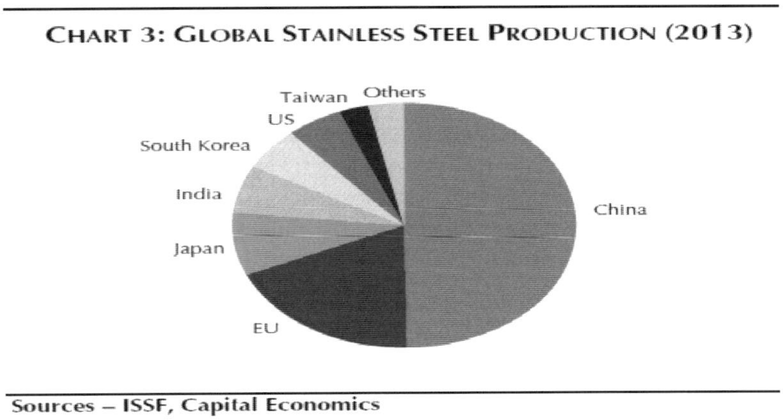

Figure 4. Production of Stainless Steel (SS) Globally [9]

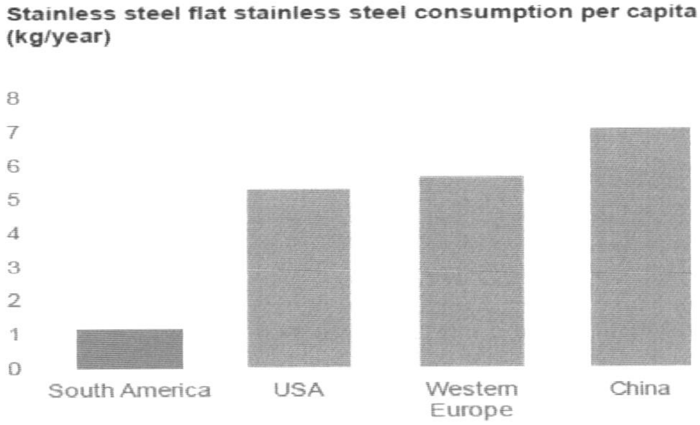

Figure 5. Consumption of Stainless Steel (SS) Per Capita In China [6]

Table 3. The Comparison Production of NPI In China Use BF, EAF, Dan RKEF [3]

	Blast Furnace (1,7 % Ni)	Blast Furnace (6 % Ni)	Shandong EAF	Inner Mongolia EAF	RKEF
Ni ore	19 %	13 %	24 %	28 %	29 %
Cargo cost	17 %	17 %	15 %	23 %	16 %
Antrasite	-	-	-	-	14 %
Coke	34 %	30 %	-	4 %	-
Fine Coke	8 %	11 %	13 %	10 %	3 %
Chalk	2 %	2 %	2 %	-	1 %
Fluorite	-	6 %	-	-	-
Energi cost	6 %	4 %	40 %	24 %	28 %
Labor cost	4 %	4 %	2 %	2 %	2 %
Etc	10 %	13 %	6 %	7 %	7 %
Cost (US $/ ton Ni)	20,300	14,500	13,900	13,000	12,600
Selling price (US $/ ton Ni)	22,700	13,900	13,900	13,900	13,900

Although the price of nickel high as shown in figure 2 above but China in producing NPI (US $ / ton Ni) was still economical. So that the production of NPI in China continues to increase from year to year as shown in table 2 above. Furthermore the products of NPI contains Ni (From the processing uses BF, EAF, and RKEF) are used to make stainless steel (SS). Stainless Steel (SS) primarily to meet the domestic demand to build infrastructure in connection with the Beijing 2008 Olympic Games. After the completion of the Olympics, stainless steel (SS) apart being used in the country also exported.As for the comparison of the production of NPI in China uses BF (Blast Furnace), EAF (Electric Arc Furnace), and RKEF (Rotary Kiln Electric Furnace), can be seen in table 3.

3. The production of NPI in China after the applicable of Mining Law in Indonesia

The entry into force of the Mining Act 2009 began January 12, 2014, automatically the supply of laterite from Indonesia to China stopped. This fact led to the production of NPI in China disrupted so that the production of stainless steel also impaired. According to experts the production of 510 kt Ni in the form of NPI in 2013 will dropped to 330-370 kt Ni after 2013 as shown in figure 6 below.

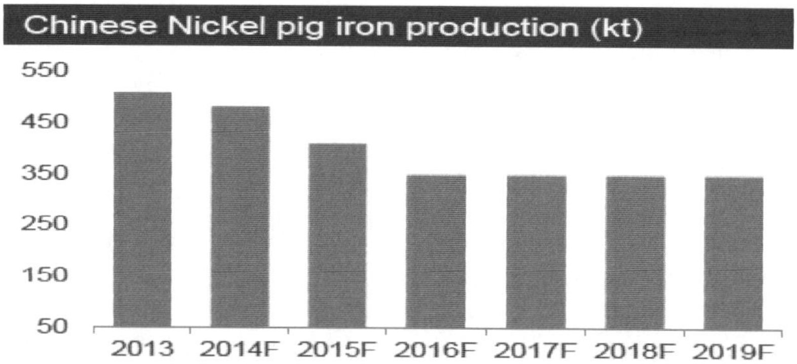

Figure 6. Estimated Production NPI (Nickel Pig Iron) In China After 2013 [10]

In order the production of NPI is not disturbed, China kept trying to get the low grade laterite especially the low grade saprolite contain Ni > 1.5% from the other country than Indonesia. Or the factories of NPI relocate out of China especially to Indonesia. According to experts exports laterite to

China from Philippine continued to rise after the Indonesian Mining Law came into effect January 12, 2014, the predictions of experts shown in figure 7 below.

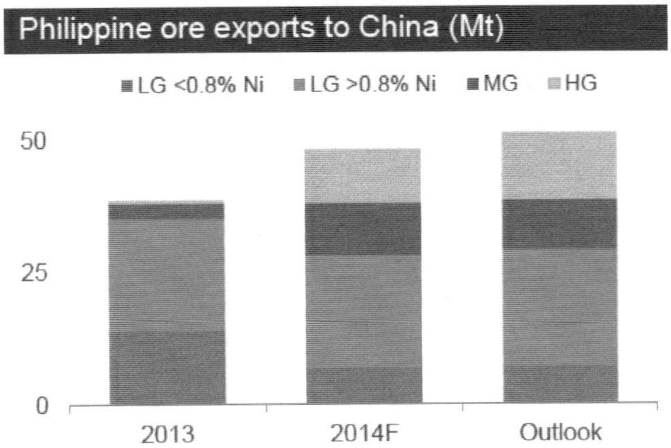

Figure 7. Exports laterite from Philippines to China [10]

4. Relocation Factories NPI (Nickel Pig Iron) from China to Indonesia

The Mining Law opens opportunities for domestic and foreign investors to process the raw materials of mineral in Indonesia. Because the mandate of Mining Law 2009 obligatory to process the raw materials of mineral in Indonesia, and stopping export the raw materials of mineral began January 12, 2014. By processing the mineral within the country then there is add value for the mineral, and the opportunities job will open for the nation's children in Indonesia. If the raw materials of mineral continuous exported, Indonesia practically get no benefit however only disadvantages from the environmental damage. Because mining is always damaging to the environment, and the environment must be restored after finished mined (post-mine).If the raw materials of mineral exported continuous, only foreigners get benefit from Indonesia's mineral which processed by them. For example laterite export to China or DSO (Direct Shipping Ore) in 2013, Indonesia only get ± US $ 1.7 billion. Whereas the value of nickel (Ni) contained in the raw materials laterite which exported ± US $ 10.5 billion if processed first [10].

With the enactment of the Minerba Act, China must relocate its NPI plant to Indonesia if China wants to survive as a stainless steel (SS) producer in the world as shown in figure 4 above. Since most of China's NPI products are high grade NPI (see table 2 above) made from imported laterite from Indonesia. Meanwhile, to produce stainless steel (SS) nationally the use of NPI reached 53%, the reality can be seen in figure 8 below [11].

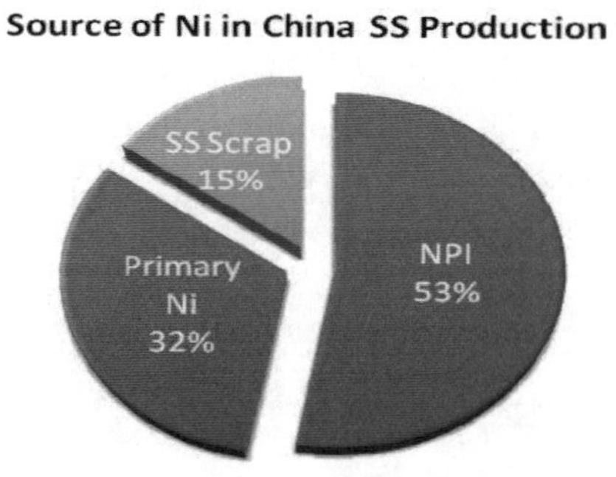

Figure 8. The use of NPI, Primary Nickel, And Scrap of Stainless Steel (SS) To Produce Stainless Steel (SS) in China

According to the authors of the Indonesian government both at the central and regional levels should be smart in anticipating the relocation of NPI factory from China. The anticipation is by making strict regulations for labor, the environment, industry and trade. Because the Chinese people in living a life known as a tough and hardworking people. So it can beat the other nations including Indonesia if the nation not smart. The defeat mainly in industry and trade. For example the United States (USA) lost to China in the world of industry and commerce. For the same television quality, television products from China circulating in the United States (USA) are cheaper than television products from the United States (USA).

5. Conclusion

- The entry into force of the Mining Law 2009 began in January 2014 affect the production of NPI in China. The production Ni ± 510 kt in the form of NPI in 2013, it is predicted the production Ni dropped to 330-370 kt Ni after 2013.
- To keep the production of Ni in the form of NPI in China remained stable then China relocate factories NPI mainly in Indonesia.
- According to the authors of the Indonesian government both at the central and regional levels should be smart in anticipating the relocation of NPI factory from China. The anticipation is by making strict regulations for labor, the environment, industry and trade. Because the Chinese people in living a life known as a tough and hardworking people. So it can beat the other nations including Indonesia if the nation not smart.

References

[1] Lennon J 2014 *The nickel market outlook: from over - supply to shortage?* Consultant to Macquire Research.
[2] Jiang K, Wang H, Li L and Liu S 2014 *The Process Research And Development For Copper, Nickel And Cobalt In BGRIMM (Beijing General Research Institute Of Mining And Metallurgy).* ALTA Free Paper
[3] Glencore 2013 *The Realities of Nickel Market*
[4] Anonim 2016 Presentasi PT INDOFERRO.
[5] Anonim 2016 Nickel Pig Iron. PT Indoferro. Origin Indonesia. Fondel Metals BV www.fondel.com
[6] Di Maulo T 2015 Golman Sachs Global Metals & Mining, Steel Conference December 01
[7] Mackenzie W 2015 Nickel Processing Falmouth UK
[8] Jansson M 2014 *From Indonesia to China and Back Again* Commodity Strategy Handelsbanken Capital Market

[9] Anonim 2015 Industrial Metals Focus Capital Economics
[10] Glencore 2014 Sell-side analyst visit *Nickel*
[11] Anonim 2016 Presentation on NAC Nickel Asia Corporation

The material from Lampung as coarse aggregate to substitute andesite for concrete-making

M Amin, Y I Supriyatna, S Sumardi

Research Unit for Mineral Technology - Indonesian Institute of Sciences (LIPI)
South Lampung, Indonesia

Email: muha047@lipi.go.id

Abstract. Andesite stone is usually used for split stone material in the concrete making. However, its availability is decreasing. Lampung province has natural resources that can be used for coarse aggregate materials to substitute andesite stone. These natural materials include limestone, feldspar stone, basalt, granite, and slags from iron processing waste. Therefore, a research on optimizing natural materials in Lampung to substitute andesite stone for concrete making is required. This research used laboratory experiment method. The research activities included making cubical object samples of 150 x 150 x 150 mm with material composition referring to a standard of K.200 and w/c 0.61. Concrete making by using varying types of aggregates (basalt, limestone, slag) and aggregate sizes (A= 5-15 mm, B= 15-25 mm, and 25-50 mm) was followed by compressive strength test. The results showed that the obtained optimal compressive strengths for basalt were 24.47 MPa for 50-150 mm aggregate sizes, 21.2 MPa for 15-25 mm aggregate sizes, and 20.7 MPa for 25-50 mm aggregate sizes. These results of basalt compressive strength values were higher than the same result for andesite (19.69 MPa for 50-150 mm aggregate sizes), slag (22.72 MPa for 50-150 mm aggregate sizes), and limestone (19.69 Mpa for 50-150 mm aggregate sizes). These results indicated that basalt, limestone, and slag aggregates were good enough to substitute andesite as materials for concrete making. Therefore, natural resources in Lampung can be optimized as construction materials in concrete making.

1. Introduction

Lampung has mineral resources which can be used as industrial materials, but mineral optimizing in Lampung is not yet optimal because minimum information for industry players related to those minerals. Natural minerals that can be used as materials for industries include limestone, granite, feldspar, basalt, and iron. These minerals, if they are optimized, can substitute split stone coming from andesite. This andesite split stone is usually used as a material for concrete making, but its availability in nature is decreasing.

Concrete making technology development is closely related to using local potentials which are needed to be improved to be more useful for local people [1]. Previously"ape" stone from North Sulawesi was used as a non-sand concrete aggregate material with the ratio of 1:6 by using cylinder object sample and it produced 7.67 MPa compressive strength, which was applied to non-sand conblock concrete [2]. Another concrete making research used fly ash, slag and limestone which were mixed with PCC cement and they produced 41.37 MPa compressive strength [3]. The concrete making by using fine aggregate from limestone (25%) and sand (100%) in West Sumba Island in the form of cylinder object sample of 15 x 30 cm at 28 days of testing age produced the compressive strength of

28.28 MPa and tensile strength of 2.92 MPa [4]. High strength concrete making by using sand and gravel aggregate from North Sulawesi in the form of cubical object sample of 150 x 150 x 150 mm at 28 days of testing age produced a compressive strength of 62.64 MPa. These research results indicate that the longer the age of testing, then the stronger is the compressive strength value [5].

The next concrete making used the following compositions. Composition A: CEM= 300 gr, water = 150 ml, sand of 0.4 mm = 603 gr, limestone 4-31 mm = 1344 gr and super plasticizer = 1.8 gr; composition B: cement= 108 gr, fly ash = 162 gr, water = 139 ml, sand of 0.,4 mm = 656 gr, and super plasticizer = 3 gr ; and composition C: cement = 108 gr, fly ash = 162 gr, water = 139 gr, sand of 0.4 mm = 400 gr, slag 4-31.5 mm = 1628 gr, and super plasticizer = 3 gr. The testing object sample was in the form of cubic of 150 x 150 x 150 mm. The compressive strength values for composition A, B, and C was 23.9 MPa, 30.7 Mpa, and 33.7 respectively. The flexural strength values for composition A, B, and C was 30.6 Gpa, 35.7 Gpa, and 39.3 Gpa respectively MPa respectively, while the specific gravity test results for composition A, B, and C were 2346 kg/m^3, 2358 kg/m^3, and 2782 kg/m^3 [6]. The coarse aggregates of river gravel and split gravel and sand fine aggregate were used for making cylindrical concrete of 10 mm and 20 mm height. The results of compressive strength values of 3, 7, 14, and 28 days showed that at 28 days, the concrete by using split gravel aggregate produced the highest compressive strength value of 30.74 MPa and 24.60 MPa for the concrete using river gravel [7].

The brick waste from demolished building could be recycled and reused as coarse aggregate material for concrete. It was made in the form of cubic of 40 x 40 x 60 mm and it was tested for compressive and flexural strength at 28 days. The results were 35.73 MPa of compressive strength and 6.63 MPa of flexural strength [8]. Types of cement and aggregates influence concrete compressive strength. A research was done on cement types of Portland type I (PC I), pozzolan Portland (PPC), composite cement (PCC) with gravel and split stone aggregates of max 20 mm. The concrete object sample was cubic of 150 x 150 x 150 mm with the testing age of 3, 7, 28, and 90 days. The compressive strength test results showed that the highest values were obtained at 90 days for gravel aggregate and cement of PC I = 49.78 MPa, PPC cement = 52.50 MPa, and PCC cement = 54.67 MPa. The compressive strength values for split stone aggregate and PC I cement = 54.36 MPa, PPC cement = 56.67 MPa and PCC cement =57.47 Mpa [9].

Volcanic ash from volcanic mountains could be used as concrete mixture based on micro silica by making the following mixture compositions. Composition A = cement paste + fine aggregate (sand) + coarse aggregate (small size gravel); composition B = cement paste + 100 mesh volcanic ash + sand aggregate + small size gravel aggregate; composition C = cement paste + micron volcanic ash + sand aggregate + small size aggregate with ratio of 1:5; composition D= epoxy resin + micron volcanic ash + sand aggregate + small size gravel aggregate with ratio of 1:5; and composition E = epoxy resin + micron volcanic ash + fine aggregate + small gravel aggregate with ratio of 1:3. All composition were made into the cylindrical concrete of 27.5 mm and 55.0 mm height. It was dried in an autoclave for 2 hours at 121°C and then heated in the oven at 100°C for 24 hours. Compressive strength test results for composition A, B, C, D, and E were 36.72 MPa, 52.23 MPa, 62.14 MPa, 801.73 Mpa, and 850.50 Mpa respectively. Standard value for normal concrete is 196.1 – 490.3 kgf/cm^2 and high quality concrete is > 490.3 kgf/cm^2[10].

Solid waste from construction in the form of small size concrete fragment could be used as coarse aggregate for concrete making. Concrete object samples of 10 x 10 x 40 cm were made with the following compositions; composition A = 50% waste, water = 7.90 kg, cement = 17.37 kg, sand 25.73 kg, split stone = 26.77 kg, waste 4.76 kg; composition B: water = 7.90 kg, cement = 25.73 kg, split stone = 26.77 kg, and waste = 9.51 kg. The compressive strength test was conducted at 3, 7, and 28 days. Compressive strength results at 3 days for composition A and B were 2.14 MPa and 1.61 MPa respectively. Compressive strength result at 7 days for composition A and B were 2.93 MPa and 2.40 MPa respectively. Compressive strength result at 28 days for composition A and B were 4.00 MPa and 3.33 MPa respectively [11]. In the concrete construction, 75% of the volume is aggregate; both rough and fine aggregates. There are two types of aggregate. Fine aggregate (sand) is less than 5 mm and coarse aggregate (gravel/split stone) is more than 5-63 mm. The higher the aggregate size, the smaller is the surface width to wet with water, so that it will reduce cement water factor and produce increasing compressive strength of the concrete (Figure 1 and Figure 2). The best maximum size of aggregate to concrete compressive strength with different cement contents is aggregate size more than 38.1 mm [12].

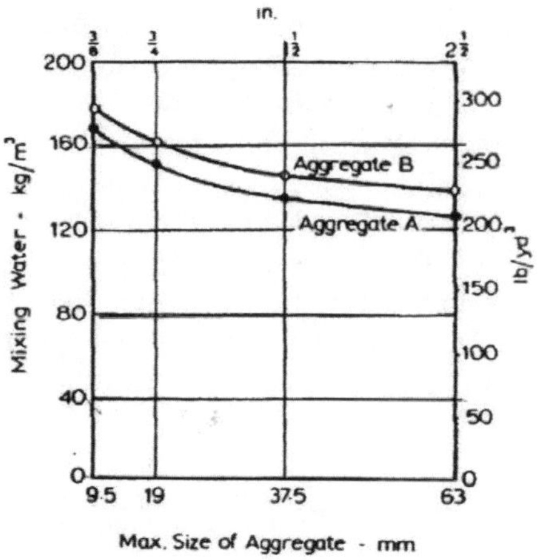

Figure 1. The influence of aggregate size to concrete compressive strength [12]

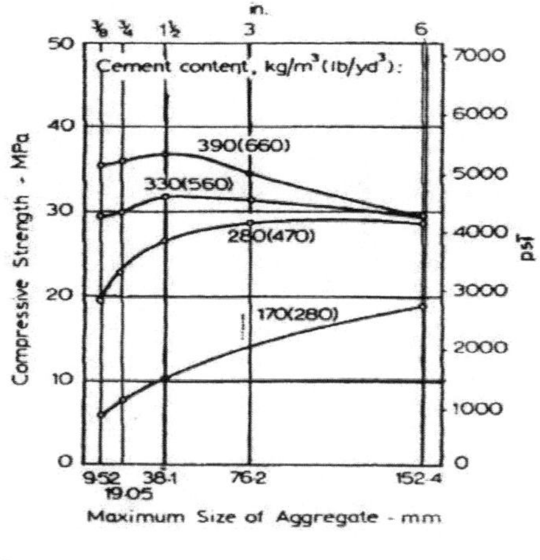

Figure 2. The influence of aggregate size to water need [12]

Conducting specific gravity test and water absorption test of the concrete and aggregate, the tests should in accordance with SNI procedure number 1969:2008 about the procedure of specific gravity test and water absorption of coarse aggregate [13]. The size of coarse aggregate influences concrete's compressive strength. Split stone sizes are divided into 5-10 mm, 10-20 mm, 20-30 mm, and combination between these three sizes. The compressive strength test results on the concrete samples of 150 x 150 x 150 mm showed that concrete mixture by using 5-10 mm split stone produced 31.63 MPa compressive strength; concrete mixture by using 10-20 mm split stone produced 32.75 MPa compressive strength; and concrete mixture by using 20-30 mm split stone produced 36.08 MPa compressive strength; and combined aggregates of split stone produced 38.29 MPa compressive strength value. The compressive strength test results which were converted into 28 days of concrete age showed improvement; where 7 days concrete improved 43%, 14 days concrete improved 13%, and 21 days concrete improved 5.1%. Sand Equivalent Test is a method to use to find out the relative

IOP Conf. Series: Materials Science and Engineering **285** (2017) 012016 doi:10.1088/1757-899X/285/1/012016

proportion of clay material containing of 4.75 mm passing filter aggregate as it is shown in Figure 3 [14].

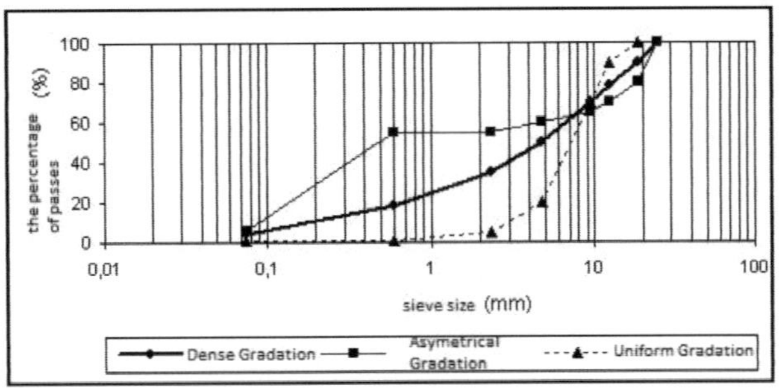

Figure 3. Types of aggregates [14]

2. Research methods

This was an experimental laboratory research by using the following parameters.

a. Coarse aggregate variations: andesite, granite, slag, limestone, feldspar, and basalt.
b. Aggregate size variations : A= 5-15 mm, B= 15-25 mm, and 25-50 mm
c. Water/cement factor, w/c=0.61
d. Compressive strength tests at 14 days
e. Cubical object sample of 150 x 150 x 150 mm (3 sample for each variation)

Materials used came from Tanjung Bintang. Aggregate materials and PCC cement came from Lampung province. Tools were one set of the sieve to analyze sand gradation, concrete mixing machine, balance, cubic molding, laboratory glass equipment, and compressive strength test machine (universal testing machine).

The research steps were as follows:

a. Material preparations included taking aggregates, milling and filtering them by using sieves of # 15 mm, # 25 mm, and # 50 mm.
b. Material testing included sand testing (tests of gradation, specific gravity, water absorption, and mud content) and coarse aggregate testing (tests of specific gravity and water absorption).
c. Making concrete object sample by weighing materials, mixing, stirring, molding, and removing from the molding by referring to the composition of K.200 and w/c = 0.61.
d. Maintaining concrete object samples in humid condition by covering the objects with a wet cloth for 14 days after they were made.
e. Testing the object samples' compressive strengths with Universal Testing Machine.
f. Analyzing testing result data to obtain conclusions from studied variables.

3. Results and discussions

3.1 Fineness Modulus of Sand

The influence of fine aggregate (sand) to resulted concrete quality is measured based on whether the concrete quality complied with requirements or not. The concrete quality is analyzed with Fineness Modulus of Sand to find out the fineness or hardness values of a particular aggregate. Fineness or hardness of aggregate will influence concrete paste adhesiveness. Too fine sand at the concrete mixture will make the fine parts will be lifted to the upper mixture and this will reduce concrete paste adhesiveness level. Data of Fineness Modulus of Sand analysis results are presented in Table 1.

Table 1. Results of fine aggregate gradation analysis

Sieve hole size (mm)	Retained weight (gr)	Retained weight (%)	Cumulative weight (%)	Passing sieve cumulative weight (%)	ASTM C 33 requirement
4.75	1.68	0.34	0.34	99.66	95-100
2.36	39.55	7.91	8.25	91.75	80-100
1.18	90.85	18.18	26.43	73.57	50-85
0.60	138.42	27.70	54.13	45.87	25-60
0.30	122.43	24.50	78.63	21.37	10-30
0.15	91.13	18.24	96.86	3.13	2-10
the rest of the sieve	15.64	3.13	-	-	
Sum	499.70	100	264.65	335.35	

The estimation of Fineness Modulus of Sand (FMS) (1)

$$FMS = \frac{\text{total of cumulative weigh}}{100} = \frac{264.65}{100} = 2.64\% \quad (1)$$

The values of Fineness Modulus of Sand (FMS) show that the gradation of sand coming from Tanjung Bintang belongs to moderate sand (FMS = 2.20 – 2.60% = fine sand, FMS = 2.60 – 2.90% = moderate sand, FMS = 2.90 – 3.20% = coarse sand). Considering the fineness level of the sand coming from Tanjung Bintang, the sand from Tanjung Bintang meets the requirement of fine gradation of ASTM with fineness modulus ≥ 2.30% and ≤ 45 % (at a particular size of sieve), so that it can be used as a fine aggregate material for concrete making. The highest value from sand gradation test results with the percentage of retained weight is 27.70% with a sieve size of 0.60 mm. Further, the fine aggregate should also meet some requirements or characteristics as they are presented in Table 2.

Table 2. The characteristics of sand from Tanjung Bintang (in %)

Characteristics	Value	Requirement
Mud content (%)	4.13	< 5
Specific gravity (gr/cm^3)	2.50	-
Water absorption (%)	9.15	-
Fineness Modulus of Sand	2.65	1.5-3.8
Water content (%)	2.09	-
Bulk Density (gr/cm^3)	1.21	-

Table 2 shows that the sand from Tanjung Bintang, South Lampung district, has mud content 4.13% and it is below SII 0052 requirement value (< 5%). The water absorption value is 9.15% and it complies with the requirement that it does not exceed 10%. These results show that the sand from Tanjung Bintang can be used as fine aggregate for concrete making.

3.2 Water Absorption

Coarse aggregates in this research come from varying minerals coming from Lampung which include granite, feldspar, andesite, basalt, limestone and slag from iron smelting waste. The coarse aggregate quality is valued based on the values of water absorption and specific gravity. The water absorption and specific gravity tests are conducted based on SNI 1969:2008. The testing results are presented in Table 3.

Table 3. The results of water absorption test of coarse aggregate mineral coming from Lampung

Types of coarse aggregate	Water absorption test (%)			Aggregate surface condition
	Size A	Size B	Size C	
Andesite	1.89	0.56	0.42	Coarse
Granite	2.07	1.52	0.95	Coarse
Slag	0.32	0.22	0.04	Slippery. glassy
Limestone	0.37	0.21	0.13	Coarse
Feldspar	1.20	1.19	0.71	Coarse
Basalt	3.62	3.05	2.63	Rough with holes

Note: Size A = 5-15mm. Size B = 15 – 25 mm. Size C= 25 – 50 mm

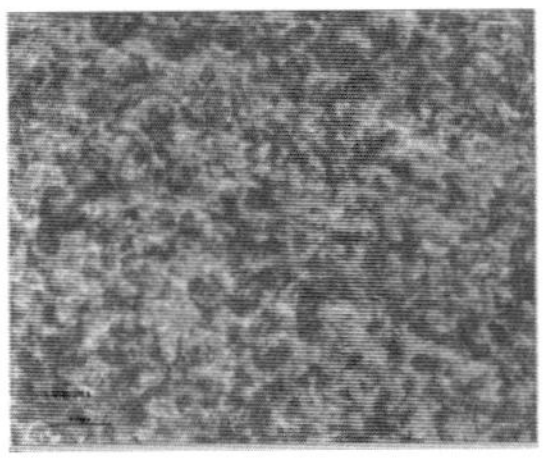
Figure 4. Andesite surface

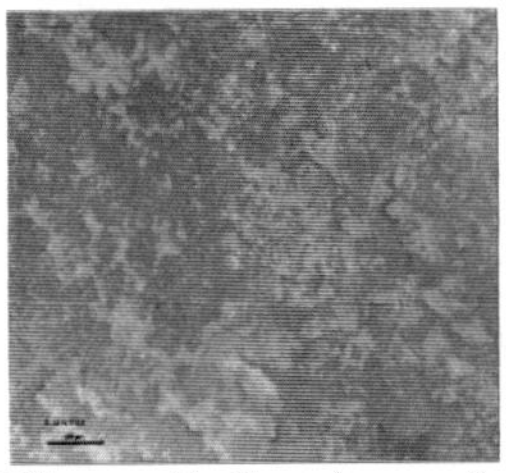
Figure 5. Granite surface

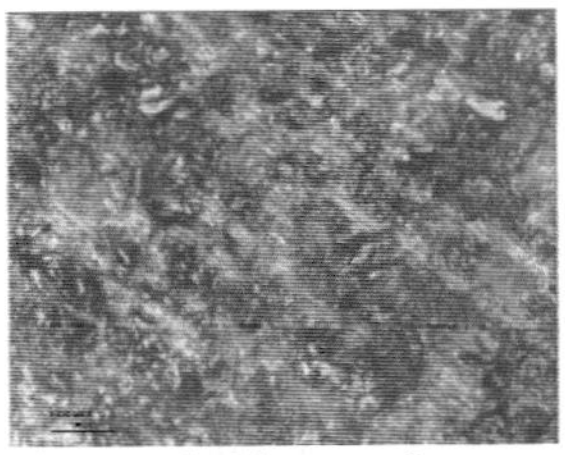
Figure 6. Slag surface

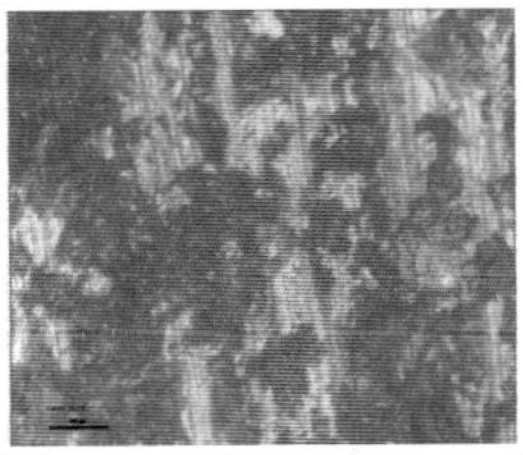
Figure 7. Limestone surface

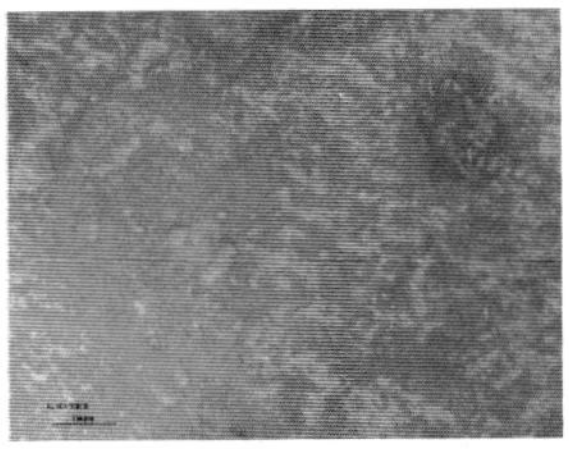
Figure 8. Feldspar surface

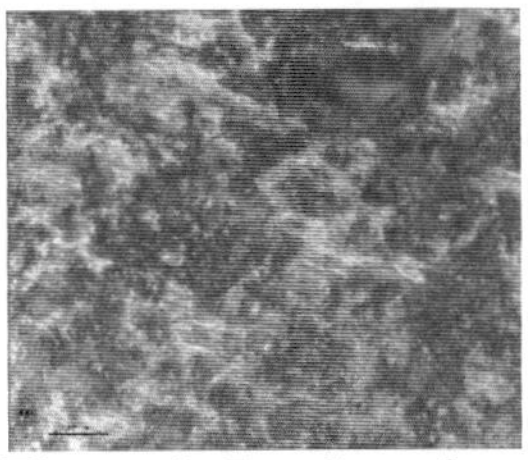
Figure 9. Basalt surface

Table 3 shows the slag water absorption value is the lowest value compared to other coarse aggregates (andesite, granite, limestone, feldspar and basalt) for all sizes of A, B, and C. This is because slag has a very smooth and glassy surface and it has small pores so that water is less absorbed into the slag. Basalt has higher water absorption compared to other types of minerals. The water absorption values of basalt for size A, B, and C are 3.62%, 3.05% and 2.63%, respectively. This is because basalt has rough surface and holes so that water is easy to enter basalt and it makes high water absorption. Water absorption on pores aggregate is 5-20% and 2% for coarse aggregate. The water absorption value will influence concrete making. The higher the pores numbers, the more water is required in concrete making. The water amount or water ratio will influence the duration of concrete drying and concrete thickness. In addition, concrete compressive strength and wear resistance will be reduced because of aggregate porosity. So that aggregate granule resistance ability against cement will be reduced. The influence of size according to Table 3 shows a tendency that the bigger the aggregate size, the smaller is absorption ability. This tendency occurs in all types of aggregates. This is because of the bigger the aggregate size in a stone form, the smaller is the surface or the lesser is the pores so that water is less absorbed. The smaller the size of a stone, the wider and the more is the surface so that more water will be absorbed by more pores. An aggregate with the coarse surface will have more absorption ability that an aggregate with a slippery surface.

The cut section images of aggregates to use in this research were taken by using a microscope to see the surface differences between these 6 types of aggregates clearer. The results are presented in Figure 4 to Figure 9.

3.3 Specific Gravity

The specific gravities of coarse aggregates are presented in Table 4.

Table 4. The results of specific gravity test of coarse aggregates from Lampung

Type of coarse aggregate	Specific gravity (gr/cm^3)	Aggregate surface condition
Andesite	2.69	Coarse
Granite	2.61	Coarse
Slag	2.72	Slippery. glassy
Limestone	2.73	Coarse
Feldspar	2.47	Coarse
Basalt	2.44	Rough with holes

Table 5. The compressive strength test results of coarse aggregate from object samples at 14 days with composition of K.200 and w/c = 0.61

Types of aggregates	Compressive strength test (MPa)			Aggregate surface condition
	Size A	Size B	Size C	
Andesite	19.69	18.93	14.39	Coarse
Granite	16.16	14.89	14.64	Coarse
Slag	22.72	14.64	13.13	Slippery. glassy
Limestone	19.69	18.68	8.58	Coarse
Feldspar	18.93	16.91	12.12	Coarse
Basalt	24.74	21.20	20.70	Coarse with holes

Note: Size A = 5-15mm. Size B = 15 – 25 mm. Size C= 25 – 50 mm

Table 4 shows that basalt mineral has the lowest specific gravity of 2.44 gr/cm^3. This is because basalt has pores and holes so that the porosity of basalt makes its light weight. Limestone and slag have higher specific gravities because they have small pores and high densities. The level of density makes bigger weight. The more mass weight increases specific gravity. Higher density makes the stone weightier. The higher the mass, the higher is the specific gravity. The higher the specific gravity of coarse aggregate used in concrete making, the weightier is the produced concrete and vice versa.

The concrete strength or the concrete compressive strength is also influenced by types and sizes of aggregates. Table 5 shows mean results of concrete compressive strength test with the material composition of K.200 and w/c = 0.61 at 14 days in the form of cubic object sample test of 150 x 150 x 150 mm by using Mykehan Farance universal machine test.

Table 5 shows that the highest compressive strength value is (24.74 MPa) obtained by the concrete using basalt aggregate with the size of 5-15 mm. The compressive strength values for concrete by using slag and limestone are 22.72 MPa and 19.69 MPa respectively. The compressive strength value for concrete by using andesite aggregate as a standard is 19.69 MPa. These results indicate that basalt, limestone, and slag aggregate with the size of 5-15 mm have higher compressive strength values than andesite, while granite and feldspar have lower values (16.16 MPa and 18.93 MPa respectively).

Table 5 shows that aggregate size A. B. and C have compressive strength value below 19.61 MPa according to standard K 200. Compressive strength values more than standard value occurs not only in basalt aggregate. The higher aggregate size, the lesser is the mixed paste coating the aggregate so that this causes lower compressive strength value. In basalt aggregate, mixed paste filling pores in aggregate so that it makes more compact and stronger concrete. The smaller the aggregate size according to the requirement of aggregate sizes, the more aggregate amount in the concrete and aggregate particles will mutually fill and bind each other. Mutually binding aggregates produce more compact, intense and stronger concrete compared to bigger aggregate sizes. Too many aggregate pores may also result in lower compressive strength value.

4. Conclusions

The best aggregate used as coarse aggregate in the concrete making is the basalt material with size 5-15 mm. Slag and limestone aggregates also have compressive strength values which are more than andesite material (19.69 MPa). Compressive strength values of feldspar and granite aggregates are lower than 19.61 MPa. Therefore, basalt, limestone, and slag materials as coarse aggregate in the concrete making have a good quality of compressive strength. The research results show that industrial materials from Lampung (especially basalt) can be optimized to be aggregate material for concrete making.

Acknowledgment

The authors would like to thank Research Unit for Mineral Technology – Indonesian Institute of Sciences that provided infrastructures and funding for this research. The authors also would like to thank the Material and Construction Laboratory in Lampung University for the cooperation to conduct concrete compressive strength test.

References

[1] Susilorini R 2015 *Pemanfatan Material Lokal Untuk Teknologi Beton Ramah Lingkungan yang Bekelanjutan* Research Report Indonesian Lybrary
[2] Trisnoyuwono D 2009 *Forum Teknik Sipil No.XIX* Poltek Kupang. Sulawesi Utara
[3] Uddin M A 2012 *Applied Mechanics and Materials* **204-208** 4105-9
[4] Kandi Y S 2012 *Jurnal Teknik Sipil* **1**
[5] Sumanjouw M D *Jurnal Ilmiah Media Enginering* **4** 215-18
[6] Papayiani I 2012 *Romanian Journal of Materials* **42** 11-5
[7] Adeputra R 2015 *Jurnal Sipil Statik* **3** 206-11
[8] Milicevic I 2013 *Recycle Clay Brik As An Aggregate For Concrete: Overview.* Tehnicky Ujesnik.
[9] Sulain A K 2009 Teknologi dan Kejuruan **32** 63-70
[10] Kurniawan C 2011 Jurnal Ilmu Pengetahuan dan Teknologi Telaah **29** 15-21
[11] Bahar A 2012 *Kajian Eksperimental Kuat Lentur Beton Yang Menggunakan Limbah Pecahan Beton Ringan Sebagai Pengganti Agregat Kasar* Jurnal Tugas Akhir Jurusan Sipil Fakultas Teknik Universitas Hasanudin Makasar
[12] Arbi M H *Majalah Ilmiah Informasi Komunikasi dan Pengkajian Iptek VARIASI* **3** 11-16
[13] Badan Standar Nasional 2008 *Cara Uji Penyerapan dan Bulk Density Pada Agregat Kasar* SNI. 19692008
[14] Zuraidah S, Wiranto H 2012 *Pengaruh Gradasi Butiran Batu Pecah Terhadap Kekuatan Beton*

IOP Conf. Series: Materials Science and Engineering **285** (2017) 012017 doi:10.1088/1757-899X/285/1/012017

The effect of sintering time on recycled magnesia brick from kiln of the cement plant

B B Aji[1], D Rosalina[2], Azhar[2] and M Amin[1]

[1]Research Division for Mineral Technology, Indonesian Institute of Sciences, Jalan. Ir. Sutami Km 15, South Lampung 35361, Indonesia
[2]Department of Chemical Engineering, University of Lampung, Jalan. Prof. Dr. Soemantri Brodjonegoro No.1, Bandar Lampung, Lampung, 35145, Indonesia

Email: bramantyo.bayu.aji@lipi.go.id

Abstract: This research aim was to investigate the effect of sintering time on reuse waste of magnesia brick from the rotary kiln of the cement plant. Reuse of the magnesia brick was carried out by mixed the kaolin as the binder. Spent refractory was used as aggregate with the composition of 85% spent refractory and 15% kaolin clay, respectively. The reuse brick then was molded with the size of 5x5x5 cm using hydraulic press under a load of 10 tons in order to forms green body. Green body then dried and sintered at 1200 ^{o}C with time variation of 2 hours, 4 hours, 6 hours, 8 hours and 10 hours, respectively. Thus, for comparison reuse brick was tested to its apparent porosity, the bulk density, and Cold Crushing Strength (CCS). The effect of kaolin addition as binder was also discussed.

1. Introduction

Refractory can be classified as ceramic materials which have ability to hold its physical and chemical properties at high temperature and extreme condition. Beside steel industry, cement industry also consume refractory in great amount such as for isolator on rotary kiln to protect kiln from high temperature, chemical reactions, and abrasion during cement production. As reported by Wiharja, refractory was used almost in all part of kiln from preheating zone, calcination, sintering, to cooling zone, respectively [1]. After service life reached, usually brick will dispose to landfill introduce to environmental problem. As reported, around 60 million tons monolithic and brick were consumed in typical cement plant [2]. Not only health problem, refractory waste also contributed to high cost to industry since refractory was disposable material.

Issues to minimize cost and environment problem related to refractory waste initiate research in reuse spent refractory. Several researches have been reported, Kwong et al reported successful recycled Mg-O refractories from EAF as raw material for slag conditioner [3], Arianpour et al [4] investigated reuse spent Magnesia brick from Steel plant as aggregate have adequate properties at composition up to 30 % and slightly better corrosion resistance than prepared sample with natural sintered magnesia, and recently, study by Avelar et al reported that recycle MgO-C refractories with dolomite addition could be satisfying industrial inquiry in order to better production [5].

Although there are already research conduct reuse spent refractories, however it still limited information provide how sintering temperature can affect to performance of recycled refractories as new brick. In ceramic manufacturing, physical properties were formed during sintering. Several factor that affect are, temperature and sintering time, particle size, composition and green body density, respectively [6]. Propitious composition, particle size, and typical temperature of sintering were the main factors to get high green body density. Green body was made by compacted the raw material with certain load pressure. Disadvantage reuse spent refractories fact that valuable mineral was declined due to oxidation and overheated. Kaolin was known for their high contain of Alumina which refers to refractory materials. When heated, kaolin will formed spinnel which is seed of mullite phase [7,8]. Mullite is refractory materials that contribute to high strength which is associated with its excellent thermal shock resistant [9]. Uses of kaolin as additive showed, it can enhance the mechanical properties of the ceramic due to changes of its crystalline phase during sintering time [7,10,11]. Therefore in this research, the aim was to investigate the effect of sintering temperature on reuse waste of magnesia brick and the effect of kaolin addition as binder.

2. Experimental

2.1. Material

Material which used was magnesia used refractory of rotary kiln from cement plant in West Java, Indonesia. Kaolin clay was added as adhesive was taken from Kemiling, Bandar Lampung City.

Table 1. Elemental analysis results of the raw materials.

Component	wt% in Used refractory	wt% in Kaolin clay
MgO	86,83	0.35
Al_2O_3	1,59	28,90
SiO_2,	1,16	42,32
Fe_2O_3	0,34	1,76
CaO	2.76	1,89

2.2. Recycle refractory brick preparation

Used refractory was grinded and shifted into two sizes aggregate. The large aggregate size was 40# to 80 # (at range 150-480 μm), smaller refractory aggregate size was -80 # (less than 180 μm) and kaolin powder was size of 100 # (150 μm), respectively. All materials were completely mixed, then molded in size 5x5x5 cm using hydraulic press under a load of 10 ton and forms green body. Total composition a refractory recycle sample (Rr) was 300 grams with 85% of used refractory and 15% kaolin clay and water. The time of sintering varies for 2, 4, 6, 8, and 10 hours, respectively. As for comparison, the refractory without kaolin clay was made as standard with composition was 100% of used refractory sintered at 1200 °C for 6 hours.

2.3. Physical properties measurement

Apparent porosity and bulk density is tested by using standard method ASTM C20-00 [12]. crushing strength or pressure strength is the maximum load when cracking occur or material destruction per unit surface area when pressured at room temperature [13]. Cold Crushing Strength test was based on ASTM C133 procedure [14]. Surface morphology was characterized using scanning electron microscopy (SEM, SU-35000, Hitachi, Japan).

3. Results and discussion

3.1. Apparent porosity and bulk density

Sintering was the main process during ceramic manufacturing and controlled its final physical properties. Diffusion of atom during particle growth, pore composition, and neck formation are among possible typical that affect to bulk density and porosity of green body.

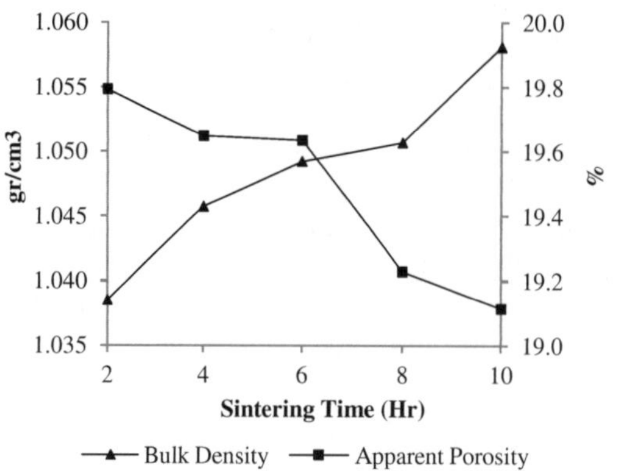

Figure 1. Bulk density and apparent porosity behaviour of recycle refractory (Rr) with increase of sintering time at temperature of 1200^0C.

As shown in figure 1, illustrated bulk density and apparent porosity of recycle refractory with the sintering time. It can be seen that, bulk density slightly increased from 1.035 gr/cm^3 2 h to1.058 gr/cm^3 at 10 hr. The result of the apparent porosity of refractory was tend to slightly decreased from 19.8 % at 2 hr to 19.6 % at 6 hr, then sharply decreased to lower level 19.18% at 10 hr of sintering time. On molding process, the given loaded pressure will shape the initial green body. The small particle filling to the empty spaces between the larger particle so the refractory will become more compact. Densification then started after heat involves. Longer time gives more time to particle growth therefore density of green body increase.

Porosity of recycle refractories decreases with increases of sintering time. During sintering, diffusion of atoms which move from particle surface to pores thus, fusion one particle to another particle occurs that called as neck. As diffusion become more intense densification occurred and denser refractory can be achieved. Discontinue of neck growth on specific temperature will promote of pore formation. The pore formed due to unbalanced diffusion and trapped gas in material [15]. The neck growth will influence the shrinkage of material pores. Therefore, as longer time of sintering densification of refractory is high and porosity composition is less.

3.2. Cold crushing strength (CCS)

Cold crushing strength (CCS) is standard of refractory which stated the strength of refractory to the given maximum load. CCS does not indicate the characteristic of refractory when it is used and more represent state the strength of refractory when it is transported. Refractory with the good strength is not easily crack or damage when it is transported. As well as with thermal conductivity, CCS also has relation with density and porosity of materials.

As illustrated in figure 2 CCS of recycle refractory (Rr) was increased from 5.25 MPa at 2 h to 6.8 Mpa at 10 hr of sintering. Similar with bulk density, longer time of sintering make green body denser and finally CCS as the state strength of refractory increase as longer time of sintering.

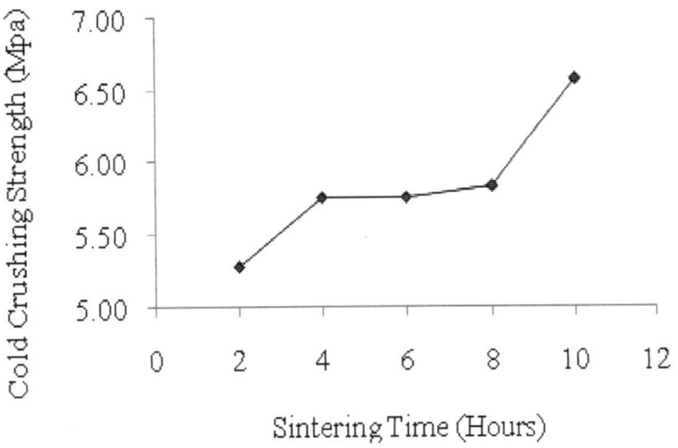

Figure 2. Behaviour of cold crush strength (CCS) with sintering time.

3.3. The effect of kaolin clay addition

Kaolin is one of clay that has good plasticity when it is mixed with water [16]. Uses of kaolin as additive expected can enhance the mechanical properties of the ceramic by exploits its crystalline phase changing during heating process. As a material with high contain of alumina, it can considerate of mullite raw material. Mullite is considered as a binding phase in most of refractory brick and it has a high resistance to melting and minimum thermal expansion as well as low thermal conductivity [17].

Table 2. The comparison of recycle refractory (Rr) and used refractory sintered at 1200^0C for 6 hr.

Sample	CCS, (Mpa)	Porosity, (%)	Bulk Density, gram/cm^3
Used kiln	48.38	5,49	2.82
Rr	5,75	19,65	1,048

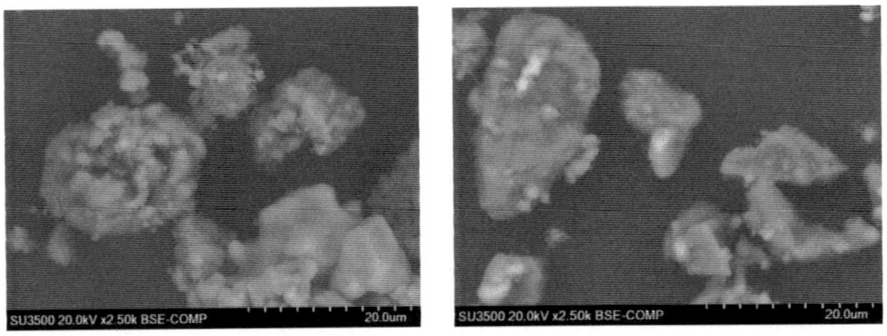

Figure 3. SEM images of prepared sample sintered at 1200 °C for 6 hours. a) recycle refractory (Rr) and b) used refractory

As presented in table 1, addition of kaolin in recycle refractory (Rr) did not give better properties compare to used refractory. CCS for Rr sample was only 7,75 Mpa much lower than used refractory sample with CCS is 48.38 Mpa. It can be explain that, Rr sample was not dense enough as represented the bulk density is only half than used kiln, were the bulk density 1.048 gram/cm^3 and 2.82 gram/cm^3, respectively. CCS value was small since density of material is low then the loaded pressure that can be accommodated is also low. This reason due to densification was not occurred during sintering. As longer time, kaolin expected can enhance densification process through mullite formation. It shown from the higher porosity of Rr than used refractory, 19.65% and 5.49%, respectively. At around temperature 450^0C, combined water in kaolin is release and leaves a large pore within refractory body, resulting in the formation of metakaolin clay. Metakaolin clay is heated around 980^0C, spinal phase is formed, an amorphous siliceous phase. As the temperature increased, the spinnel is transformed into mullite and, simultaneously SiO_2 is expelled to form of glassy phase. Then glassy silica promotes sintering, which densification initiated, and formation of third shrinkage phase at temperature above 1200^0C [18,19]. However, recycle refractory (Rr) has higher porosity than used refractory which means rather than densification another phenomenon could occur such as coarsening.

As illustrated in figure 5(a), SEM image result show that Rr has larger particle than used refractory. It is proves that rather than particle has grow and then densified, coarsening occurred resulting in imperfect grains and its various size. The growth of very large grain will eliminate the bond between particles which located at the edge of grain borders (grain boundaries), thus will decrease the strength of refractory, and it influence the refractory to become weak. Besides, the imperfect grains growth will affect formation of pores because and contribute to high percentage of refractory porosity. Meanwhile, in figure 5 (b), it can be seen that particle has more uniform in shape and size, however recycle brick still has lower bulk density than new magnesia brick [20]. It can be explained due extreme condition, used refractory has decline its properties by oxidizing or mechanical contact during processes.

4. Conclusions

Investigation variations of sintering times have been done in order to recycle waste magnesia brick into new refractory. As increasing of the sintering time, bulk density of recycle refractory (Rr) was slightly increased. However, as addition of kaolin that used as binder did not successfully enhance the final properties such as increase of CCS number. High porosity percentage of Rr indicates that may be from preparation of green body, the load pressure is not only high enough but also sintering temperature should be higher to ensure mullite formation from kaolin and increase of recycle refractory properties. Therefore, future research utilization of waste brick with kaolin addition with higher load pressure and sintering temperature could be applied in order to achieve excellent properties.

Acknowledgement

The authors acknowledge for the technical and financial support from Research Division for Mineral Technology, Indonesian Institute of Sciences and Civil Engineering Laboratory Lampung University for Strength Test Equipment.

References

[1] Wiharja 2006 *J. Tek. Lingkung.* 17–22
[2] Magnesita 2015 *A look back at the industry & pathway for a healthy succesful year* (Pennsylvania)
[3] Kwong K-S and Bennett J P 2002 *J. Miner. Mater. Charact. Eng.* **1** 69–78
[4] Arianpour F, Kazemi F and Fard F G 2010 *Miner. Eng.* **23** 273–6
[5] Avelar T C, Veiga A F, Gueguen E and Oliveira J R 2012 *Trans. Ecol. Environ.* **163** 259–70
[6] Rahaman M N 2007 *Solid state and viscous sintering* (CRC Press)
[7] Chen C Y, Lan G S and Tuan W H 2000 *Ceram. Int.* **26** 715–20
[8] Chen Y F, Wang M C and Hon M H 2004 *J. Eur. Ceram. Soc.* **24** 2389–97
[9] Schneider H, Fischer R X and Schreuer J 2015 *J. Am. Ceram. Soc.* **98** 2948–67
[10] Mostafa N Y, Shaltout A A, Abdel-Aal M S and El-maghraby A 2010 *Mater. Des.* **31** 3677–82
[11] Azeredo G and Diniz M 2013 *Constr. Build. Mater.* **38** 515–23
[12] ASTM 2015 *Standard Test Methods for Apparent Porosity, Water Absorption , Apparent Specific*

Gravity , and Bulk Density of Burned Refractory Brick and Shapes by Boiling Water Astm C20-00 0 1–3
[13] Schact C A 2004 *Refractories Handbook* ed L . Faulkner (New York, New York, USA: Marcel Dekker, Inc)
[14] ASTM C133 2014 *Standard test methods for cold crushing strength and modulus of rupture of refractories* Annu. B. ASTM Stand 97 1–6
[15] Rahaman M N . 2007 *Sintering of Ceramics* ed T & F Group (New York: CRC Press)
[16] Practical Action 2015 *Clay as a binder* Aurov. Earth Inst. 44
[17] Al-Amaireh M 2006 *J. Appl. Sci.* **6** 2605–10
[18] Venturelli C and Paganelli M 2007 *Sintering Behaviour of Clays for Process Eng.* **84** 5–8
[19] Duval D J, Risbud S H and Shackelford J F 2008 *Ceramic and Glass Materials: structure, properties and processing* ed J F Shackelford and R H Doremus (California: Springer US)
[20] Harbison and Walker 2005 *Harbison-Walker Handbook of Refractory Practice*

IOP Conf. Series: Materials Science and Engineering **285** (2017) 012018 doi:10.1088/1757-899X/285/1/012018

The effect of various sintering temperature on used refractory towards its physical properties

Sudibyo[1], Y R Wulandari[2], M Amin[1] and Azhar[2]

[1]Research Unit for Mineral Technology, Indonesian Institute of Sciences, Lampung, Indonesia
[2]Department of Chemical Engineering, Lampung University, Lampung, Indonesia

Email: sudibyo@lipi.go.id

Abstract. The used magnesia refractory from the kiln of cement industry was successfully recycled to new refractory using Kaolin as an adhesive. In this work, the temperatures of sintering were varied from 1000°C to 1500°C. The result shows that the increment temperature effects in sintering process will enhance refractory physical properties such as bulk density, cold crushing strength or pressure strength and thermal conductivity. Meanwhile, the porosity was decreased as the increase of the sintering temperature.

1. Introduction

The increasing of global refractory demand enhances the cost of raw materials and freight and environmental pressures. Searching the cheaper mineral for refractory applications creates the problem in the refractory industry. One of a cheaper refractory mineral is from recycling uses bricks. However, the mineral form used brick has many contaminants hence need to process and add some mineral to improve the physical property such as bulk density, porosity, cold crushing strength or pressure strength and also thermal conductivity [1 - 4].

The refractory production use sintering process at a certain temperature. Sintering is one of the methods to produce material with controllable density from a metal component or ceramic powder by a thermal application. This technique also able to design microstructural control that is grain size, density post-sintering (sinter density), sizes and phase distribution included pores. The binder material is needed to produce the high quality of refractory. Kaolin is one of clay that usually used as a binder material or adhesive in ceramic and refractory production due to the more stable and low cohesion at high temperature [2, 3]. This work was used Kaolin as a binder in refractory production from used brick from the cement factory. This used brick was magnesite or magnesia refractory. This brick, in general, is made from natural dead-burn magnesite which containing 90% MgO or less, and for purified magnesite, it is containing 96% MgO or more [4, 5].The temperature effect in sintering process towards refractory physical properties such as bulk density, porosity, cold crushing strength or pressure strength and thermal conductivity were studied also in this research.

2. Experiment Methods

The used magnesia bricks were from kiln unit in cement factory (Jui Shin Co. Ltd, West Java, Indonesia). The used refractory was crushed into powder. The powder than were screened to get desired size particles. New refractory were made from 300 gram material which consists 55.76% of

IOP Conf. Series: Materials Science and Engineering **285** (2017) 012018 doi:10.1088/1757-899X/285/1/012018

used refractory powder with size particle 40 -80 mesh, 29.24% of used refractory powder with size particle 80-100 mesh, 15% of kaolin size particle higher than 100 mesh. All material were mixed and then were pressed into solid (as known as green body) using a hydrolytic device with 10 ton of pressure. The dimension of green body was 5 cm x 5 cm x 5 cm (length x width x height).After that, the density of green body was measured. Finally, the green body was sintered in a furnace at various temperatures (1000°C, 1100°C, 1200°C, 1300°C, 1400°C and 1500°C) for 6 hours.

The characterisation of the product was studied in this work were apparent porosity, bulk density, cold crushing strength (CCS), and thermal conductivity. The measure of apparent porosity and bulk density was conducted based on a standard test from ASTM C20-00. CCS test was conducted using Wykeham Farrance equipment. Meanwhile, the thermal conductivity of specimen is tested by using standard guarded hot plate method, ASTM C177-04.

3. Result and Discussion

In this work, the used refractory was recycled to produce new refractory using kaolin as adhesive. The characteristic of product this experiment which were processed at various sintering temperature were listed in table 1. Meanwhile, the correlations between CCS with sintering temperatures were shown in figure 1. The figure shows that the CCS increase as the increment of sintering temperature. The increment of sintering temperature caused the enhancement number particles which were combined with the other particle making the single solid part, hence the value of CCS is increasing. In this process, the temperature was increased lower than material the melting point temperature. At this temperature, the atoms in the materials disperse cross ways the borders of the particles, blending the particles jointly and making a compact hard piece.

The effect of sintering temperature toward the percentage of porosity was shown in figure 2. The figure shows that the increasing of sintering temperature caused the decreasing of porosity percentage. This phenomenon due to the increasing of sintering temperature will increase the number of fusing of particles, hence the porosity between the particles will decrease.

Table 1. Physical characteristic from experiment product of refractory using kaolin adhesive

Temperature (°C)	CCS (Mpa)	Porosity (%)	Bulk Density gr/cm^3	Conductivity (W/m.K)
1000	1.44	20.6428	1.0358	11.6368
1100	2.06	20.4472	1.0420	11.6368
1200	5.76	19.5199	1.0503	11.8577
1300	9.56	17.8041	1.0706	12.0784
1400	13.36	16.0883	1.0909	12.2991
1500	15.5	14.9952	1.1234	12.3872

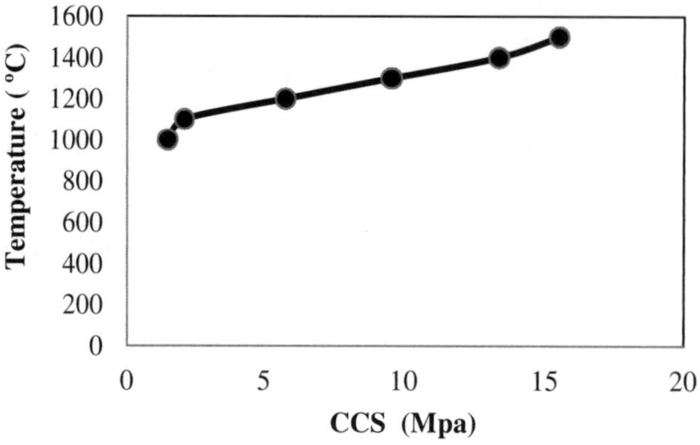

Figure 1. The effect of the sintering temperature towards cold crushing strength (CCS)

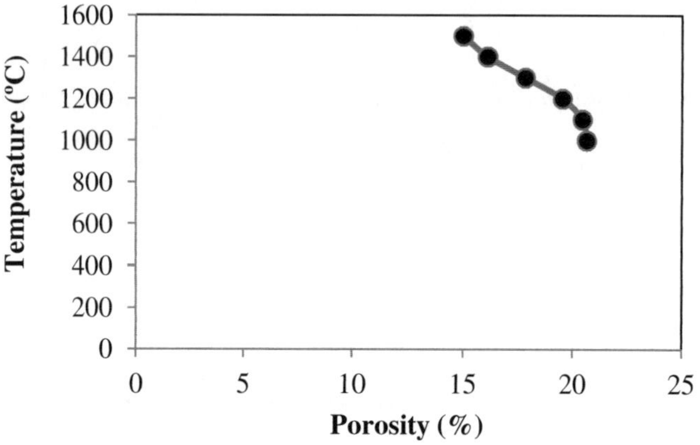

Figure 2. The effect of the sintering temperature towards porosity

Figure 3 shows the effect of the sintering temperature towards density. The figure shows that the increment of sintering temperature will enhance the density of refractory. The increasing of temperature will increase the number of fusing particles in refractory, it caused the decreasing the number of pores [5]. The density of refractory was increase as the decreasing the number of the pore in refractory.

The effect of the sintering temperature towards thermal conductivity is shown figure 4. The figure shows that the enhancement of thermal conductivity as the increment of sintering temperature. This increasing of thermal conductivity is due to the decreasing of the number of pores when the sintering temperature is increased.

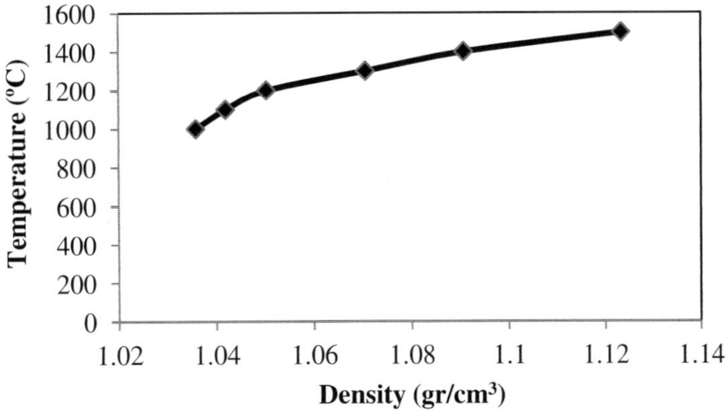

Figure 3. The effect of the sintering temperature towards density

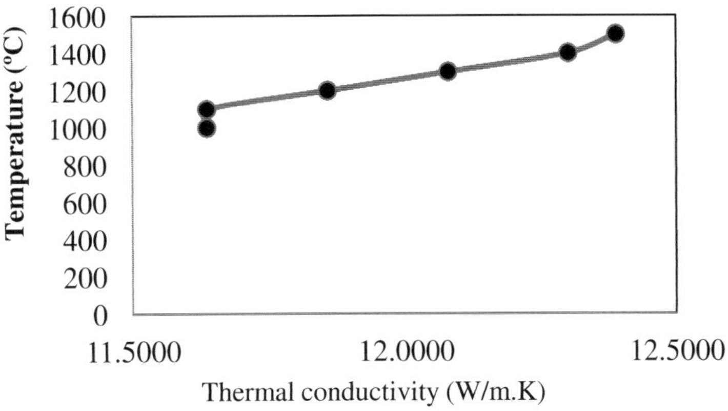

Figure 4. The effect of the sintering temperature towards thermal conductivity

Table 2. The characteristics of refractory

Sample	CCS (Mpa)	Porosity (%)	Bulk Density gr/cm^3	Conductivity (W/m.K)
Used refractory from Kiln	48.2	6.7	2.8	11.2
Recycled refractory (refractory from used refractory)	15.5	15.0	1.1	12.4
New magnesia refractory (standard)	34.5-55.2	15.5-19.0	2.8-2.9	10.6

The physical characteristics of used refractory, recycled refractory (refractory from used refractory) and new magnesia refractory [3]. The value of CCS show that recycled refractory (refractory from used refractory) is smaller than the standard new magnesia refractory, meanwhile, the value of thermal conductivity was higher than the value of the standard new magnesia refractory. These different values were due to the process of recycled refractory using 34.47-55.16 Mpa and 1500°C, meanwhile, the production of standard Magnesia refractory used 192 Mpa and 1800°C.

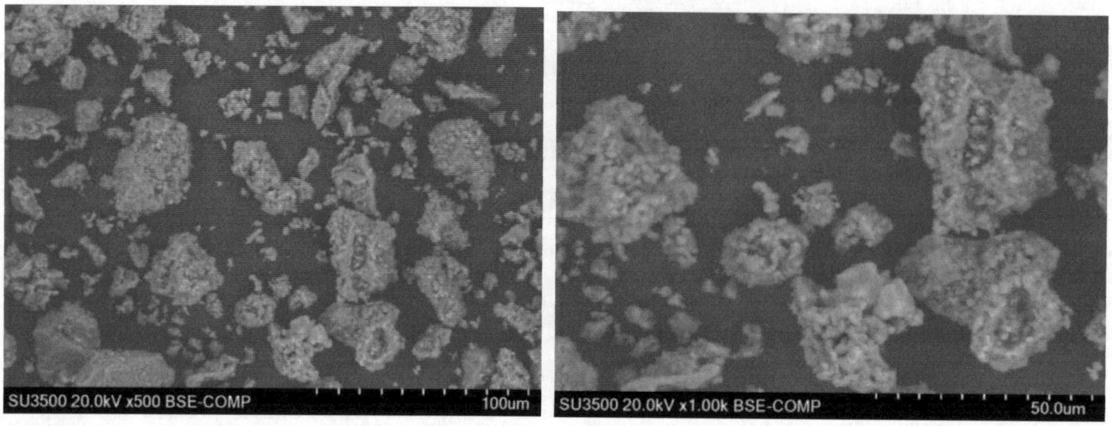

Figure 5. SEM result from sample for size of 50 μm (left), SEM result from sample for size of 100 μm (right)

SEM images from the sample which was sintered at a temperature of 1500°C were showed in figure 5a and 5b. The figures show that the growth of particle grains and the fusing of particles which show the evidence of sintering process.

4. Conclusions

The used magnesia refractory from the kiln of cement industry was successfully recycled to new refractory. Kaolin was used as adhesive in this magnesia refractory. The product of this work shows that the refractory has different physical properties than the standard of magnesia refractory due to the different pressure of green body development and the temperature of the sintering process. This differentiates of operation condition is caused by the limitation of the processing unit in a laboratory.

Acknowledgment

This work was supported by Ministry of Research, Technology and Higher Education - the Republic of Indonesia through INSINAS research grant no. RT-2016-0227 is greatly acknowledged.

References

[1] Efendy H 2008 *Studi Antioksidasi pada Refraktori MgO-C Monolitik denagn Bahan Pengikat Tar-Resin* Bandung: Institut Teknologi Bandung
[2] Schacht CA 2004 *Refractories Handbook.* New York: Marcel Dekker, Inc
[3] Carniglia S C and Barna G L 1992 *Handbook of Industrial Refractories Technology* Westwood, New Jersey: Noyes Publication
[4] Rahaman M N 2007 *Sintering of Ceramics* Boca Raton: Taylor & Francis Group, LLC
[5] Harbinson and Walker 2005 *Handbook of Refractory Practice* Moon Township: Harbinson-Walker Refractories Company

IOP Conf. Series: Materials Science and Engineering **285** (2017) 012019 doi:10.1088/1757-899X/285/1/012019

The effect of immersion time to low carbon steel hardness and microstructure with hot dip galvanizing coating method

A A Hakim[1], T O Rajagukguk[1], S Sumardi[2]

[1,2]Department of Mechanical Engineering, Faculty of Engineering, University of Malahayati, Bandar Lampung, Indonesia
[3]Reseach Unit for Mineral Processing Technology, Indonesian Insitute of Sciences, Indonesia

Email : slam009@lipi.go.id

Abstract. Along with developing necessities of metal materials, these rise demands of quality improvements and material protections especially the mechanical properties of the material. This research used hot dip galvanizing coating method. The objectives of this research were to find out Rockwell hardness (HRb), layer thickness, micro structure and observation with Scanning Electron Microscope (SEM) from result of coating by using Hot Dip Galvanizing coating method with immersion time of 3, 6, 9, and 12 minutes at 460°C. The result shows that Highest Rockwell hardness test (HRb) was at 3 minutes immersion time with 76.012 HRb. Highest thickness result was 217.3 μm at 12 minutes immersion. Microstructure test result showed that coating was formed at eta, zeta, delta and gamma phases, while Scanning Electron Microscope (SEM) showed Fe, Zn, Mn, Si and S elements at the specimens after coating.

1. Introduction

Current industry developments demand more materials for a product. Demands of metal materials in varying product components are growing. Steel is a common metal used by industries of constructions, machineries, and handcrafts. Steel uses for machinery components and constructions often demonstrate damage sooner than time use had been estimated before. Some of common causes are corrosion, metal fatigue, brittle fracture, overload, high temperature corrosion, stress corrosion cracking and wear. Effects of metal damage in a component or critical part is sometimes beyond expectation, and even it may cause disaster. Consider a steel wire for bridge drag is getting damaged because of corrosion or dynamic loads from vehicles running on the bridge that this wire receives. This may fail and this means more weight loads for other wires. This further makes the bridge fail. Continuous vibration and dynamic movement will cause metal to get fatigue. Another example is industrial pipes in mining, water pipes, oil pipes, etc. If there is no protection to these pipes, these pipes will get damaged over time caused by natural factors, chemicals, air, etc. the most common cause of metal damage is corrosion[1]. Therefore, protection and control efforts are required to prevent or minimize unexpected damage because of corrosion and other mechanical damages. Surface metal coating method is one of protection efforts. This can be done by electroplating[2], spraying or *Hot Dip Galvanizing*[3,4,5]. This research used hot dip galvanizing method and the objectives of this research were to find out the hardness value, layer thickness and microstructure of lo carbon steel material (0.02% C) after being coated with zinc.

2. Experimental

Steps to conduct this research were as follows:

1. Preparing equipment and research materials. They were grinding cutter, drilling machine, measuring cup, hot plate, Beaker flask, stopwatch, digital scale, Rockwell hardness tester, OES, metallurgy microscope, furnace machine, oven, polishing machine, crucible (galvanized container), low carbon steel (0.023%C), NaOH, resin, clean water/aquades, H_2SO_4, Zinc Ingot, Zinc Ammonium Chloride.
2. Making specimens by cutting low steel carbon material with dimension of 10 cm length, 3.5 cm wide, and 3.5 mm thickness by using grinding cutter. The specimen was drilled on each of middle tip as it is shown in Figure 1. These holes helped to facilitate specimen assembly.
3.

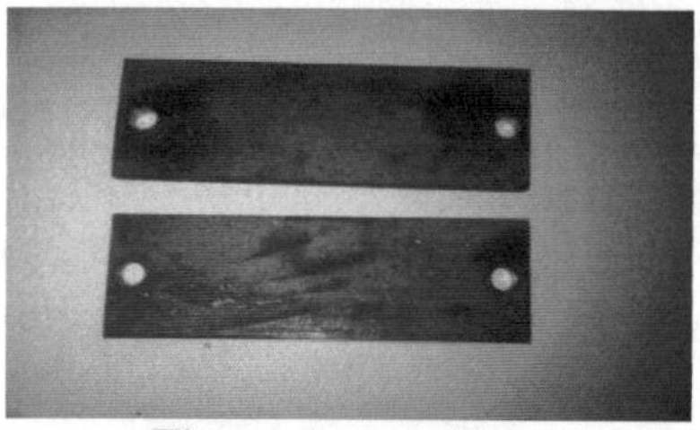

Figure 1. Specimen

Figure.2 Crucible (galvanized container)

4. Making crucible (galvanized container) by using steel pipe which was cut into two pieces and then their lower back were welded. The steel pipe was cut with dimension of 35 cm length and 20 cm diameter, and it was split into two pieces by using *bluner*. After spilt, each of lower back was connected with welding, and the edges of the split pipe were closed to form a container (Figure 2)
5. Weighing before coating process to obtain weight comparison after coating.
6. Testing chemical composition to find out chemical composition contained in a particular material or percentage of each forming element; for example, C, Si, Fe, Cu, Mg, Al, or other elements.
7. Physical cleaning by sand papering and chemical cleaning by *degreasing, rinsing I, pickling, rinsing II, fluxing*. Hot dip galvanizing process was conducted with immersion time variations of 3, 6, 9 and 12 minutes at 460^0C.
8. Weighing after coating to find out additional weight at the specimen and conducting Rockwell B hardness test, microstructure test by using *metallurgy microscope,* thickness test by using *Thickness Meter Gauge* and Conducting test with *Scanning Electron Microscope* (SEM)

3. Results and Discussions

3.1. Chemical Composition Test Result

The chemical compositions of basic material are listed in table 1.the specimen in this research is low carbon steel with 0.023% C.

Table 1. The major chemical compositions of the sample (wt.%)

Fe	C	Ni	Mn	Si	S	Al	Cr
98.6	0.023	0.038	0.80	0.199	0.001	0.045	0.01

Table 2 shows that Zn content in Zinc ingot is 98.65% and it complies SNI standard number 07-135331989 where Zinc to use for coating must be G.O.B (Good Ordinary Brand) qualified and containing 98.5% Zn

Table 2. Chemical compositions of Zinc Ingot (wt.%)

Fe	Si	S	Ni	Zn
0.18	0.71	0.43	0.039	98.65

3.2. Difference of initial (W_o) and final (W_a) Weights

Figure 3 shows that specimen weighing before and after coating suggests that the longer immersion time, then, the higher is Zn (W_{zn}) additional weight attached or coating the specimen. This can be seen from increasing specimen weight at each of time variation. 3 minutes immersion results in lowest weight difference (W_{zn}) (8.216 gram). 6 and 9 minutes immersion result in 11.596 gram and 11.818 gram weight difference (W_{zn}) respectively. 12 minute immersion shows significant difference by producing 13.656 gram weight difference (W_{zn}). These are results of continuous attaching zinc into metal surface during the process. In the *hot-dipped galvanizing*, attached zinc to base metal surface is the formation and growth of inter-metallic layer. This zinc layer is principally containing of pure zinc layer which is drawn when the working material is removed from *galvanize* and produces mixing layer between zinc and base metal (Fe-Zn).

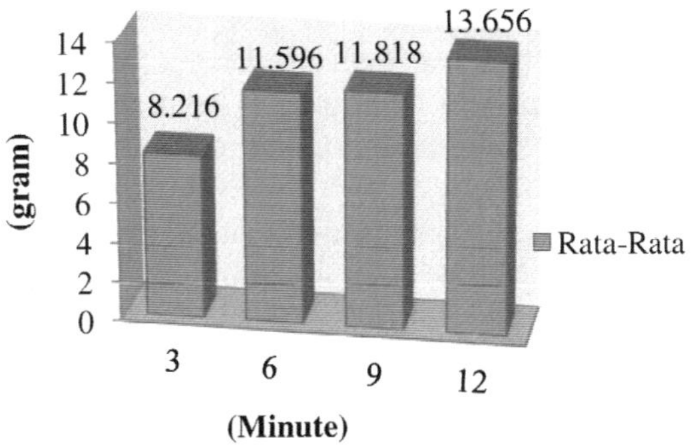

Figure 3. Graphic of difference of initial (Wo) and Final (Wa) weights of specimen based on time variations

3.3. Raw Material Rockwell Harness (HRb) test Result Before and After Coating with Immersion time Variation

Figure 4 shows that the average hardness value of the coated raw material is 86.96 kg/mm^2. After raw material is coated with zinc (Zn) by using Hot Dip Galvanizing for 3 minutes, the variations of hardness values are not significant between them. The average hardness value of 5 specimen is 76.01 kg/mm^2. The hardness value of raw material seems to decrease. It is because the raw material has been protected with Zn metal with lower hardness value (42.04 kg/mm^2) than raw material, so that indenter of hardness testing equipment only touches Zn layer and only a little touches base material; only on *Zeta* layer containing of 96.257% Zn and 3.743 Fe, so that hardness value becomes smaller than initial hardness value of raw material. 6, 9, and 12 minutes immersion show average hardness values of 74.10 kg/mm^2, 71.824 kg/mm^2, and 65.604 kg/mm^2 respectively. These hardness values are more decreasing compared with 3 minute immersion. This is caused by thicker Zn metal will coat base material along with longer immersion time so that indenter only touches Zn layer and presses only a little the base material. Obtained data show that the longer the immersion time, the smaller is the hardness value.

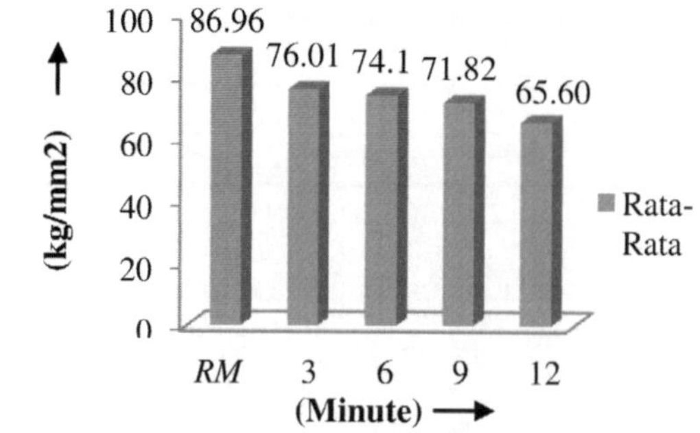

Figure 4. Raw material Rockwell hardness (HRb) test results before and after coating with immersion time variations

3.4. Layer Thictness Test Result

Figure 5 shows layer thickness result at each of immersion time variation. At 3 minutes, the average of layer thickness is 85.742 μm, and this is proportional to the weighing result difference (W_{Zn}) before and after coating and the thinnest layer is obtained with 3 minutes immersion.

6 minutes immersion produces 176.40 μm layer thickness and this is proportional to the weighing result difference (W_{Zn}) between 6 and 3 minutes immersion, where 6 minutes immersion produces thicker layer than 3 minutes immersion. 9 minutes immersion produces average layer thickness of 183.52 μm. This shows a little difference with 6 minutes immersion but it is proportionally thicker compared to 6 minutes immersion. 12 minutes immersion produces average layer thickness of 217.30 μm, and it demonstrates significant difference than lesser immersion time. This is proportional to previous weighing where 12 immersion time produce thicker layer than other less time immersion variations.

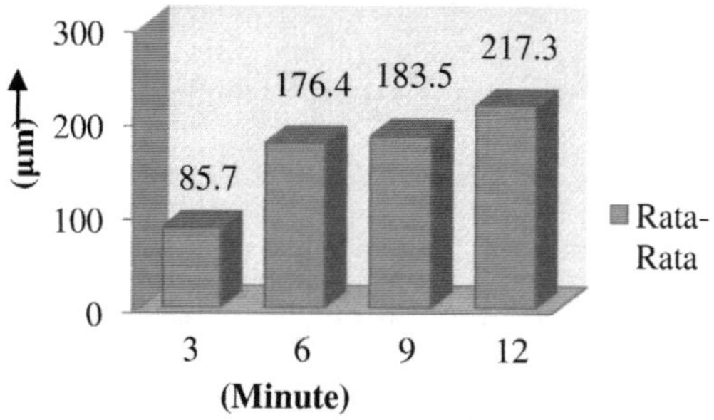

Figure 5. Layer thickness test result

3.5. Microstructure Test Results

Figure 6. Microstructure of base material

Figure 6 demonstrates microstructure image of raw material where domination of *ferrite* crystal appears in white bright color toward dark color *pearlite* crystal. *Ferrite* phase is called as alpha (α). Inter atom space is small and dense so that it will accommodate lesser Carbon atom. The maximum carbon dissolve is 0.025% at 723^0C and its crystal structure is BCC (body center cubic). At room temperature, its carbon degree is 0.008% so that it is considered as a pure iron. Ferrite is magnetic until 768^0C and it has low toughness, high tenacity, medium corrosion resistance and it has most smooth structure amongst Fe_3C diagram. Meanwhile *pearlite* phase is mechanical mixture containing of two phases; *ferrite* with 0.025% carbon content and *cementite* in form of *lamellar* (layer) with 6.67% carbon content with interstitial forms. Thus *perlite* is the microstructure of *eutectoid lamellar* reaction. This domination shows that the raw material is a metal which is not too hard and it is in fact a low carbon steel

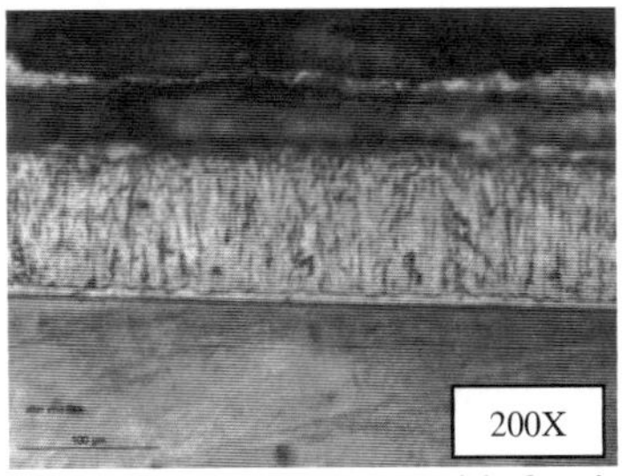

Figure 7. Micro structure with 3 minute time variation

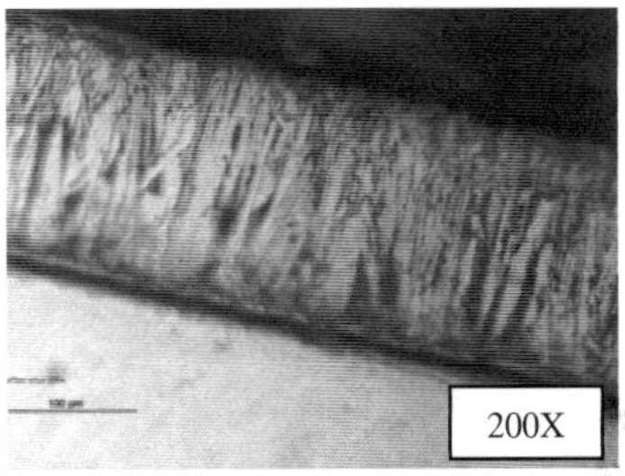

Figure 8. Micro structure with 6 minute time variation

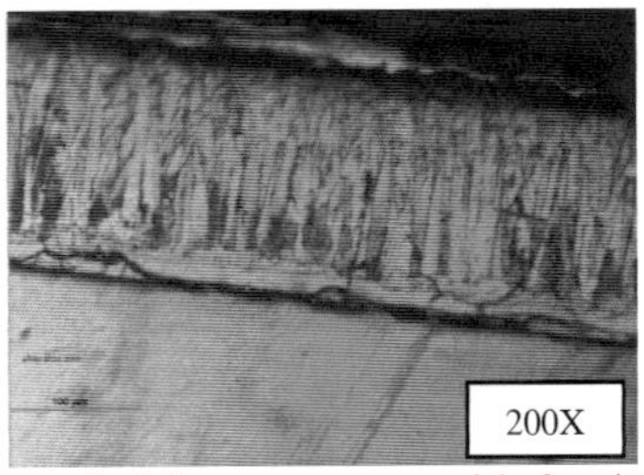

Figure 9. Micro structure with 9 minute time variation

Figure 10. Micro structure with 12 minute time variation

The microstructures of the images above demonstrate result of Zinc (Zn) coating to low steel carbon surface and it is shown that the zinc coating is not evenly distributed. Specimen having pre-treatment is immersed into a container with liquid zinc at 460^0C temperature for 3 (Figure 7), 6 (Figure 8), 9 (Figure 9) and 12 minutes (Figure 10) so that mixture Fe-Zn layer is formed. The formed Fe-Zn mixture in the outer layer is Eta (η) phase, Zeta (ζ) phase and Delta (δ) phase, while layer in steel interface is Gamma (γ) phase. The raw material does not undergo microstructure change. Eta (η) phase is the outer layer which is composed from pure 100% of zinc (Zn) with hexagonal structure and it has soft and ductile mechanical characteristics. Zeta (ζ) phase is the second outer layer after Eta (η) phase which is composed of 94% Zn and 6% Fe ($FeZn_{13}$), and it has monoclinic structure with hard and brittle mechanical characteristics. The next layer is Delta (δ) layer composed of 90% Zn and 10% Fe ($FeZn_7$) with hexagonal structure and it has ductile mechanical characteristic. The last layer is

Gamma (γ) layer composed of 75% Zn and 25% Fe (Fe_3Zn_{10}) with BBC structure and it has thin, hard and brittle mechanical characteristics.

4. Conclusions

a) The longer immersion time, the smaller is the hardness value. Highest hardness value obtained after *hot dip galvanizing* process at 3 minutes immersion time produces 76.012 kg/mm^2 Rokwellhardness (HRb).
b) The longer time immersion, the longer is layer thickness. The highest layer thickness value after *hot dip galvanizing* process obtained after 12 minutes immersion is 217.3 μm.
c) Microstructure testing result shows that raw material composed of *pearlite* and *ferrite*, where *ferrite* crystal dominates more than *pearlite* crystal. Specimen of *hot dip galvanizing* at 3 minutes immersion shows that Zn coating is not evenly distributed and it is formed in stratifying form. Zn layer will be thicker along with longer time immersion until 12 minutes immersion time variation.
d) 3, 6, 9, and 12 immersion time do not change microstructure of specimen after *hot dip galvanizing* process.

Acknowledgements
This work was financially supported by Research Unit for Mineral Processing Technology and Department of Mechanical Engineering, Faculty of Engineering, University of Malahayati

References

[1] Lu W K, Elsenbaumer R L and Wessling B 1995 *Synthetic Metals* **71(1-3)** 2163 – 6
[2] Schlesinger M and Paunovic M 2000 *Modern Electroplating* 4th ed.Wiley New York 13-15
[3] Culcasi J D, Sere P R, Elsner C I and Di Sarli A R 1999 *Surface and Coating Technology* **122(1)** 21 – 3
[4] Wang P B, Xuping J H S, Zhi L and Fucheng Y 2008 *Surface and Coating Technology* **122(9)** 1785 – 8
[5] Wang P B 2007 *Effect of Zinc Bath Temperature on the Coatings of Hot-Dip Galvanizing*, China

IOP Conf. Series: Materials Science and Engineering **285** (2017) 012020 doi:10.1088/1757-899X/285/1/012020

The effect of aluminium on the microstructure and hardness of high austenitic manganese steel

F Bahfie[1], B B Aji[1], F Nurjaman[1], A Junaedi[1] and E H Sururiah[2]

[1] Research Division for Mineral Technology, Indonesia Institute of Science, Ir. Sutami Km 15 South Lampung, Lampung 35361, Indonesia
[2] Physic Department, Faculty of Mathemathics and Natural Sciences, Lampung University, Prof. Dr. Soemantri Brodjonegoro No. 1 Bandar Lampung, Lampung 35145, Indonesia

Email : fathan.bahfie@lipi.go.id

Abstract. High-strength manganese aluminium and austenitic steels have the reason of their perfect combination of high mechanical properties and good plasticity for the structural elements. They have the microstructure stability, and goodstrength properties because of the addition of aluminium. The effect of aluminium on high austenitic manganese steel were investigated. The samples wereexamined with several tests such as microstructure, chemical content, and hardness. The effect of aluminiumhad been influenced the formation of the Fe-Al-Mn-C phase and the lower hardness of samples.

1. Introduction

In recent new, the structural elements (automotive and railway industries) are used high-strength manganese-aluminium austenitic and austenitic-ferritic steels by reason their perfect combination of high mechanical properties and good plasticity. The application of construction materials is relied on improvement of their casting properties and deformability under conditions of plastic working by their development and commercialization. A proper selection of chemical composition, modification of the initial microstructure, grain refining, and application of proper parameters of the thermo-mechanical treatment may be accepted by its, completing an optimal combination of strength and plastic properties[1-9].

In various industries,the demand for high strength steels with excellent ductility has been increasing and automobile manufacturing in particular, aimed at not only improving the productivity, safety and comfort levels, but also reducing the weight of automobile body to decrease fuel consumption and exhaust emission as well [7-10]. On other hand, the newest high-manganese steels, such as transformation induced plasticity (TRIP) and twinning induced plasticity (TWIP), have the microstructure stability, stacking fault energy value and good strength properties [10-14].Material TWIP (twining induced plasticity) is represented by Fe-Mn-C chemical composition with low aluminium content, finally, even with poor silicon, subsequently. Even though, material TRIPLEX (beside iron three elements) is established on the origin of Fe-Mn-C-Al with aluminium content higher than 8 wt % and without silicon content. Depending on high manganese type and on carbon content manganese reaches higher level than 19 wt. % usually and in this way recognizes the basic austenite microstructure of FCC type, consequently [15-17].

Therefore, the main of this research was to investigate the effect of low weight percent aluminium on the microstructure and hardness of high austenitic manganese steel. Whereas high carbon in high austenitic manganese steel can affect on its microstructure.

2. Experimental

The high carbon austenitic manganese steel with aluminium and non-aluminium was investigated. The chemical composition of the sample is presented in table 1. Steel, aluminium scrap, and ferromanganese was melted at temperature 1600°C and then it was poured in the mould. The effects of aluminium on the hardness and microstructure of austenitic manganese steel were examined. The measurement of hardness and density was conducted by using Rockwell hardness testing machine and Archemedes's teory, while the microstructure was analyzed by an optical microscope (Nikon Eclipse MA 100, Japan).

Tabel 1. Chemical composition

	Mass fraction of element, wt. %				
	C	Mn	Si	Ni	Al
Sample 1	1.15	15	0.73	0.26	0.002
Sample 2	1.06	20	0.7	0.23	2.78

3. Result and Discussion

3.1 The characterization of high carbon austenitic manganese steel

The microstructure of high carbon austenitic manganese steel in as-cast condition comprised austenitic and some of the carbides $(Fe,Mn)_3C$ (black) in austenite (white) and grain boundaries, as shown in figure 1. According [18],The carbides were formed and precipitated along austenite grain boundaries due to the slow cooling in solidification process. This carbide network resulted a negative effect on toughness. From figure 1, the matrix was austenite. The hardness of high carbon austenitic manganese steel was only 19 HRC.

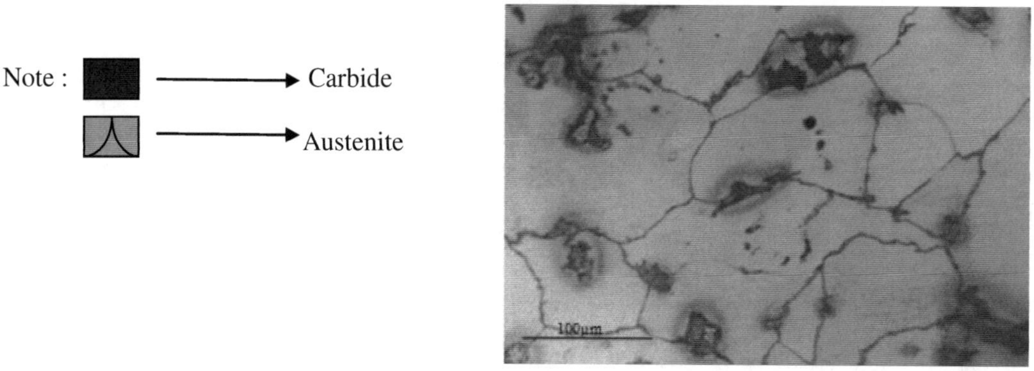

Figure 1. Microstructure of sample 1

3.2 The effect of aluminium on the microstructure and hardness of high carbon austenitic manganese steel

The addition of aluminium was conducted to dissolve the carbides that were found in sample 1. In figure 2 shows the microstructure of austenitic manganese steel after adding aluminium. It was found that Fe-Mn-Al-C phase appeared on its structure. According [4], the characteristic of high aluminium manganese alloys with high content of carbon was the presence in the structure of carbide variable composition $(Fe,Mn)_3AlC_{1-x}$ with FCC lattice in which atoms of iron or manganese are arranged on the faces, and a carbon atom is in the center . Its matrix will be dominated by dendrite of Fe-Al-C structure and austenite matrix. This Fe-Al-C structure resulted the low hardness of this austenitic manganese steel. All carbides were dissolved into the dendrite Fe-Al-C by adding aluminium. Carbon will be decreased in the austenitic matrix in table 1 and figure 2.The hardness of sample 2 was only 11,5 HRC.The hardness of sample 2 was lower than sample 1.

Note : → Fe-Al-C Structure.

→ Austenite.

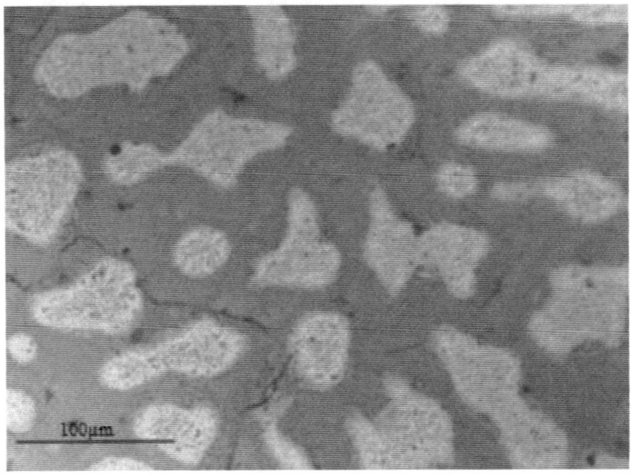

Figure 2. Microstructure of sample 2

4. Conclusion

The microstructure of material 1 consisted the austenitic matrix with carbide along its grain boundaries. However the material 2 founded the Fe-Al-C dendrite in austenite matrix was caused the addition of aluminium. The hardness have the small effect of aluminium. At higher weight percent of aluminium would be affected larger the dispersion of Fe-Al-C dendrite, and lower hardness. Adding aluminium hadn't.

Acknowledgment

We would like to thank the Indonesian Institute of Sciences for the financial support, Mrs. Nurbaiti Marsas and Mr. Zulkifli Ilyas for the technician support.

References

[1] Hamada A S, 2007 *Manufacturing, Mechanical Properties and Corrosion Behaviour of High-MnTWIP Steels* (Czech Republic:Universitatis Ouluensis) 1 - 56.
[2] Kuc D, Hadasik E, Niewielski G, Schindler I, Mazan-cová E, Rusz S and Kawulok P 2012 *Archives of Civil and Mechanical Engineering* **12** 312 - 7
[3] Lasek S and Mazancova E 2013 Metalurgija **52(4)** 441 - 4
[4] Mazancova E, Ruziak I and Schindler I 2012 *Archives of Civil and Mechanical Engineering* **12** 142 - 7
[5] Wiewiórowska S 2013 Archives of Metallurgy and Material **58** 573 - 8
[6] Jabłońska M and Śmiglewicz A 2013 *Defect and Diffusion Forum* **334-335** 177 – 81
[7] Bleck W and Phiuon K 2007 *Freiberg: Säsiche Fachtagung Umformtechnik, Werkstoffe und Komponen-ten für Fahrzeugen* **14** 38 - 55
[8] Jab łońska M, Śmiglewicz A, Niewielski G and Hetmańczyk M 2011 *IOP Conference Series - Materials Science and Engineering* **22** 1 - 5
[9] Farahata A I Z, Hamedb O, El-Sisi A and Hawash M 2011 *Materials Science and Engineering A* **530** 98 – 106
[10] Jabłońska M B 2014 *Arch. of Metal. and Mater.* **59(3)** 1193 - 7
[11] Grajcar A, Różanski M, Kaminska M and Grzegorczyk B 2014 *Arch. of Metal. and Mater.* **59 3** 1164 - 9
[12] Yang W S and Wan C M 1990 *Journal of Materials Science* **25** 1 - 3

[13] Kuziak R, Kawalla R and Waengler S 2008 *Archives of Civil and Mechanical Engineering* **2** 103 - 18
[14] Chen L, Zhao Y and Qin X 2013 *Acta Metall Sin.* **26** 1 - 15
[15] Scott C, Allain S, Farel M and Guelton N 2006 *Rev. de Met.* **103** 307 – 17
[16] Schumann V H 1972 *Neue ütte* **17** 605 – 9
[17] Grässer O and Frommeyer G 1998 *Material Sci. and Technology* **14** 1213-121
[18] Nurjaman F, Bahfie F, Widi A and Shofi A 2017 *IOP Conf. Series: Journal of Physics: Conf. Series* **817** *012063*

IOP Conf. Series: Materials Science and Engineering **285** (2017) 012021 doi:10.1088/1757-899X/285/1/012021

Influences of magnetic field on the fractal morphology in copper electrodeposition

Sudibyo[1], M B How[2] and N Aziz[2]

[1]Research Unit for Mineral Technology, Indonesian Institute of Sciences, Lampung, Indonesia
[2]School of Chemical Engineering, Universiti Sains Malaysia, Pulau Pinang, Malaysia

Email: sudibyo@lipi.go.id

Abstract. Copper magneto-electrodeposition (MED) is used decrease roughening in the copper electrodeposition process. This technology plays a vital role in electrodeposition process to synthesize metal alloy, thin film, multilayer, nanowires, multilayer nanowires, dot array and nano contacts. The effects of magnetic fields on copper electrodeposition are investigated in terms of variations in the magnetic field strength and the electrolyte concentration. Based on the experimental results, the mere presence of magnetic field would result in a compact deposit. As the magnetic field strength is increased, the deposit grows denser. The increment in concentration also leads to the increase the deposited size. The SEM image analysis showed that the magnetic field has a significant effect on the surface morphology of electrodeposits.

1. Introduction

Electrodeposition plays a very important role in various types of industries. Electrodeposition is not only used to enhance the face value of metal articles by improving the appearance aesthetically but electrodeposition is also used to improve contact resistance, reflection properties of materials and to impart friction properties [1]. The problem of obtaining a uniform, dense and compact deposition is the problem in electrodeposition. There are numerous studies that had been carried out to reduce it [3. 7 8]. One of the methods of tackling this problem is MED. It had been found that different strength of magnetic field will affect the growth pattern and its morphology [7, 8]. Moreover, in the MED technology, the electrolyte concentration is also a crucial factor in determining the growth pattern of the deposited metal [9]. The objectives of this work are to study the effect of magnetic strength on growth fractal and its morphology of copper electrodeposits. The influence of electrolyte concentration towards growth fractal also has been studied.

2. Experiment Procedure

The copper fractal electrodeposits were developed in a flat circular cell with a copper wire cathode (1 mm in diameter) at the center and a copper ring anode (thickness 0.5 mm, outer diameter 70 mm and inner diameter 50 mm) as shown in Figure 1. The distance between anode and cathode was 49.5 mm. Copper sulfate solutions of various concentrations were made up using $CuSO_4 \cdot 5H_2O$ in distilled water. The $CuSO_4 \cdot 5H_2O$ concentration was varied in the range from 0.1 to 0.75 M in 0.5 M H_2SO_4. The depth of the aqueous film can be varied by regulating the total volume of the $CuSO_4$ solution. The copper was electrodeposited with a voltage of 6 V. A Ferrite and Neodymium permanent magnets were used to provide a weak (300 gauss) and strong (900 gauss) magnetic field, respectively. The resulting fractal patterns were photographed using a digital camera and then analyzed using Matlab image processing. A mass microbalance was used to measure the mass of fractal electrodeposits. The maximum diameter of fractals was measured using a ruler. The surface morphology of fractal electrodeposits was analyzed and characterized using scanning electron microscopy (SEM).

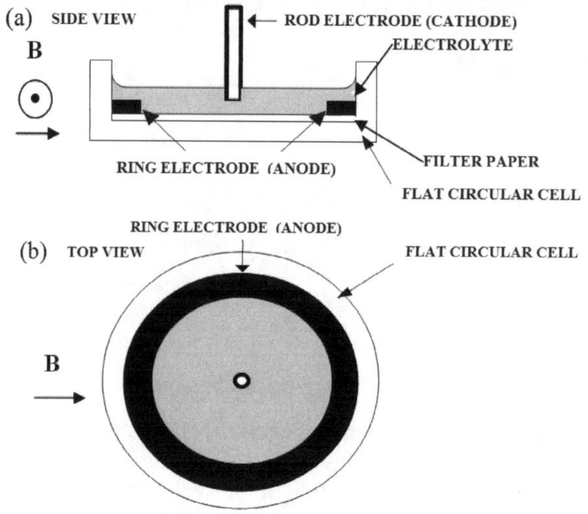

Figure 1. Schematic diagram of a flat circular electrochemical cell to investigate the growth fractal produced via MED: (a) side view (b) top view

3. Result and discussion

3.1. Effect of Electrolyte Concentration on Magnetic Electrodeposition

From figure 2a to figure 2e, we found that the dendritic structure grows larger and more compact as the concentration of $CuSO_4$ increases. As the concentration increases, the tiny dendritic structure in Figure 2a branched out further to a state as shown in figure 2e. Then, the dendritic structure becomes denser when the concentration was increased to 0.2 M. At 0.5 M, the dendritic structure already achieved a dense branching morphology (DBM) state [12]. Since the concentration keeps increasing, the dense branching morphology will inhibit more copper from depositing within it. Thus it will cause the aggregate structure to spread out. Consequently, as the concentration used is increased to 0.75 M, the electrodeposited copper had DBM as well as protruding dense branches from the DBM structure as shown in figure2e.

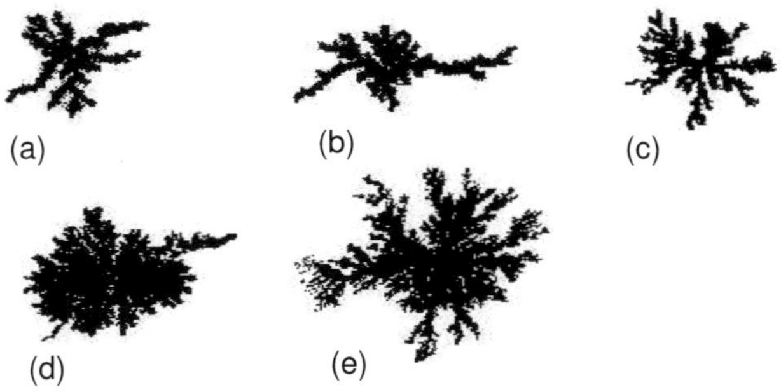

Figure 2. Copper electrodeposits (applied voltage of 6 V, neodymium magnet, time duration 20 minutes) : (a) $CuSO_4$ 0.1 M, (b) $CuSO_4$ 0.15 M, (c) $CuSO_4$ 0.2 M (d) $CuSO_4$ 0.5 M (e) $CuSO_4$ 0.75 M

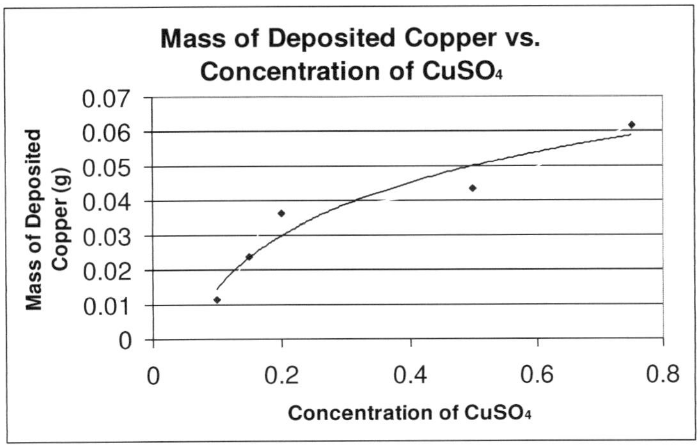

Figure 3. Mass copper electrodeposits in variation concentration of $CuSO_4$

Besides that, when the concentration is increased, the mass of the deposited copper will also increase accordingly as seen in Figure 3. This increment of growth pattern and the mass deposited can be linked to the increasing concentration of the $CuSO_4$ in the electrolyte. As the concentration increases, so do the free cations (Cu^{2+}) available in the electrolyte. These free cations are more ready to be deposited at the deposition area to form the deposited metal [14]. As the quantities of these free cations increases so do the deposition intensity per unit area. This will result in a higher rate of copper deposition as well as produce an aggregate with higher mass and larger in size.

The increment in concentration also will increase the collision rate of the species within the electrode. It is due to the increasing quantity of substances within a unit area. As the collision rate increases, it will also induce a turbulent flow of the electrolyte. In return, this will increase the flux of the species [13]. As a result, the thickness of the Nernst diffusion layer will gradually be reduced and this will decrease the screening effect too. The magneto-hydrodynamic (MHD) effect caused by the magnetic field will also decrease the screening effect at the deposition site [2, 12]. When the screening effect of the Nernst diffusion layer decreases, more same charged cations are able to converge at any single point at the deposition area without being repulsed away [15, 16, 17]. Thus, more cations are able to discharge at the deposition site which will results in a denser, larger and heavier growth as shown from figure. 2a to figure 2e and figure 3.

However, based on figure 3, when the concentration of the $CuSO_4$ in the electrolyte is increased, the mass of the deposited copper increases not in a linear fashion. The increment tends to be greater when the concentration range is lower and decreases as the concentration increases. This observation occurs as a further increment of concentration will result in the increase of cations. The high amount of cation will increase the cathodic polarization and decreases the anodic polarization. The polarization occurred will reduce the difference in potential between anode and cathode to a minimum [1]. Consequently, it tends to resist the flow of the current to the cathode. Thus, the increment in the mass of deposited copper is reduced. Similar observations are made when the experiment is conducted with variation in the electrical potential supplied.

3.2. Effect of Magnetic Field on Surface Morphology

When conventional electrodeposition is carried out without the magnetic field influence, it is observed that the copper fractal formed a diffusion limited aggregation (DLA)-like structure (figure 4a). The fractal pattern becomes denser and compact as the increase of magnetic field as shown in figure 4b. This phenomenon is due to the presence of MHD effect (figure4b). This MHD effect is actually generated by the magnetic force and as the magnetic field strength increases, the strength of the MHD effect increases as well. This MHD effect acts by reducing the thickness of the Nernst diffusion layer. When the thickness is reduced, the screening effect within the layer is reduced as well [2, 12]. Then, more copper ions are able to release at a given point in the deposition area. Hence, the overall aggregate will become more compact. Consequently, the mass deposited will also increase.

The MHD effect also will cause a turbulent flow near the deposition area which will enhance the ionic mass transfer. The additional metal ion introduced by the turbulent flow will also encourage more side branches to develop beside the main branch.

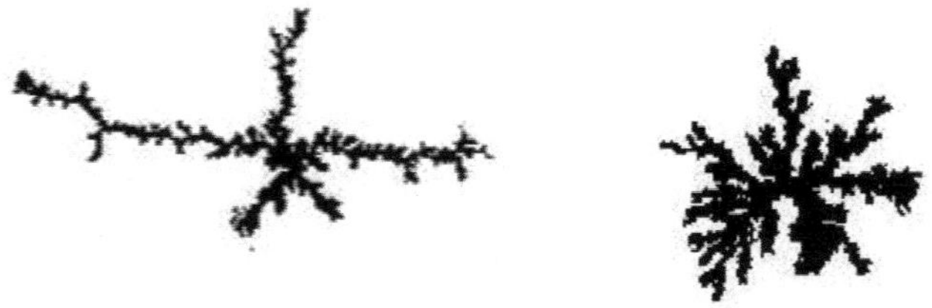

(a) (b)

Figure 4. Copper electrodeposits (applied voltage 6 V, $CuSO_4$ 0.2 M, time duration 20 minutes): (a) without magnet, (b) in strong magnetic field [18]

Figure 5a shows the SEM image of copper electrodeposited with the absence of a magnetic field. The long cylindrical structures in the diagram are the fiber strands of the filter paper. Each of the tiny clusters represents a new nucleation point and when aggregated together, it forms the domain of the electrodeposited copper as in the figure 5a. Even at a microscopic level, the overall domain of the deposited copper is sparse. Each of the domains is situated far away from each other and it also seems to be coarse. At a higher magnification (figure 5b), the tiny cluster of deposited copper within the domain is found to be highly inconsistent in its size and shape. Some clusters are big and some are small. The shapes of the cluster are also highly irregular. Even as the surface of a tiny cluster may seem to be smooth; however, when aggregated together, the overall domain surfaces are found to be coarse.

Figure 6a shows the SEM picture of the copper deposited in the presence of an external applied magnetic field. It can be seen that the overall electrodeposited copper is much compact and uniform in shape. Most of the aggregate is also spherical shaped and it contains several large domains of 1-3.5μm in size. At a higher magnification (figure 6b), these domains are actually made up of tiny clusters of

IOP Conf. Series: Materials Science and Engineering **285** (2017) 012021 doi:10.1088/1757-899X/285/1/012021

copper nanoparticles. These tiny clusters of copper nanoparticles indicate that fresh nucleation occurs and the growths of the particles take place uniformly under the influence of the magnet. The overall surface morphology of the domains and the clusters are also observed to be much smoother and uniform. Thus, the process of electrodeposition in the presence of an externally applied magnetic field is found to be assisting and increasing the fresh nucleation as well as the growth process. The cluster structure also tends to aggregate together and the surface is much smoother and regular. This observation can be explained via the magneto-hydrodynamic effect generated by the magnetic field.

The MHD effect will increase the ionic mass transfer near the cathode surface by creating a convection flow near the electrodeposition area. Thus, the metal grains tend to grow uniformly and have a smooth surface in a magnetic field since the convection provide sufficient metal ions to each grain during the growing process.

The MHD convection also reduces the thickness of Nernst diffusion layer and the screening effect within it. As the screening effect is reduced, more metal ions are able to approach the electrodeposition side unhindered [15, 16, 17]. Fresher nucleation will occur and the growth of the new nucleation point also will be more uniform and fast as the result of the increasing quantity of copper ions.

(a) (b)

Figure 5. SEM photograph of copper deposited : (a) without magnet at 10.00 K X, (b) without magnet at 20.00 KX (higher magnification)

(a)

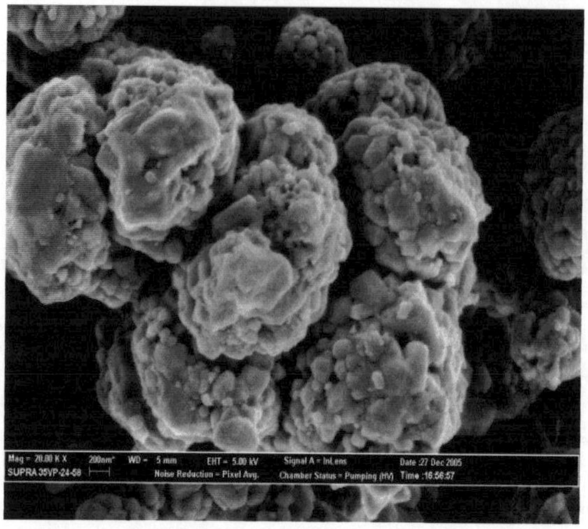

(b)

Figure 6. SEM photograph of copper deposited: (a) under strong magnetic field effect (900 gauss) at 10.00 K X, (b) under strong magnetic field effect (900 gauss) at 20.00 KX (higher magnification).

4. Conclusions

Based on the result and discussion above, the increment of bath concentration as well as the magnetic field may result in a heavier deposit but one must bear in mind that the deposit may be large in size. Meanwhile, the increasing of the magnetic field will increase more compact and dense electrodeposits. The surface morphology of copper electrodeposits also shows that electrodeposits under magnetic effect more compact and uniform and shape. Hence, we can conclude that the introduction of magnetic field on electrodeposition able to improve the surface morphology of electrodeposits.

Acknowledgment

This work was supported by FRGS under grant No. 607113, Malaysia, and by Ministry of Research, Technology and Higher Education- the Republic of Indonesia through INSINAS research grant no. RT-2016-0227 is greatly acknowledged.

References

[1] Sudibyo, How M B, Basir N I and Aziz N 2008 *Study of magnetic field effects on copper electrodeposition*. RSCE – SOMCHE, Kuala Lumpur, Malaysia, 961
[2] Krause A, Uhlemann M, Gebert A, and Schultz L 2004 *Electrochimica Acta* **49** 4127
[3] Matsushima H, Bund A, Plieth W, Kikuchi S, and Fukunaka 2007 Y *Electrochimica Acta* **53** 161
[4] Sudibyo, Ismail A B, Uzir M H, Idris M N and Aziz N 2009 *International Conf. on Green Tech. and Eng*. 52
[5] Sudibyo, Ismail A B, Uzir M H, Idris M N and Aziz N *2009 International Conf. on Green Tech. and Eng*. 57
[6] Matsushima J T, Trivinho-Strixino F and Pereira E C 2006 *Electrochimica Acta* **51** 1960
[7] Mogi I and Kamiko M 1996 *J. Crystal Growth* **166** 276
[8] Mhiochain T R N, Hinds G, Martin A, Chang Z Y E, Lai A, Costiner I and Coey J M D 2004 *Electrochimica Acta* **49** 4813
[9] Coey J M D and Hinds G 2001 *J.Alloys Comp*. **326** 238
[10] James A M 1984 *Electrochemistry Dictionary* (John Wiley & Sons, Ltd)
[11] Mansur Filho J C, Silva A G, Carvalho A T G and Martins M L 2005 *Physic A* **350** 393
[12] Bund A, Koehler S, Kuehnlein H H and Plieth W 2003 *Electrochimica Acta* **49** 147
[13] Ganes V, Vijayaraghavan D and Lakshminarayanan V 2005 *Appl. Surf. Sci* **240** 286

[14] Leventis N, Chen M, Gao X, Canalas M and Zhang P 1998 *J. of Phys. Chem.* B **102** 3512
[15] Chopart J P, Douglade J, Fricoteaux P and Olivier A 1991 *Electrochimica Acta* **36** 459
[16] Fahidy T Z 2001 *Prog. Surf. Sci.* **68** 155
[17] Hinds G, Spada F E, Coey J M D, Mhiochain T R, and Lyons M E G 2001 *J. Phys.Chem.* B **105** 9487
[18] Sudibyo, How M B, Basir N I and Aziz N 2008 *Study of magnetic field effects on copper electrodeposition* RSCE – SOMCHE Kuala Lumpur Malaysia 961 - 4

IOP Conf. Series: Materials Science and Engineering **285** (2017) 012022 doi:10.1088/1757-899X/285/1/012022

The ferromanganese production using Indonesian low-grade manganese ore using charcoal and palm kernel shell as reductant in mini electric arc furnace

Y I Supriyatna[1], Z Zulhan[2], and Y Triapriani

[1]Research Unit for Mineral Technology, Indonesian Institute of Sciences, Lampung 35361, Indonesia.
[2]Department of Metallurgical Engineering, Bandung Institute of Technology, Bandung 40132, Indonesia

Email: yayat_iman@yahoo.com

Abstract. The series of ferromanganese production have been conducted using charcoal and palm kernel shell as a reducing agent to replace of cokes. The experiment was preceded by the characterization of raw materials. Manganese ore and limestone were analyzed using XRF and XRD. Charcoal and palm kernel shells were analyzed proximate to determine its carbon content. Based on the analysis of raw materials then calculated mass balance to determine the raw material requirements for each experiment. The variations are the use of reductant ratio (1, 1.5 and 2.1 stoichiometric ratio). Products and slag are analyzed using OES and AAS to determine its chemical composition. The results showed that the use of palm kernel shell as a reductant better than charcoal for all use ratio (1, 1.5 or 2.1 stoichiometric ratio). The highest percentage of manganese extraction using palm kernel shells as a reducing agent is 49.91% (75.58% Mn, 15.75% Fe, 2.12% C, 5.23% Si, 0.08% P) with the product of FeMn is 6,6 kg. The highest percentage of manganese extraction using charcoal as reductant is 44.16% (72.35% Mn, 18.44% Fe, 1.93% C, 5.69% Si, 0.02% P) with the product of FeMn is 6,1 kg. The results showed that palm kernel shell and charcoal could potentially be used as a reductant in the production ferromanganese.

1. Introduction

Indonesia has the potential of manganese reserves contained in several locations such as Sumatra, Java, Nusa Tenggara Timur and a small portion in Kalimantan, Sulawesi, and Papua. Reserve of manganese in Java Island based on data from Center for Geological Resources (PSDG) – The Ministry of Energy and Mineral Resources (ESDM) is in West Java (Sukabumi, Tasikmalaya), Central Java (Kebumen, Magelang and Wonogiri) and East Java (Blitar, Jember, Malang, Pacitan, Trenggalek, Tulungagung)[1]. Manganese resources of which there are fairly large in East Java province. Complete data about the mineral resources of manganese in East Java province is shown in Table 1.

Until the end of 2013, more manganese is exported in raw material form without unprocessed. On January 12, 2014, Act No. 4 of 2009, on the obligation for the holder of Mining Business License (IUP) and Special Mining Business License (IUPK) to increase value-added products of mineral and coal through processing and refining, formal applicable. In Article 103 paragraph 1 states that "Holders of Production Operation IUP and IUPK are required to perform processing and refining of mining production in the country" [2]. Associated with added value activities of minerals through processing and refining are regulated and clarified through the regulation No 7 of 2012 Minister of Energy and Mineral Resources. The regulation amended several times. Recent changes are regulation of ESDM RI No 8 of 2015 concerning mineral added value through Activities of Mineral Processing and Purification whose set a minimum limit for the product of processing and/or purification for 12 metallic minerals. Manganese minerals in accordance with Annex I ESDM No. 8 of 2015 abaout minimum limit of processing and refining within the country in form ferroalloys is ferromanganese with Mn contents $\geq$ 60% [3].

A ferromanganese is an alloy form between the iron and manganese elements in which manganese is the dominant element. The content of manganese in ferromanganese generally ranges between 70% - 90% Mn. The content of the other elements in these alloys, among others, carbon (C) up to 7.5%, Si up to 3%, P up to 0.4% and S up to 0.05%, the rest consisting of elemental iron (Fe) [4].

Another type of iron and manganese alloy with manganese contents lower (10-25%) is known as Spiegel Eisen. Classification this ferromanganese showed in Table 2.

Table 1. Manganese mineral resources in East Java Province [1]

No	District / Town	Locations	Sub-district	Resource (ton)	Information
1.	Tulungagung	Trenggalek	Basuki, Watulimo	214	Hypothetical
		Sukerejo	Rejotangen	530	Hypothetical
		G. Cemenung	Rejotangen	54,000	Mn content : 47,9% (measured)
2.	Blitar	Daerah Tambakrejo, Wlingi	Tambakrejo, Wlingi	2,000	Measured
		G. Jimbe, G. Puncukasem, Wlingi	Wlingi	22,000	Mn content : 34,24% - 41,43% (measured)
3.	Jember	Karangbale, G. Marcondon, Sekunir	Puger	489,563	Designated
4.	Pacitan	Tegalombo dan Bandar	Tegalombo, Bandar	18,315	Measured
5.	Trenggalek	Daerah Kec. Bandung	Bandung	45,000	Measured
		Desa Melis dan Sukorame	Gandusari	3,000,000	MnO_2 content : 79,4% (reserve)
6.	Lumajang	Ds. Kalirejo	Donomulyo	30,000	Mn content : 50-90% (Designated)

Table 2. Classification of ferromanganese according to ASTM standards [5]

Elements (%)	Low Carbon Ferromanganese		Medium Carbon Ferromanganese				High Carbon Ferromanganese (standard)		
	Grade A	Grade B	Grade A	Grade B	Grade C	Grade D	Grade A	Grade B	Grade C
Mn	85-90	80-85	80-85	80-85	80-85	80-85	78-82	76-78	74-76
C (max)	0.75	0.75	1.50	1.50	1.50	1.50	7.50	7.50	7.50
Si (max)	2.00	5 – 7	1.50	1.00	0.70	0.53	1.20	1.20	1.20
P (max)	0.20	0.30	0.30	0.30	0.30	0.30	0.35	0.35	0.35
S (max)	0.02	0.02	0.02	0.02	0.02	0.02	0.05	0.05	0.05

Raw materials for both methods were manganese ore. Reducing agents are cokes or coal and as a flux used limestone. The reduction reaction that occurs in the element of Mn and Fe in the ferromanganese smelting process is [6]:

$$MnO_{2\,(s)} + \tfrac{1}{2}\,CO_{(g)} = \tfrac{1}{2}\,Mn_2O_{3(s)} + \tfrac{1}{2}\,CO_{2(g)} \qquad \Delta H_{298} = -99{,}4\ kJ \quad (1)$$

$$Mn_2O_{3(s)} + 1/3\ CO_{(g)} = 2/3\ Mn_3O_{4(s)} + 1/3\ CO_{2(g)} \qquad \Delta H_{298} = -66{,}5\ kJ \quad (2)$$

$Mn_3O_{4(s)} + CO_{(g)} = 3\ MnO_{(s)} + CO_{2(g)}$ $\Delta H_{298} = -51{,}2$ kJ (3)

$1/3\ C_{(s)} + 1/3\ CO_{2(g)} = 2/3\ CO_{(g)}$ $\Delta H_{298} = +57{,}5$ kJ (4)

$½\ Fe_2O_{3(s)} + 3/2\ CO_{(g)} = Fe_{(s)} + 3/2\ CO_{2(g)}$ $\Delta H_{298} = -11{,}59$ kJ (5)

$½\ Fe_2O_{3(s)} + 3/2\ C_{(s)} = Fe_{(s)} + 3/2\ CO_{(g)}$ $\Delta H_{298} = +247{,}1$ kJ (6)

$MnO_{(l)} + C_{(s)} = Mn_{(l)} + CO_{(g)}$ $\Delta H_{298} = +236{,}4$ kJ (7)

$SiO_{2\ (l)} + 2C_{(s)} = Si_{(l)} + 2\ CO_{(g)}$ $\Delta H_{298} = +725{,}9$ kJ (8)

Pyrolusite is the general manganese compounds contained in manganese ore, are relatively unstable and easily reduced in solid form in the presence of CO gas or under reductive conditions [7]. Currently, the main technology in pyrolusite reduction process uses coal and/or coke as a reductant and fuel. However, this reductant causes in pollution and greenhouse gas emissions. The use of fossil fuels such as coke and coal led to an increase in greenhouse gases in the atmosphere from 280 ppm to 390 ppm, which has serious impact on global warming [8].

In addition to environmental concerns, the price of coal and coke which are relatively high is another obstacle that must be faced by businessmen in order to increase production and reduce the cost of production. Therefore, manganese producers are under pressure large enough to change the existing processes and/or develop a new process that is more environmentally friendly to meet the challenge of climate change while remaining economical.

Recently, the use of biomass as a fuel and reductant in the reduction process of manganese ore has attracted a lot of attention. For example, research by Tian et al [7] in 2010 using corn cobs as a reducing agent to extract manganese from low-grade manganese ore. Research conducted Zhao et al [8] in 2010 showed manganese oxide ore can be reduced by roasting using a straw at a temperature below 650 ° C. Straw is a renewable biomass resource and can generally be found as waste in the food and agriculture industry.

Another biomass that can be used as a carbon source is palm kernel shell which is a waste of the palm oil industry. Indonesia is one of the largest producers of palm oil is about 16-17 million tons/year. Approximately 20-25% (3.5-4 million tons / year) palm kernel shells obtained from palm oil production [9]. Palm kernel shells can be converted into palm kernel shell coal that can serve as a reductant such as coal and coke. Besides palm kernel shells, carbon sources derived from biomass that can be used as a reducing agent is charcoal.

Based on these facts, these researchs was conducted to ferromanganese production using manganese ore obtained from the sub-district Puger Jember - East Java. The smelting process carried out in a mini electric arc furnace with palm kernel shells (PKS) and charcoal as a reductant. The processing time, the ratio of reductant addition and the ratio of the addition of limestone as a flux using the research results with coke as a reductant [10]. This study was a preliminary research in assessing the potential of palm kernel shells and charcoal as a reducing agent in the smelting of Indonesia manganese ore in a mini electric arc furnace.

2. Experimental Procedures

The research includes several stages as follows :

1. Preparation and analysis of manganese ore, limestone, palm kernel shells (PKS), and charcoal as a raw material for the ferromanganese production. Cokes and charcoal are crushing to obtained a similar size that is 2-4 cm. Manganese ore and limestone were crushing to obtained a size of 1-3 cm. Then, manganese ore and limestone were grinding using ball mill up to -200 mesh for analyzed using XRF and XRD to determine contents of elements and compounds present in the raw material. Palm kernel shells and charcoal was proximate analyzed to determine the carbon content.
2. Subsequent experiments are ferromanganese production using PKS and charcoal as a reductant in mini electric arc furnace (Figure 1). The experiment process of smelting manganese ore in an electric arc furnace begins by dividing the raw material equally into 6 parts according to its composition. The first and second enters for the beginning of the

process. Then the three, four, five and six sections are entered at an interval of 5 minutes. Temperature measurements using immersion thermocouple (NSP 203R C / AL model) were performed prior to the pouring process of metal and slag. The melted metal and slag are poured into the mold that has been provided.

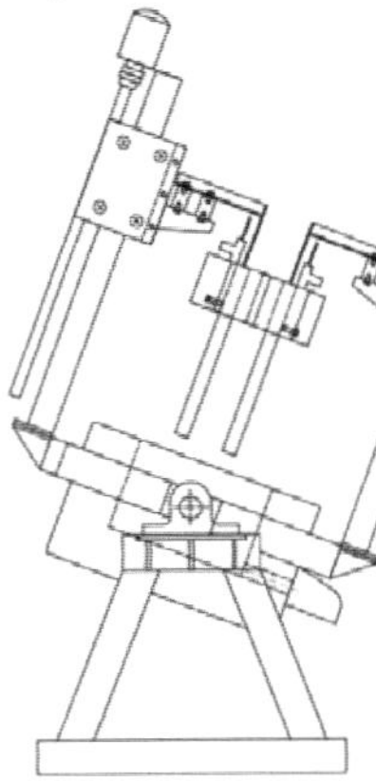

Figure 1. Design of mini Electric Arc Furnace

3. Analysis of the product and the slag from an experiment using OES and AAS to obtain the effect of variations reductant into the percentage of manganese extraction.

3. Results and Discussion

3.1. Characteristics of Raw Materials

The chemical composition of raw materials can be seen in Table 3 to Table 5.

Table 3. Chemical composition of Limestone

Compound	Amount (%)
CaO	55.65
SiO_2	0.13
Al_2O_3	0.22
Fe_2O_3	0.20
MgO	0.01
MnO	0.63
Na_2O	0.01
K_2O	0.04
TiO_2	0.03
P_2O_5	0.02
SO_3	0.00
LOI	43.06

Table 4. Chemical composition of Manganese Ore

Compound	Amount (%)
MnO_2	52.7
SiO_2	29.11
Al_2O_3	2.46
Fe_2O_3	8.17
TiO_2	0.12
K_2O	1.06
CaO	2.67

MgO	0.57
Cr_2O_3	1.36
P_2O_5	0.16
CuO	0.071
BaO	0.59
SO_3	0.003
Co_2O_3	0.001
NiO	0.002
PbO	0.59
ZnO	0.24
SrO	0.084
Y_2O_3	0.001

Table 5. Result of Proximate Analysis

Compound	Palm Kernel Shell	Charcoal
Fixed Carbon (wt%)	21.11	76.85
Volatile Matter (wt%)	67.43	8.75
Ash (wt%)	2.30	4.37
Moisture (wt%)	9.16	10.03

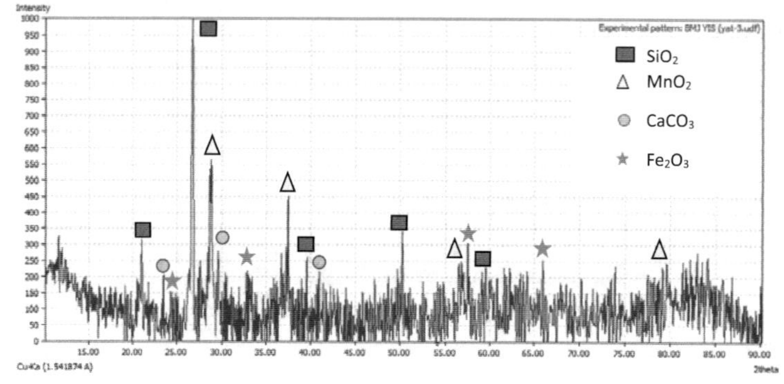

Figure 2. XRD pattern of manganese ore

The results of XRF analysis (Table 1) and XRD (Figure 2) show that the most dominant compound in the manganese ore is Pyrolusite (MnO_2). Other detected compounds are SiO_2, Fe_2O_3, and $CaCO_3$.

3.2. Result of Smelting Process

In these experiments, the processing time and the addition of limestone, using data from previous studies conducted Yayat et al [10]. The addition of limestone is 7 kg and processing time is 70 minutes. Manganese ore is used 30 kg.

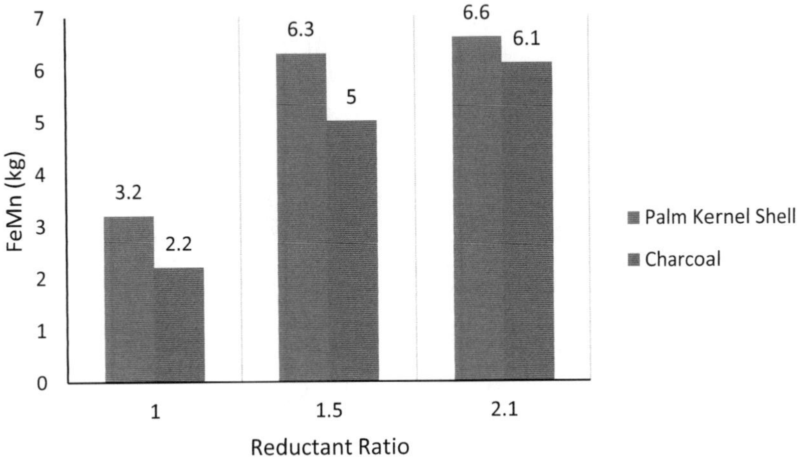

Figure 3. Effect of reductant type to the ferromanganese product in the composition of the 30 kg of manganese ore, 7 kg of limestone and 70 minutes of the processing time

Ferromanganese products using palm kernel shells and charcoal as reductant indicate the same trend. The trend is in line with the increasing used of reductant then the ferromanganese product increased (Figure 3). This could be due to the presence of excess carbon which would keep the carbon in the system so that it can produce CO gas is needed to reduce MnO_2 to be MnO and ensuring the availability of carbon to reduce MnO in slag to metallic Mn. Based on Figure 3, when added a reductant 1 stoichiometry, 1.5 stoichiometry and 2.1 stoichiometry experimental results show the highest ferromanganese products generated using palm kernel shells as a reductant (3.2 kg FeMn for 1 stoichiometry, 6.3 kg FeMn for 1.5 stoichiometry and 6.6 kg FeMn for 2.1 stoichiometry).

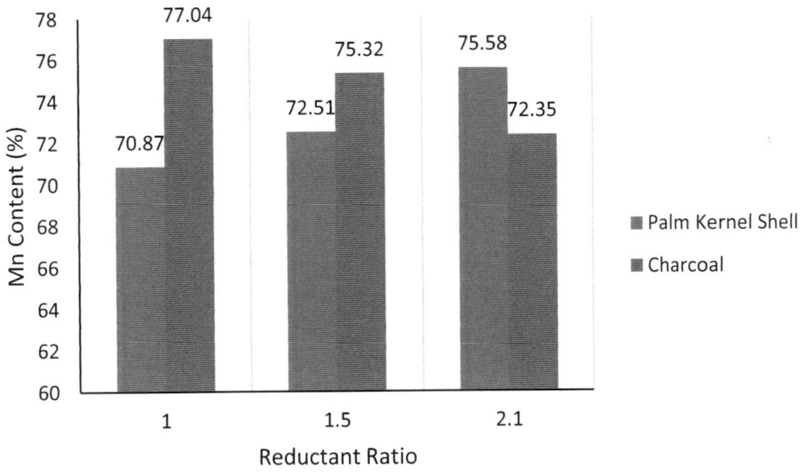

Figure 4. The influence of reductant type on Mn content in the product on the composition of 30 kg of manganese ore, 7 kg of limestone and 70 minutes of the processing time

The manganese content results of the experiment showed different trends for different types of reductant. Palm kernel shells indicate increased usage of reductant then the manganese content is also increasing. The increasing use of charcoal reductant then the manganese content in the product obtained a slight decline (Figure 4).

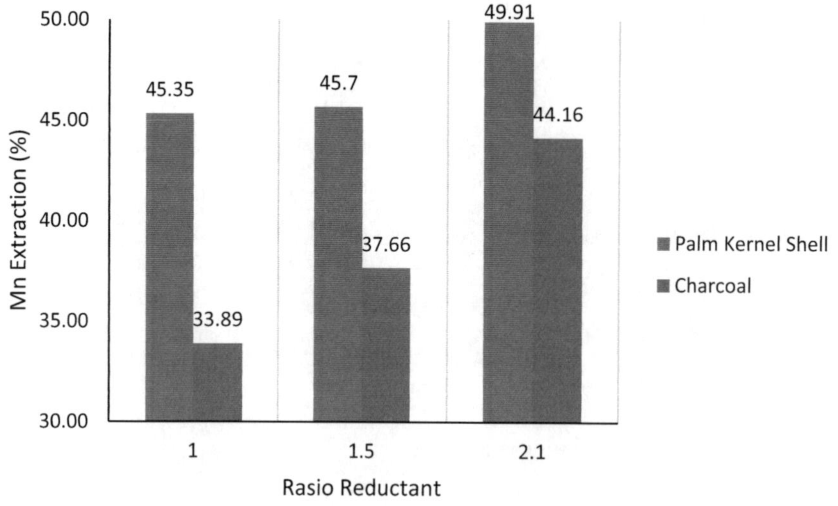

Figure 5. The influence of reductant type on Mn extraction percentage on the composition of 30 kg of manganese ore, 7 kg of limestone and 70 minutes of the processing time.

The highest percentage of manganese extraction for 1 stoichiometry, 1.5 stoichiometry and 2.1 stoichiometry obtained using palm kernel shells as reductant (Figure 5). The highest percentage of manganese extraction for reductant palm kernel shells or charcoal obtained in 2.1 stoichiometry.

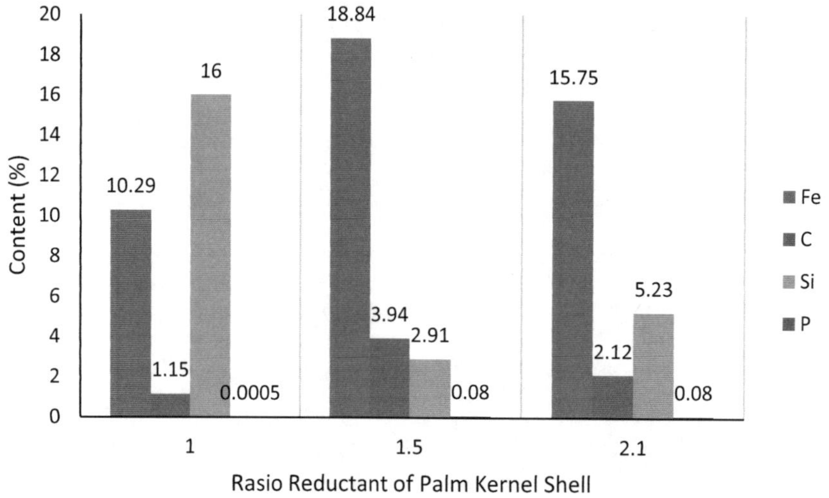

Figure 6. The effect of adding palm kernel shells as a reductant on the composition of the product composition on the 30 kg of manganese ore, 7 kg of limestone and 70 minutes of the processing time

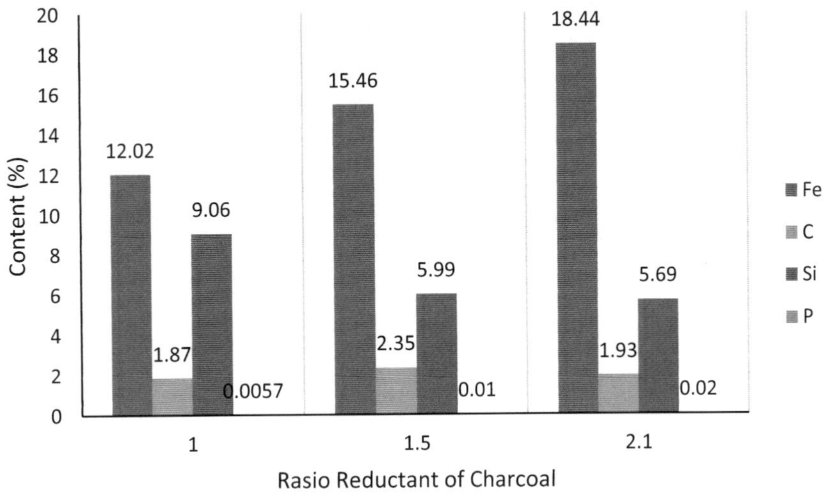

Figure 7. The effect of adding charcoal as a reductant on the composition of the product composition on the 30 kg of manganese ore, 7 kg of limestone and 70 minutes of the processing time

The results showed a trend with increasing use of a reductant (palm kernel shells and charcoal) effect to decreased of Fe content, increased of C content, increased of Si content, decreased of P content (Figure 6 and Figure 7).

4. Conclusion

The results showed that palm kernel shell and charcoal could potentially be used as a reductant in the production ferromanganese. Thc highest percentage of manganese extraction using palm kernel shells as a reducing agent is 49.91% (75.58% Mn, 15.75% Fe, 2.12% C, 5.23% Si, 0.08% P) with the product of FeMn is 6.6 kg. The highest percentage of manganese extraction using charcoal as a reductant is 44.16% (72.35% Mn, 18.44% Fe, 1.93% C, 5.69% Si, 0.02% P) with the product of FeMn is 6.1 kg.

Acknowledgments

The authors would like to thank Research Unit for Mineral Technology – Indonesian Institute of Sciences that provided infrastructures and funding for this research.

References

[1] http://webmap.psdg.bgl.esdm.go.id/geosain/neraca-total-prov,php, accessed on 9-10-2015 at 11.32
[2] Law of the Republic of Indonesia No 4, 2009, About Mineral and Coal Mining, Obligations for Holders of Mining Business Licenses (IUP) and Special Mining Business Licenses (IUPK) to increase value added of mineral and coal mining products through processing and purification activities 21
[3] Regulation of the Ministry of Energy and Mineral Resources of the Republic of Indonesia Number 8, 2015, abaout Increase of Mineral Added Value through Minerals Processing and Purification Activity, Attachment I (Minimum Limitation of Processing and Purification of Metallic Minerals Commodity) 4
[4] Supriyatna Y I 2013 *tekMIRA Journal* **9** 50-60
[5] American Society for Testing and Materials (ASTM) 1993 *Annual Book of ASTM Standards - Ferrous Castings; Ferroalloys* **1-2** 59-61
[6] Olsen S E, Tangstad M, Lindstad T 2007 *Production of Manganese Ferroalloys* Trondheim - Norwegia Tapir Academic Press pp 205-213

[7] Zhao Y, Zhu G, Cheng Z 2010 *Hydrometal.* **105** 96-102
[8] Zhang M, Yuan Y C, Liu Y Z *J. Energy Research and Information* **21** 15-19
[9] Japan Steel Plantech Co. 2010 *Report for Improving the Efficiency of Carbonizing Plant for PKS Charchoal in Malaysia* pp. 2-17
[10] Supriyatna Y I, Zulhan Z, Firdiyono F *tekMIRA Journal* **12** 105-15

IOP Conf. Series: Materials Science and Engineering **285** (2017) 012023 doi:10.1088/1757-899X/285/1/012023

Effect of basicity and reductant amount in the nickel pig iron (NPI) production from Indonesian limonite ore in submerged electric arc furnace (SAF)

W Astuti[1], R Andika[2], F Nurjaman[1]

[1]Research Division for Mineral Technology, Indonesian Institute of Sciences (LIPI) Jl. Ir. Sutami Km. 15, Tanjung Bintang, Lampung Selatan, Indonesia
[2]Department of Physics, Lampung University, Indonesia

Email: widi004@lipi.go.id

Abstract. The effect of basicity and reductant amount on the nickel and iron recovery of the nickel pig iron (NPI) production from Indonesian limonite ore was investigated in the experimental study using submerged electric arc furnace (SAF). Indonesian limonite ore used in this study originated from Sulawesi Island with the composition of Ni (1.26%) and Fe (43%). Metallurgical coke was applied as the reductant. This study showed that the the highest nickel and iron recovery as well as metal yield can be resulted from the basicity of 0.8 and reductant amount of 0.23 kg coke/kg limonite ore. Nickel content in the NPI produced was around 3 – 4%. It was concluded that this experiment can produce medium grade NPI.

1. Introduction

Nickel is an important metal used in the production of stainless steel and alloys [1]. Approximately 65% and 12% of the overall nickel consumption worldwide are primarily consumed for stainless steel production and for super alloy or nonferrous alloy manufacturing, respectively [2]. Nickel laterite ores, which comprise73% of the world's proven nickel reserves (approximately 160 million tons), are the dominant source of nickel [3]. Indonesia is one of the largest producers of nickel ore in the world and it has approximately 16% of total global nickel resources in the form of lateritic ores. They are widely distributed and spread among the islands of Kalimantan, Sulawesi, Halmahera, and Papua [4].

Nickel laterite ores are difficult to beneficiate by physical methods because of their complex mineralogy [5]. Thus, metallurgical technology is usually applied to extracting nickel from nickel laterite ores. In recent years, the rapid growth in stainless steel demand has dramatically increased the production of nickel metal by approximately 4.7% per year. Therefore, the recovery of nickel from laterite ores using pyrometallurgy has been extensively investigated.

In this study, the production of nickel pig iron (NPI) from an Indonesian low-grade nickel limonite ore was performed in submerged electric arc furnace (SAF) using metallurgical coke as a reductant.The effects of basicity and reductant amount on the metal yield and recovery of nickel and iron were investigated.The present study aims to provide a basis for the utilization of nickel laterite ores to produce nickel pig iron (NPI) by pyrometallurgical method.

2. Materials and Method

2.1. Materials

The low-grade limonite ore used in this study was collected from Sulawesi Island, Indonesia. The chemical composition of the limonite ore that is analyzed by X-ray fluoresence (XRF) and X-ray diffraction (XRD) pattern is given in Table 1 and Figure 1, respectively. Metallurgical coke was employed as a solid reductant in the smelting process. The proximate analysis of the coke is listed in Table 2. The coke is a good reductant because of its high fixed carbon and volatile matter, low ash, and relatively low harmful element (S, P) content. The ore and coke were ground to 100% passing screen aperture of 74 μm to increase the contact between them before they were mixed and agglomerated into pellets with diameter 10-20 mm. Limestone was used as a flux to adjust the basicity of material during the smelting process of lateritic nickel ore.

2.2. Experimental Method

Reduction and melting process of low-grade limonite ore to produce nickel pig iron (NPI) were conducted in SAF with capacity of 30 kg/heat. The effect of basicity on the recovery of nickel and iron was conducted by adjusting the composition of limestone in the raw material with the basicity used in this experiment is 0.8; 1; and 1.4. The effect of reductant amount was also studied by varying of coke amount in the raw material. The composition of raw material in this smelting process is listed in Table 3. The pouring temperature of slag and hot metal was 1246-1578 ^{o}C. Metal that produced from the experiments were analyzed using optical emission spectrometry (OES) to get their chemical composition. Recovery of metal particularly iron (Fe) and nickel (Ni) as well as the yield of metal can be calculated from the iron and nickel content in the nickel pig iron produced.

Table 1. Chemical composition of limonite ore from Sulawesi Island, Indonesia

Parameter	Ni	Fe	Si	Mg	Ca	Al
wt. %	1.26	43.10	3.23	6.29	0.34	2.83

Table 2. Proximate analysis of metallurgical coke

Parameter	Moisture	Volatile matter	Ash	Fixed carbon
wt. %	0.20	9.65	27.18	62.97

Table 3. The composition of raw material

Trial	Laterite ore	Coke (kg)	Limestone (kg)	Basicity
1	30	7 (stoichiometry)	-	0.8
2	30	7 (stoichiometry)	0.90	1
3	30	7 (stoichiometry)	3.12	1.4
4	30	10.5 (excess + 50%)	-	0.8
5	30	5.6 (lean - 25%)	-	0.8

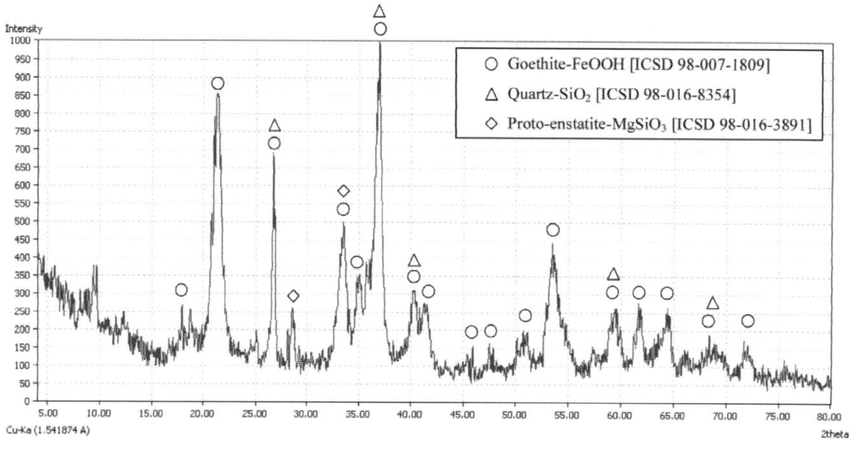

Figure1. X-ray diffraction pattern of limonite ore

Table 4. Rietveld analysis of nickel laterite ore

Compound	Geothite-FeOOH	Quartz-SiO2	Enstatite-MgSiO3
wt. %	76.5	7.9	15.6

3. Result and Discussion

As listed in Table 2, the nickel laterite ore has high in Fe content and it was in form of goethite, as shown in Figure 1. Quartz (SiO_2), and proto-enstatite ($MgSiO_3$) also found in this lateritic ore. From the Rietveld analysis, as listed in Table 4, it was 76.5%, 7.9%, and 15.6% for goethite, quartz, and enstatite, respectively.

3.1. Effect of Basicity

The quarternary basicity, which is calculated as the ratio of CaO and MgO to SiO_2 and Al_2O_3, as expressed in Equation 1, is used in this experiment to investigate its effect on metal yield and metal recovery. From the Figure 2 and Figure 3, the 0.8 basicity showed the optimum metal yield and metal recovery, both for Ni and Fe recovery. The metal yield and recovery tend to decrease when it was 1.0 basicity and it tend to increase at 1.4 basicity.

$$B = \frac{CaO + MgO}{SiO_2 + Al_2O_3} \quad (1)$$

From other research, as reported by Li *et al.*, the optimum basicity for smelting the nickel laterite ore was 1.0 [6]. Zhu et al., also reported the optimum basicity was 1.0, nevetheless the basicity was calculated fom binary basicity as the ratio of MgO to Al_2O_3 [7]. The lower basicity in this experiment probably due to the higher Fe content, i.e. 43.1%, than the previous research.

3.2. Effect of Reductant Amount

From Figure 4 and Figure 5, the optimum metal yield and metal recovery was obtained by the addition 7 kg of coke as stoichiometry calculation. Nickel oxide and ferrous oxide needs carbon as reductant to transform into metal, Ni and Fe. The less amount of coke than stoichiometry resulted in low metal yield and recovery. It means there was no-suifficient coke to reduce the metal oxides. Nevertheless, the same result was also found in excess coke of stoichiometry. The metal yield and metal recovery was lower than stoichiometry. Coke contains S and P. More coke means in more harmful element in smelting process. S and P inhibit the reduction of oxides in smelting process, thus it lowering the recovery of metal .

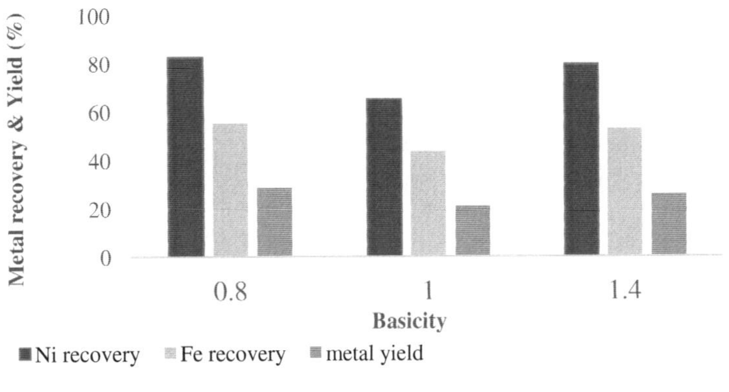

Figure 2. The effect of basicity on the metal yield

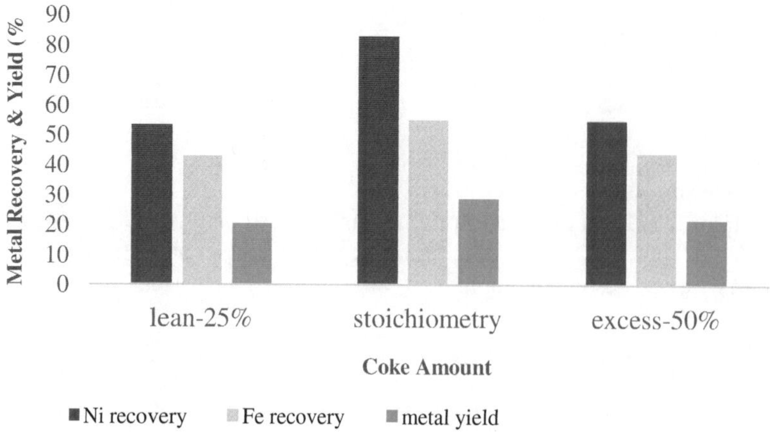

Figure 3. The effect of coke amount on the nickel and iron recovery

Table 5. Smelting proces result

Trial No.	NPI Composition (wt.%)		Pouring Temperatur (C)	Metal yield (wt.%)	recovery Ni (wt.%)	recovery Fe (wt.%)
	Fe	Ni				
1	82,81	3,64	1480	28,67	82,81	55,21
2	88,9	3,91	1485	21,00	65,17	43,42
3	87,3	3,86	1246	26,00	79,65	52,79
4	87,1	3,19	1435	21,67	54,85	43,89
5	90,7	3,31	1578	20,33	53,42	42,89

4. Conclusion

This study showed that the the highest nickel and iron recovery as well as metal yield can be resulted from the basicity of 0.8 and reductant amount of 7 kg of coke as stoichiometry calculation. Nickel content in the NPI produced was around 3 – 4%. It was concluded that this experiment can produce medium grade NPI.

Acknowledgement

We would like thanks to Ministry of Research, Technology and Higher Education of The Republic of Indonesia through the scheme of Incentive Research Program for The National Inovation System with the grant number 04/INS-2/PPK/E/E4/2017 and also acknowledge to Indonesian Institute of Sciences from for financial supporting in publication through Material Development, Energy and Manufacturing Engineering Program 2017.

References

[1] Chang Y F, Zhai X J, Fu Y 2008 *Trans. Nonferrous Met. Soc. China* **18** 969-73
[2] Johnson J, Reck B K, Wang T, Graedel T E 2008 *Energy Policy* **36** 181-92
[3] Li B, Wang H, Wei Y G 2011 *Miner. Eng.* **24** 1556-62
[4] Supriyadi A 2008 *Pengolahan bijih nikel yang sesuai dengan keanekaragamankarakteristik bijih Kolokium Pertambangan* (Bandung: Puslitbang Tekmira)
[5] Dalvi A D, Bacon W G, Osborne R 2004 *International Convention. North Carolina, USA: Trade Show & Investors Exchange* 1-27
[6] Li G, Zhi Q, Rao M, Zhang Y, Cai W, Jiang T 2013 *Powder Technol.* **249** 212-19
[7] Zhu D, Zhou X, Luo Y, Pan J, Bai B 2016 *High Temp. Mater. Proc.* 1-6

IOP Conf. Series: Materials Science and Engineering **285** (2017) 012024 doi:10.1088/1757-899X/285/1/012024

Study on Sumbawa gold ore liberation using rod mill: effect of rod-number and rotational speed on particle size distribution

A Prasetya, A Mawadati, A M R Putri and H T B M Petrus

Center of Advanced Material and Sustainable Mineral Processing
Department of Chemical Engineering, Universitas Gadjah Mada,
Jl. Grafika 2 Yogyakarta 55281, Indonesia

E-mail: bayupetrus@ugm.ac.id

Abstract. Comminution is one of crucial steps in gold ore processing used to liberate the valuable minerals from gaunge mineral. This research is done to find the particle size distribution of gold ore after it has been treated through the comminution process in a rod mill with various number of rod and rotational speed that will results in one optimum milling condition. For the initial step, Sumbawa gold ore was crushed and then sieved to pass the 2.5 mesh and retained on the 5 mesh (this condition was taken to mimic real application in artisanal gold mining). Inserting the prepared sample into the rod mill, the observation on effect of rod-number and rotational speed was then conducted by variating the rod number of 7 and 10 while the rotational speed was varied from 60, 85, and 110 rpm. In order to be able to provide estimation on particle distribution of every condition, the comminution kinetic was applied by taking sample at 15, 30, 60, and 120 minutes for size distribution analysis. The change of particle distribution of top and bottom product as time series was then treated using Rosin-Rammler distribution equation. The result shows that the homogenity of particle size and particle size distribution is affected by rod-number and rotational speed. The particle size distribution is more homogeneous by increasing of milling time, regardless of rod-number and rotational speed. Mean size of particles do not change significantly after 60 minutes milling time. Experimental results showed that the optimum condition was achieved at rotational speed of 85 rpm, using rod-number of 7.

1. Introduction

Indonesia is a country rich in natural resources. One of those resources is gold. Production of gold in Indonesia reaches 100 ton/year (https://www.orori.com/). In nature, gold is not found in its pure form, this exists in rocks which is called ore. Purification process of rock containing gold is commonly done in Indonesia, both traditionally in small scale and production of bigger scale with modern technologies [1]. Gold processing by traditional method known as amalgamation method. The principle of this method is to bind the gold contained in the rock by using mercury. Amalgamation method consists of several processes, such as size reduction process, mixing process, collecting and settling process, panning and washing process, and heating process [2, 3]. Gold purification method by amalgamation is harmful to the environment. Panning and washing process are usually done in the river where the remaining mercury will be carried by the water flow and thus contaminate the river [4]. Meanwhile, the mercury that evaporates during the heating process will be carried by the air. This will harm the health of the surrounding community. Based on that fact, gold purification method of amalgamation needs to be replaced by other methods which are more environmentally friendly.

IOP Conf. Series: Materials Science and Engineering **285** (2017) 012024 doi:10.1088/1757-899X/285/1/012024

An example of an environmentally friendly gold purification method is the gravity method. This method separates gold from its impurities by using gravitational force. This process utilized a tool called sluice box. The working principle of this tool is to drain the water on the crushed slurry ore where the other minerals with the lower density than gold will be carried by the water while the gold itself will be left and settled. This separation process will work well if the particles (non gold particles) have uniform size (homogeneous) [5-7].

In order to obtain particles of small size and homogeneous, the ore will go through comminution stage by grinding process (milling). In the mining world, this comminution process plays a role in liberating precious minerals (gold) from the impurities. In this research, the selected milling process is the process done by rod mill.

Rod mill is a grinding process using a rotating cylindrical spindle. The grinding media is usually made of steel and arranged parallel to the mill. The result of this grinding process using rod mill is flat rock with flat surface. This flat surface of gold is desirable, as it will increase its effective size during sieving process and ease the separation process by gravity method.

This study uses variations in the number and the speed of rotary rod mill to see which milling process results in the most homogenous particle size distribution. From this study, the most optimum milling process for gold purification using gravity method will be obtained.

2. Methodology

The ore tested in this study is gold ore from Sumbawa, West Nusa Tenggara, Indonesia. Gold ore were crushed using hammer and then screened to get desired size (-2.5/+5 mesh). This condition was taken to mimic real application in artisanal gold mining. The diameter and length of the rod mill are measured using a roll meter. Then the critical speed of rod mill can be calculated by using the critical velocity equation by B.A.Wills. The speed variations used in this study were 60 rpm, 85 rpm, and 110 rpm.

$$N_c = \left[\frac{42{,}3}{\sqrt{D-d}}\right] \text{rpm} \quad (1)$$

N_c : critical speed, rpm
D : inner diameter, meter
d : rod diameter, meter

The observation on effect of rod-number and rotational speed on comminution by using rod mill was conducted by variating the rod number of 7 and 10. The comminution kinetic was applied by taking 1.5 kg sample at 15, 30, 60, and 120 minutes for size distribution analysis. Particle results of each experiment for each rotational speed and each interval of time put into the sieve shaker with sieve size 70, 100, 115, 140, 170, 200, 230 and 270 mesh. The results of the top and bottom of each sieve are weighed using a digital analytical scale. Then by using the data of weighing results at any time could be made a cumulative distribution graph. The change of particle distribution of top and bottom product as time series was then treated using Rosin-Rammler distribution equation.

$$Y_d = \exp(-A\,(d/d_{avg})^n) \quad (2)$$

Y_d : mass fraction of the particles retained in the sieve of size d
d : sieve size, mm
d_{avg} : average diameter, mm
n : size distribution parameters
A : size distribution parameters

3. Result and Discussion

3.1 The Effect of rod mill rotary speed on particle size distribution of Sumbawa gold ore particle size distribution analysis with Rosin-Rammler equations: Rosin Rammler's A parameters

Experiment conducted by using rod mill with rod number of 7 on rotational speed of 60, 85, and 110 rpm. The value of A is a parameter in the Rosin-Rammler equation which serves to align the retained mass fraction curve of the experimental results to the outcome of the equation and does not significantly affect to the shape of the curve which describes the homogeneity of the particle size distribution. Based on Figure 1, the value of A is only slightly affected by milling time and is not influenced by the rotational speed of the rod mill. Its value does not significantly affect the homogeneity of the particle size distribution. Based on this experiment, obtained that the value of A is getting bigger with the increasing time of milling. It can be seen on Figure1 that on the same milling time, the value of A at each rotational speed tends to be same. Therefore it can be concluded that the value of A is not influenced by the size of the rotation speed of the rod mill.

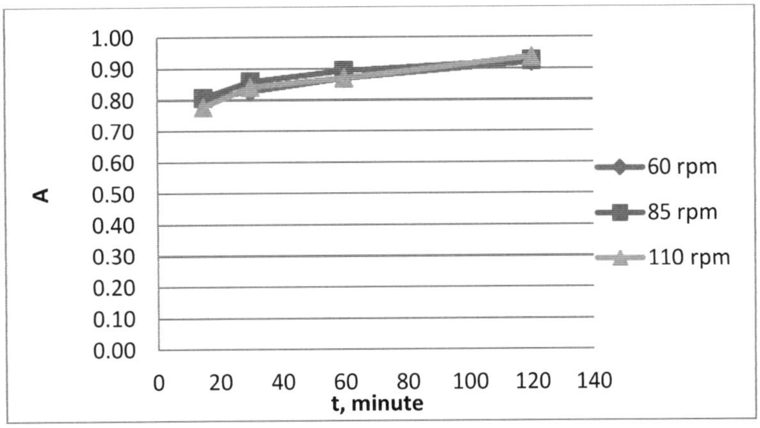

Figure 1. Time relation chart of value A with number of rod 7 at various rotational speed

3.2 Particle size distribution analysis with Rosin-Rammler equations: Rosin Rammler's n parameters

The value of n is a parameter in the Rosin-Rammler equation which describes the homogeneity of the particle size distribution. It is affect on cumulative distribution curve. Based on Figure 2, value of n is affected by the milling time and the rotational speed of the rod mill, so that the value of n affects the homogeneity of the particle size distribution.

Based on the experiments, obtained that the value of n is getting smaller as the milling time increases. On the same milling time, there are different values of n at each rotation speed of the rod mill. However, at milling time of 120 min, n values tend to be the same at every rotational speed. Therefore it can be concluded that the value of n is influenced by the rotational speed, but the effect of rotational speed on the value of n decreases with increasing milling time. It can be seen at milling time 120 minutes, the value of n at each rotation speed tend to be the same.

The smaller value of n will result in an increasingly non-homogeneous distribution of particle sizes, so the cumulative distribution curve generated is getting wider. In Figure 2 shows that at almost any point of time, the smallest n value is obtained at a rotational speed of 85 rpm, so that at a rotation speed of 85 rpm is obtained the best particle size distribution, since it is spread in various sizes or not homogeneous.

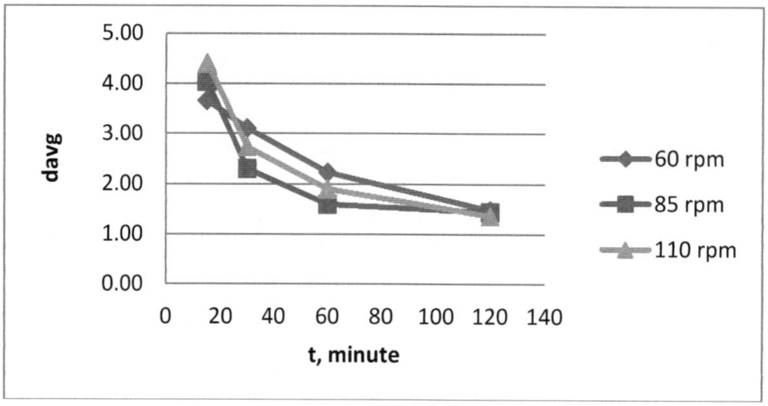

Figure 2. Time relation chart of value n with number of rod 7 at various speeds

3.3 Particle size distribution analysis with Rosin-Rammler equations: diameter average

In the experiment using rod number of 7, it indicate that the increasing milling time effect on the smaller diameter average value. The particle size is getting smaller as the milling time increases. In the graph above shows that at the same time milling, d_{avg} values obtained at each rotation speed is not much different. The particle size is influenced not significantly by the rotational speed.

Figure 3 shows that at almost any point of time, the smallest d_{avg} value is obtained at 85 rpm rotational speed. It can be conclude that the comminution process is fastest at rotational speed of 85 rpm than the two other rotational speeds.

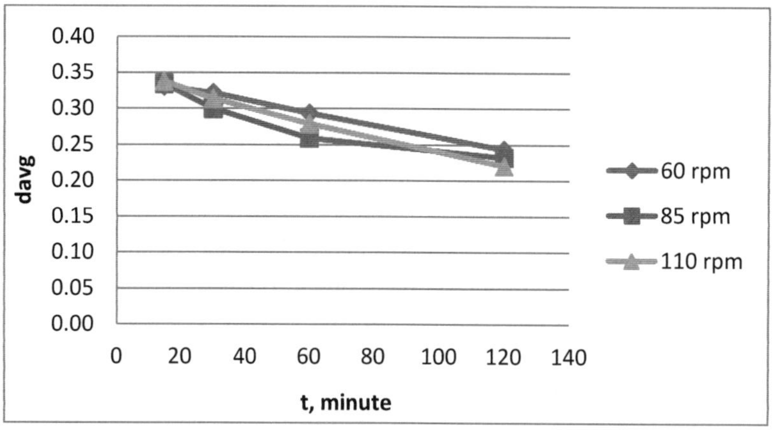

Figure 3. Time milling relation chart (t) with average diameter (d_{avg})

Figure 4 shows at speeds of 60, 85, and 110 rpm with the number of rod 7, as milling time increase, the d / d_{avg} value is greater. This is due to the longer milling time, the particle diameter of the milling results will be even smaller, so the diameter average (d_{avg}) value will be even smaller. Because the smaller the value of d_{avg}, the greater the value of d / d_{avg}. Thus experiments at speeds of 60, 85, and 110 rpm are in accordance with the theory.

At 60, 85, and 110 rpm rotational speeds, the fractional data of the experimental mass (Yd experiments) on the 15 and 30 minutes much less deviate from the mass fractional data of equation (Yd equation), this is probably due to the milling process which has not been stable in the early time. Meanwhile, in the 60 and 120 Yd data the experiment can be said to be in accordance with the data Yd equation.

As the milling time increases, the curve widened as seen on the larger d/d_{avg} value. It shows that the particle size distribution is increasingly dispersed (not homogeneous). This increasingly distributed size distribution is influenced by the smaller value of n as the milling time increases as shown in Figure 4.

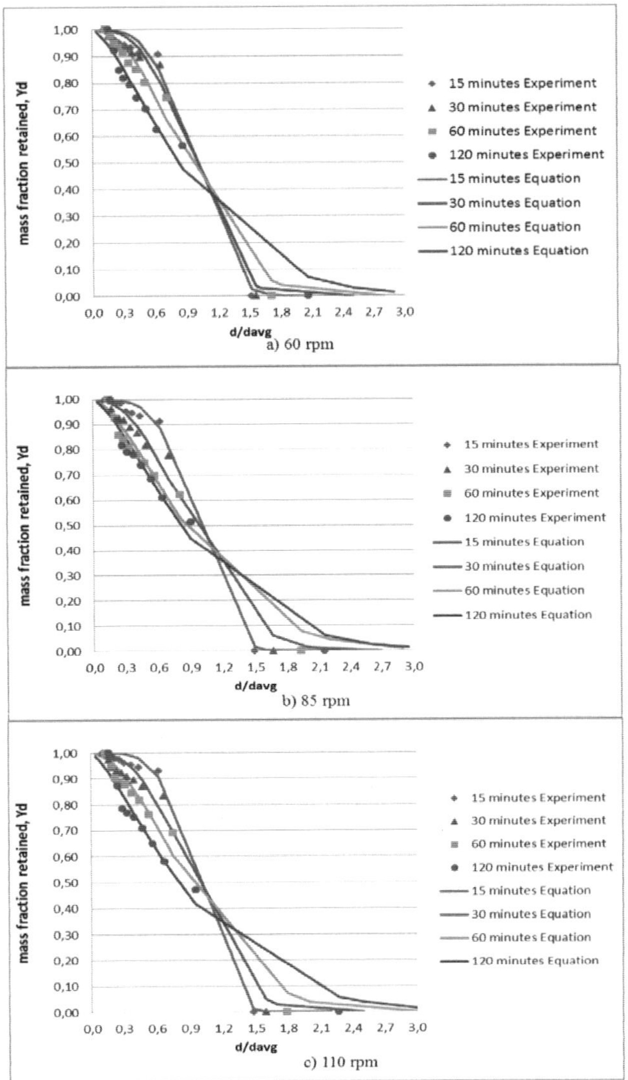

Figure 4. d / d_{avg} relation chart with Yd at various rotational speed with number of rod 7

3.4 The effect of number of rod on Sumbawa gold particle size distribution Sumbawa gold particle size distribution analysis with Rosin-Rammler equations: Rosin Rammler's A parameters

The experiment using rod mill with number of rod 7 and 10 with a rotational speed of 85 rpm obtain the value of A which increases with the addition of milling time. However, increasing the value of A is not significant with increasing milling time. So it can be concluded that the value of A is slightly influenced by milling time. It can be seen on Figure 5 that with the same milling time, the value of A on each number of rods tend to be the same. However, at minute 120, a significant difference of value A of each number of rods is obtained. Therefore it can be deduced that the value of A is not influenced by the number of rods, but the effect of the number of rods against this A value will increase with increasing milling time.

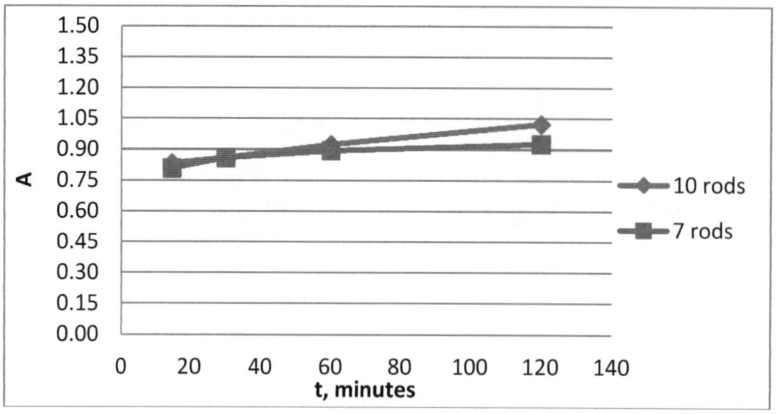

Figure 5. Time relation graph of value A with number of rod 7 and 10 at rotational speed 85 rpm

3.5 Particle size distribution analysis with Rosin-Rammler equations: Rosin Rammler's n parameters
In the experiments with the number of rods 7 and 10 at a rotational speed of 85 rpm obtained the value of n is getting smaller as the milling time increases. Meanwhile, with the same milling time, we get different values of n on each number of rods. However, at milling time of 120 min, we get n values which tend to be the same in both rod numbers. Therefore it can be concluded that the magnitude of the value of n is influenced by the number of rods, but the effect of the number of rods on the value of n decreases with increasing milling time and the smaller particle size, it is seen at milling time 120 min, Already tend to be the same.

Figure 3 shows that at all time points, the smallest n value is obtained on the number of rods of 10 pieces but only slightly different with number of Rod 7 with increasing time of milling process. The number of rods of 10 is obtained the best particle size distribution, because it is spread in various sizes or not homogeneous but the number of rods 7 become more effective compare to number of rod 10.

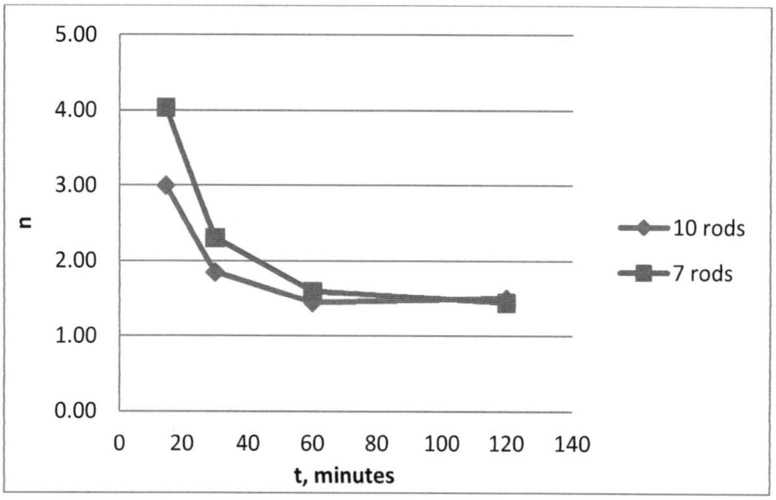

Figure 6. Time relation chart of value n with number of rods 7 and 10 fruits at rotational speed 85 rpm

3.6 Particle size distribution analysis with Rosin-Rammler equations: diameter average (d_{avg})
The experiments with rotational speed of 85 rpm, indicate that the increasing milling time effect on the smaller the d_{avg} value. This indicates that the particle size is getting smaller as the milling time

increases. At the same milling time, the d_{avg} value obtained at each rod number is not much different, but with the increasing milling time, the difference in d_{avg} value is more significant. It can be concluded that the particle size is influenced by the number of rods, and the influence of the number of rods is getting bigger along with the increasing time of milling.

Figure 7 shows that at all time point, the smallest d_{avg} value is obtained on the number of rods of 10 pieces. It can be concluded that comminution process on the number of rod 10 effects on the smaller d_{avg} value than the number of rod 7 but the differences is not too significant.

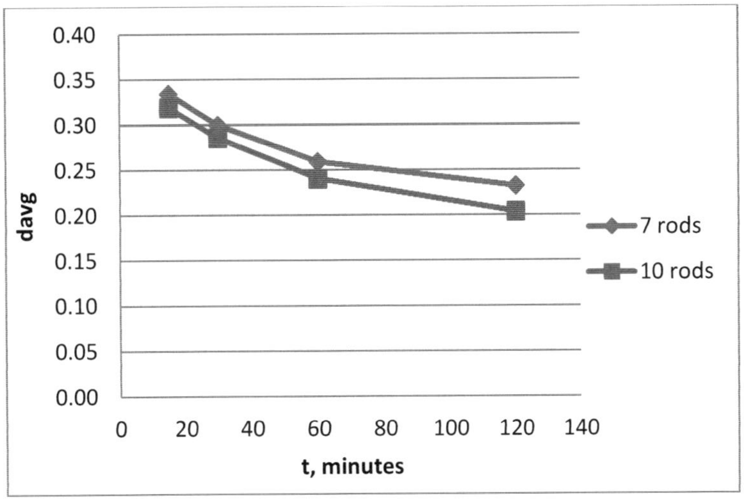

Figure 7. Time milling relation chart (t) with average diameter (d_{avg})

The experiments with a rotational speed of 85 rpm with the number of rod 7, the longer milling time effect on the greater of d / d_{avg} value. It is because the longer milling time, the diameter particles will be smaller so the value d/d_{avg} will be greater. In both graphs on Figure 7, the mass fractional data retained by the experimental results (Yd experiments) on the 15 and 30 minutes are much too distorted from the mass fractioned data of the equation (Yd equation), similar to the previous experiments, probably due to unstable milling processes at the beginning of the time. Meanwhile, in the 60 and 120 minutes, Yd data the experiment is in accordance with the data of Yd equation. Figure 7 show that as the milling time increases, the curve widened as seen on the larger d / d_{avg} value. The widening curve illustrates the increasingly dispersed (not homogeneous) particle size distribution. This increasingly distributed size distribution is influenced by the smaller value of n as the milling time increases as shown in Figure 8.

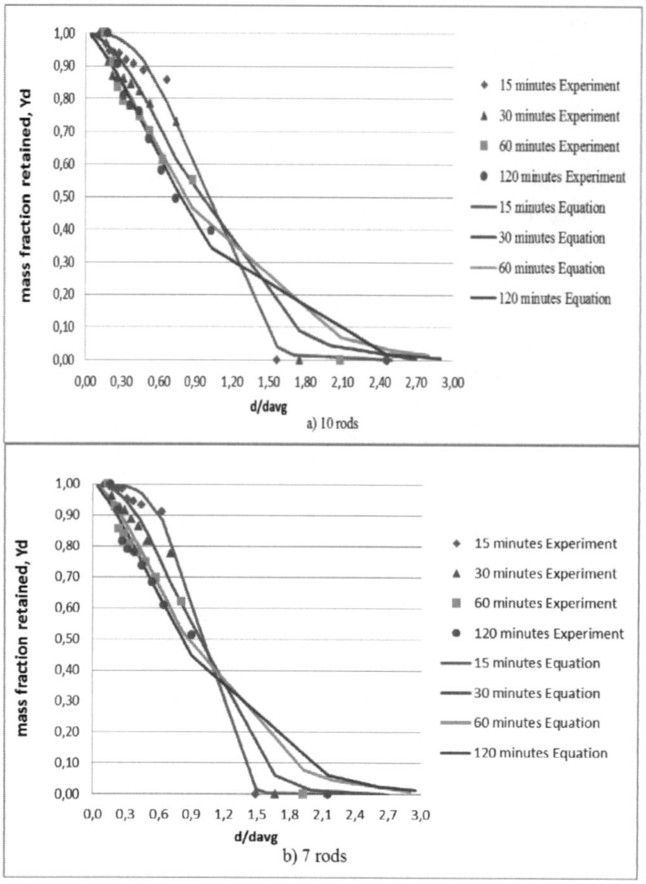

Figure 8. Relation chart d/d_{avg} with Yd at rotational speed 85 rpm with 7 and 10 rods

4. Conclusions

From this experiment, it can be concluded that the longer the milling time, the less homogeneous the particle size distribution. The higher the rotational speed and the more number of rod will provide higher size reduction. The homogeneous distribution can be obtained in a condition at which the rotational speed is 85 rpm and the number of rods is 10 rods. While Rosin-Rammler could fit and predict the phenomenon occured in this comminution experiment.

References

[1] Aye M T, Pramumijoyo S, Idrus A, Setijadji L D, Imai A, Araki N, Arif J 2011 *J. SE Asian Appl. Geol.* **3** 12-22
[2] Kelly E G, Spottiswood D J 1982 *Introduction to Mineral Processing* Michigan: John Wiley and Sons pp 132-62
[3] Mular A L, Halbe D N, Barrat D J 2002 *Mineral Processing Plant Design, Practice, and Control* Littleton: Society for Mining, Metalurgy, and Exploration Inc. **1** p 712
[4] Eugene W W L, Mujumdar A S 2009 *Gold Extraction and Recovery Processes* Singapore: National University of Singapore pp 5-7
[5] Wills B A 1985 *Mineral Processing Technology* Cornwall:Camborne School of Mines pp 14-20
[6] Zhou J, Jago B, Martin C 2004 *SGS Mineral* **3** 1-3

[7] Pedersen H K, Andersen A 2007 *Geoviden* **2** 5-6

IOP Conf. Series: Materials Science and Engineering **285** (2017) 012025 doi:10.1088/1757-899X/285/1/012025

A study of bauxite tailing quality improvement by reverse flotation

W Wulandari[1], M Purwasasmita[1], E Sanwani[2], W Malatsih[1], F Fadilla[1]

[1]Chemical Engineering Department, Faculty of Industrial Technology, Institut Teknologi Bandung, Indonesia
[2]Metallurgical Engineering Department, Faculty of Mining and Petroleum Engineering, Institut Teknologi Bandung, Indonesia

Email: winny@che.itb.ac.id

Abstract. The pre-treatment of bauxite ore from Tayan, West Kalimantan includes washing and screening fine bauxite particles (-2mm) prior as the feed to the Bayer process for producing alumina. These fine particles are believed to have high content of silica which is detrimental to the process. This washed bauxite tailing still has a significant amount of alumina content. Previous research has indicated that bauxite ore can be upgraded by applying reverse flotation method to reduce its silica content in the ore. Therefore, this study is aimed to utilize reverse flotation method to recover alumina content from washed bauxite tailing. The reverse flotation experiments were carried out at pH of 6 and 8; while the particle sizes were varied at -140+270 mesh and -270 mesh, using a batch and circuit configuration. The result of this study shows that the batch reverse flotation can recover alumina in the tailing up to 81.4%, however the silica content is still significant. The complexity of silica-alumina minerals in the tailing prevents a complete separation of the ores by only using reverse flotation.

1. Introduction

Bauxite is the raw material for producing alumina, both as Metallurgical Grade or Chemical Grade via the Bayer process. West Kalimantan has large amount of bauxite reserve, and has been mined by PT Aneka Tambang to be processed by PT Indonesia Chemical Alumina. Bauxite from West Kalimantan is categorized as lateritic bauxite. Typical bauxite from Tayan area has alumina content ranging from 41 to 46%, with silica content of 5 % to 17%, giving alumina to silica ratio of 5 to 9 [1].

Silica in bauxite comprises of two types, reactive silica and nonreactive silica [2]. Reactive silica is the clay type minerals, including kaolinite ($Al_2[Si_2O_5](OH)_4$), haloisite ($Al_2[Si_2O_5](OH)_4.2H_2O$), and zussmanite ($K(Fe^{2+},Mg,Mn)_{13}[AlSi_{17}O_{42}]$, while non-reactive silica includes quartz (SiO_2). Silica combined as clay and other silicates dissolves in the digestion process and forms sodium aluminium silicate, which precipitates as Desilication Precipitates (DSP), and therefore responsible for the loss of caustic soda as well as reduces the alumina recovery from the ore [3]. Silica in the form of quartz is not attacked during the digestion process, but it reduces the production capacity of bauxite processing.

The current practice to process bauxite ore from Tayan, West Kalimantan, by PT Aneka Tambang is to wash and screen bauxite ore. The washed bauxite that has particle size +2 mm is believed to have low silica content and suitable for processing in the next process step, which is the digestion in the Bayer process. The bauxite that has particle size -2+1 mm also can be processed to the Bayer process as the alumina content for this fraction is about 40%. The washed bauxite tailings that have particle size less than 1 mm has not been utilized yet. This results in a big dump of washed bauxite tailings. The alumina content in the washed bauxite tailing is still significant, at about 36.3 wt%.

One of the methods to physically separate gangue materials is reverse flotation. The term "reverse flotation" refers to flotation that collects unwanted material at the froth phase, while the economic material is collected in the base phase. This is beneficial for high silica bauxite as the collector chemicals collect rich silica phase in the froth phase and the upgraded bauxite can be utilized for further process.

Reverse flotation has been applied to process high silica bauxite in order to increase the bauxite quality, particularly for Chinese diaspore [4]. Xia *et al* [5] used Gemini cationic surfactant, BDDA, as a collector for reverse flotation to separate illite, pyrophyllite and kaolinite from diaspore bauxite. The bench-scale separation was reported to obtain Al to Si ratio of 9.72 with 81.25% of Alumina recovery. Wang *et al* [6] used quartenary ammonium salt (DTAL) to obtain good separation from similar diasporic bauxite with Al to Si recovery ratio of 10 and alumina recovery of 86%. Recent research by the authors reported that reverse flotation can be applied to separate silica from bauxite ores from Tayan. The ratio Al to Si can be increased from 7 up to 14, by using dodecylamine as the collector [7].

While reverse flotation has been successfully applied to separate major silica from bauxite ores, no literature is available for reverse flotation method that applied to the tailing. This paper aims to understand the reverse flotation of washed bauxite tailing, thus attempts to increase the quality of washed bauxite tailing by such method. In particular, the operating conditions that is suitable for the reverse flotation is examined, and the product from the concentrate and tailing are characterized in order to study the phenomena for better understanding.

2. Experimental

This study used washed bauxite tailing obtained from PT Aneka Tambang. About 68 kg of these tailings with particle sizes of -2 mm were dried and were reduced to 17 kg and 2 kg by using coning and quartering sampling technique (ASTM D346). The tailings were then screened to obtain 3 kg of -2+1 mm fraction, while the latter was further milled using ball mill to obtain a fraction of -140+270 mesh (-0.105+0.053 mm) and -270 mesh (-0.053 mm).

The experiments were carried out in two configurations, batch and circuit. For batch configuration, every experiment used 455 gram of fine sample which has been obtained by using sampling splitting method (ASTM D2013). The reverse flotation experiments were conducted in a 1.5 liter of Denver flotation equipment with 1 liter of aqua dm. It used 2 ml of dodecylamine 98% as the collector agent, 9 gram of sodium hexametaphosphate (SHMP) as the dispersant, food grade starch as the depressant, and Dowfroth 1016 as the frothers. To adjust the acidity/basicity, sulfuric acid and sodium carbonate was used. During the flotation process, the concentrate (at the froth) was scrapped and collected from the top, while the tailing was collected at the end of experiment. Both products were dried by using oven at temperature of 110 °C for 16 hours. The products were characterized by using X-Ray Fluorescence (XRF) to determine the oxide content and XRD to determine the major phase.

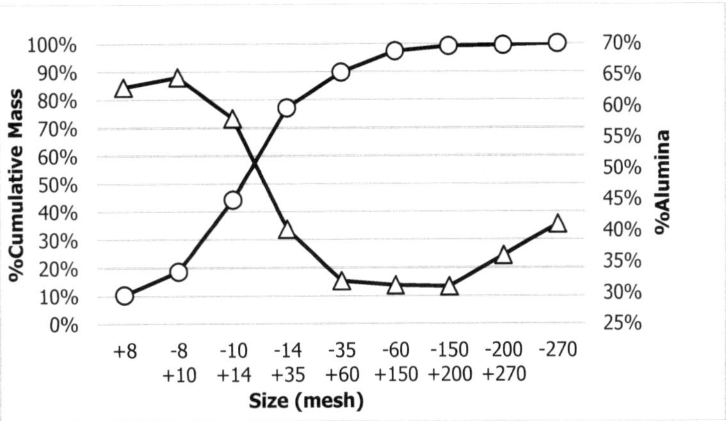

IOP Conf. Series: Materials Science and Engineering **285** (2017) 012025 doi:10.1088/1757-899X/285/1/012025

Figure 1. Mass cumulative of washed bauxite tailing particle distribution

3. Result and Discussion

3.1. Feed Characterization

About 2 kg of feed was characterized by using sieving analysis and XRF analysis. Figure 1 shows the mass cumulative of washed bauxite tailing and its particle size distributions. It is obtained that the bauxite feed with particle size of +8 mesh (+1 mm) has cumulative mass of 45% while the particle size -14 mesh (- 1mm) has cumulative mass of 55%, respectively. Figure 1 also shows alumina content of each fraction by using XRF analysis. As the particle size of bauxite is smaller, the alumina content is also decreased. For typical fraction size of -1mm, XRF analysis of the feed shows that washed bauxite tailing comprises of 36.80 wt% of Al_2O_3, 24.05 wt% of SiO_2, 34.15 wt% of Fe_2O_3, 3.47 wt% of TiO_2, 0.76wt% of SO_3, 0.25 wt% of V_2O_5, and others.

Figure 2 shows the mineralogy of washed bauxite tailing. It is shown that the washed bauxite tailing consists of Gibbsite, quartz, hematite, kaolinite, zussmanite, and pyrophyllite. Overall, the washed bauxite tailing still contains gibbsite as the source of alumina, quartz as the non-reactive silica, and kaolinite, zussmanite and pyrophyllite as the reactive silica minerals. Hematite also presents as the impurities, with the XRF indicates the presence of hematite as high as 34.15%.

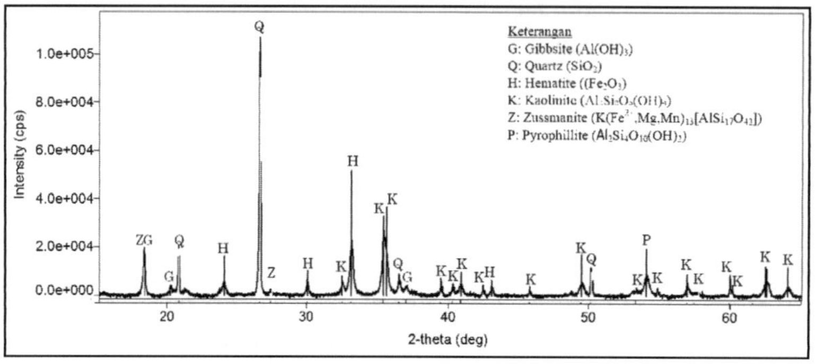

Figure 2. XRD analysis of washed bauxite tailing feed

3.2. Effect of Operating Condition to Reverse Flotation

Batch reverse flotation was carried out at pH 6 and 8 at two different particle sizes: -140+270 mesh and -270 mesh. Figure 3 shows the alumina recovery at the concentrate (the depressed section) from the reverse flotation. For the particle sizes -140+270 mesh, the alumina recovery obtained was 81.4% at pH 6 condition, and 80.6% at pH condition. The recovery is similar with other work reported in the literatures, such as Wang *et al* reported alumina recovery of 78% from reverse flotation of diaspore bauxite (6) .

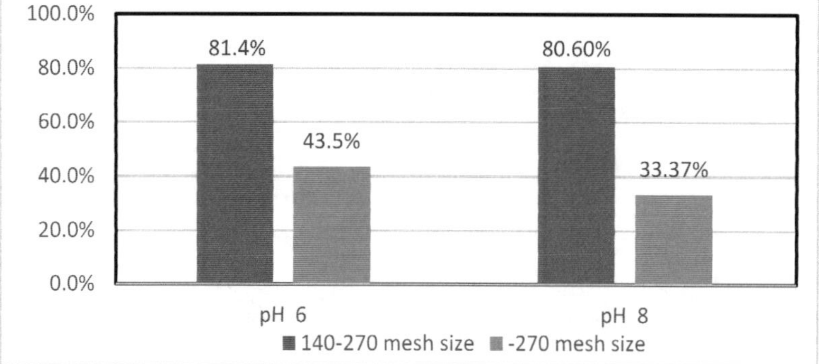

Figure 3. Alumina recovery at pH 6 and 8 for different particle sizes

The reverse flotation has been found not effective in recover alumina, as in Figure 3, the alumina recovery of alumina in at the concentrate is only 43.5% at pH 6 and 33.37% at pH 8. This can be explained by the settlement phenomena that may happen very fast as well as the probability of entrainment of very small particles for those -270 mesh particles to the froth phase, thus reduce alumina recovery at the concentrate.

3.3. Mineralogy analysis of Reverse Flotation

Figure 4 shows the XRD pattern of (a) concentrate and (b) tailings of batch reverse flotation of washed bauxite tailings. Overall, both of concentrate and tailings contain similar minerals, including gibbsite, quartz, kaolinite, haloisite, and zussmanite. Pyrophillite is present at the tailing but not at the concenrtate, The absence of pyrophillite at the concentrate shows that the pyrophillite at the feed has been completely liberated at the experiment.

The similar minerals at the concentrate and tailings was caused by the silica and alumina are bounded and not completely liberated. In addition, as the clay minerals also contains alumina, such as kaolinite, pyrophillite, the tailings at the experiments still contains alumina with fraction of 35%. Table 1 shows XRF analysis of concentrate and tailing from washed bauxite tailing reverse flotation. From all 8 runs, it is shown that the alumina and silica content in the concentrate is similar, ranging from 34-36 % for alumina and 23-26% for silica. This composition is similar to the feed composition, thus implying that the reverse flotation is not effective enough to be applied to the tailings of washed bauxite.

4. Summary

Reverse flotation method has been applied to increase to separate silica from washed bauxite tailings. While reverse flotation method has been successfully applied to raw bauxite ore from Tayan, Kalimantan, this method is not effectively separate silica and increases the quality of its washed bauxite tailing. The good condition of reverse flotation is obtained at pH 6 condition, with particle size of -140+270 mesh, with alumina recovery of 86.20%. However, the alumina content in the concentrate is still similar, only 35.78% from 36.80% from the feed. The silica content in the concentrate is also similar at 25% from 24.05 % from the feed. The complexity of silica-alumina minerals in the tailing prevents a complete separation of the ores by only using reverse flotation.

IOP Conf. Series: Materials Science and Engineering **285** (2017) 012025 doi:10.1088/1757-899X/285/1/012025

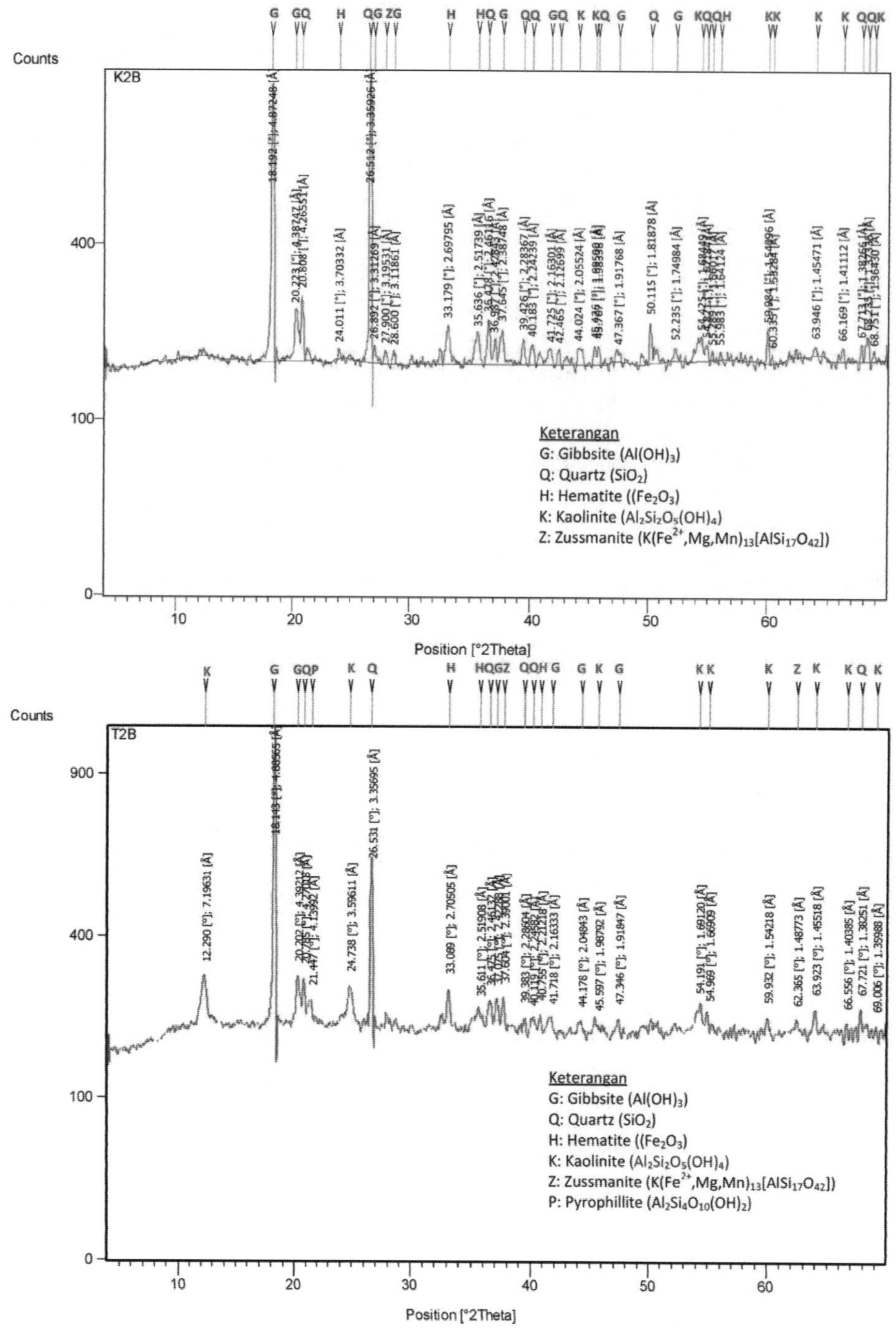

(a)

(b)

Figure 4. XRD Analysis of (a) Concentrate and (b) Tailings of Batch Reverse Flotation of Washed Bauxite Tailing

Table 1. XRF Analysis of Washed Bauxite Tailing Reverse Flotation

Run	Condition	Concentrate		Tailings	
		Al_2O_3	SiO_2	Al_2O_3	SiO_2
1	-140+270 mesh particle size	35.72%	25.13%	43.71%	17.14%
2	pH =8	35.57%	25.28%	41.68%	19.17%
3	-140+270 mesh particle size	35.58%	25.27%	43.61%	17.24%
4	pH =6	35.78%	25.07%	41.81%	19.04%
5	-270 mesh particle size	36.93%	23.92%	36.75%	24.10%
6	pH =8	35.65%	25.20%	37.62%	23.23%
7	-270 mesh particle size	34.20%	26.65%	39.44%	21.41%
8	pH =6	34.08%	26.77%	38.89%	21.96%

References

[1] Surata M, Suksiano O, Pratomo M, Supriyadi 2010 *Kalimantan Coal and Mineral Resources Seminar*

[2] Roach G I, White A J 1988 *Dissolution Kinetics of Kaolin in Caustic Liquors Light Metals 1988* ed Boxall LG p 240

[3] Adamson E J, Bloore E J, Carr A R 2013 *Basic Principles of Bayer Process Design Essential Readings in Light Metals: Alumina and Bauxite.* ed Donaldson D, Raahauge BE (John Wiley and Sons, Inc) p 100

[4] Xu Z, Plitt V, Liu Q 2004 *Miner. Eng.* **17** 1007

[5] Xia L, Zhong H, Liu G, Huang Z, Chang Q. 2009 *Int J Miner Process* **92** 74

[6] Wang Y, Hu Y, He P, Gu G 2004 *Miner. Eng.* **17** 63

[7] Wulandari W, Purwasasmita M, Sanwani E, Pixelina AA, Maulidan A 2017 *AIP Conf. Proc.* **1805** 050004

Physical-chemistry of Nawangan's phyropyllite and its prospective as environmental friendly geopolymer materials

S Mutrofin, T Setyaningsih , F N Wati, D Purwonugroho

Department of Chemistry, Brawijaya University Malang, Indonesia

Email: mutrofin@ub.ac.id

Abstract. The chemical composition and thermal behaviour of Nawangan-phyropyllite have been studied using XRF, powder XRD and FTIR. The fourier transfornation infrared was applied to analyze the phyropyllite after treating by calcinating at variuos temperature. Initial investigation has also been carried out by adding sodium hydroxide and potassium hydroxide to study the possiblity of phyropyllite as geopolymer materials. The phyropyllite contains Si(57.7\%), Al(16,7%), K (20.6%), Fe (2.47%) Ti (2.33%) and Cu (0.088%). Based on the XRD difractogram, peaks at 2 theta (9°, 20°, 21°, $26\ ^o$, $34\ ^o$, $36\ ^o$ and $39\ ^o$) were characteristic for phyropyllite. While, infrared study showed that at 3630 cm^{-1}, 756 cm^{-1} and 938 cm^{-1} are responsible for phyropyllite's peaks. The hydroxyl bonded to alumina still existed under heating up to 400 ^oC and disappered at 600 ^oC. It indicted that covalent bond of Al-OH was broken. By heating at 600 ^oC, the peak at 1021 cm^{-1} splitted into two peaks, 990 cm^{-1} and 1049 cm^{-1}. It may be due to the displacive transition. By adding NaOH 10 M , the peak intensity of Al-OH (3630 cm^{-1}) reduced to 17% but the peak intensity of Al=O (1661 cm^{-1}) incresed to 14% and the new peak (5%) emerged at 1387 cm^{-1}(O-Al-O). The most reactive phyropyllite was`obatained by adding KOH 5 M. The present of reactive functional groups (Al=O, O-Al-O and Al-OH) indicates that the local phyropyllite has a good change as geopolymer materials.

1. Introduction

Phyrophillite is one among a lot of minerals in Indonesia that has not been explored due to the limitation of its physical-chemistry properties. It is found at East Java, East Nusa Tenggara, Sumatera and West Java. Phyropyillte in Nawangan–Pacitan East Java spreads out in about 37 ha [1,2].

It is categorized as methamorphic rock, which is the hardness of 1-2 mosh scale and in general, is isomorphous to talc. Its structure is a dioctahedral alumina between two tetrahedral silicas, as shown in Figure 1.

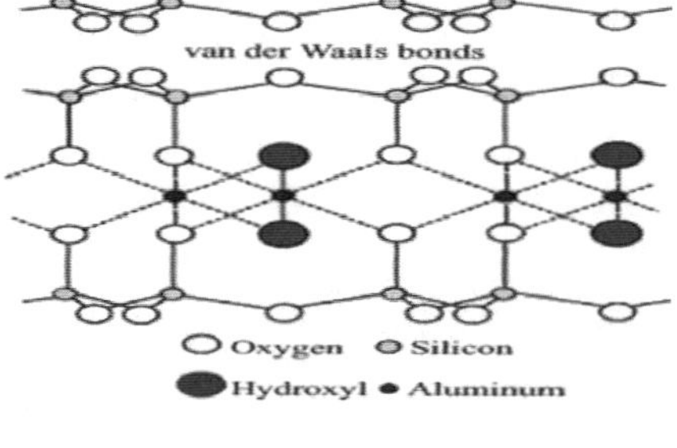

Figure 1. The structure of phyropyllite [3]

It has a lamellar structure which the van der Waals bond connected between the tetrahedral SiO_4 layers and terminal hydroxyl bonded covalently to the octahedral alumina. Phyropyllite is classified as 2:1 aluminosilicate based on the Si:Al ratio. It is also can be applied as fire and heat resistance materials (refractory products), filler and adsorbent. Mutrofin and Anggraeni (2007) had reported that Phyropyllite taken from Sumbermanjing Malang Selatan-Indonesia work significantly as filler when it was applied to concrete [4,5]. Adding 15% phyropyllite of total concrete mixture increases the compression stress up to 42 % from the characteristic compression stress of concrete without phyropyllite (25 MPa). Phyropyllite as filler and fine aggregate added to asphalt was able to increase the stability of hot roller aspalt toward acid [5].

In this coming work, initial study is conducted to investigate the possibility of phyropyllite as geopolymer materials. According to Joseph Davidovits, there are five types of geopolymer materials based on the Si:Al ratio. They are: 20:1 < Si : Al < 35 : 1; Si : Al > 3 : 1 ; Si : Al = 3 :1; . Si : Al = 2 : 1 and Si : Al = 1 : 1. As the ratio of Si : Al of phyropyllite is 2:1, it has a possibility as a geopolymer material candidate and can be applied as cements, concrete and fire/heat resistance materials [6].

The term geopolymers is introduced as inorganic polymers chain or network consisting of aluminosilicate which is bonded covalently, low calcium content and heat or fire resistant. One of the benefits of geopolymer materials is its low calcium concentration that is greener chemistry when it is used as eco-green cement's materials compared to the Portland cement that produces a big amount of CO_2 emission to the space.

2. Experimental details

2.1 Materials

Nawangan phyropyllite, NaOH AR (Sigma-Aldrich), KOH AR , KBr and aquadest.

2.2 Characterizations

The chemical composition of Nawangan-phyropyllite was analyzed using XRF PANanalytical Type Minipal 4, its phase or crystallinity was characterized using powder XRD PANalytical (PHILIPS) expert pro. Both analyses were done at Centre of Mineral Analysis, State University of Malang. While, the analysis for sodium oxide was carried out using AAS/AA6800 and the infrared study was performed using FTIR Shimadzu 8400 LS, using KBR powder, the number of scanning was 10 and the resolution was 4.0. Both analyses were done at analysis laboratory, Chemistry Department, Brawijaya University. Prior to use, phyropyllite ground to fit the size of 120 mesh. Heat treatment was performed by putting the mineral in the furnace (Carbolite) in various temperatures (200 °C, 400 °C, 600 °C and 800 °C) for 5 hours.

3. Result and Discussion

The chemical composition of Nawangan-phyropyllite analyzed using XRF is tabulated in Table 1 and the picture of phyropyllite are presented in Figure 2. The percentage of silica is the highest and the ratio of silica and alumina is 3:1, but it does not mean that the mineral is not phyropyllite. The present of quarz and kyanite are common in natural mineral. Iron oxide and titanium oxides are detected also and it supposes as hematite and rutile.

Potassium oxide, vanadium oxide (karelianite) and copper oxide (cuprite) are found as impurities that fulfill the space in between the phyropyllite structure.

Table 1. Chemical composition of Nawangan phyropyllite.

The Elements	Si	Al	K	Fe	Ti	Cu
% wt.	57.7	16.7	20.6	2.47	2.33	0.088

X-ray powder diffraction analysis is carried out for further characterization. Phyropyllite is amorphous which shows specific sharp peaks at 2θ (9°, 20°, 21°, 26 °, 34 °, 36 ° and 39 °) based on the JCPDS data. There are 3 identified peaks responsible for phyropyllite with the significant intensity, that are at 2θ (9°, 21° and 26 °) [7]. The present of quarz, rutile and hematite are also detected. The difractogram of Nawangan-phyropyllite is exhibited in Figure 3.

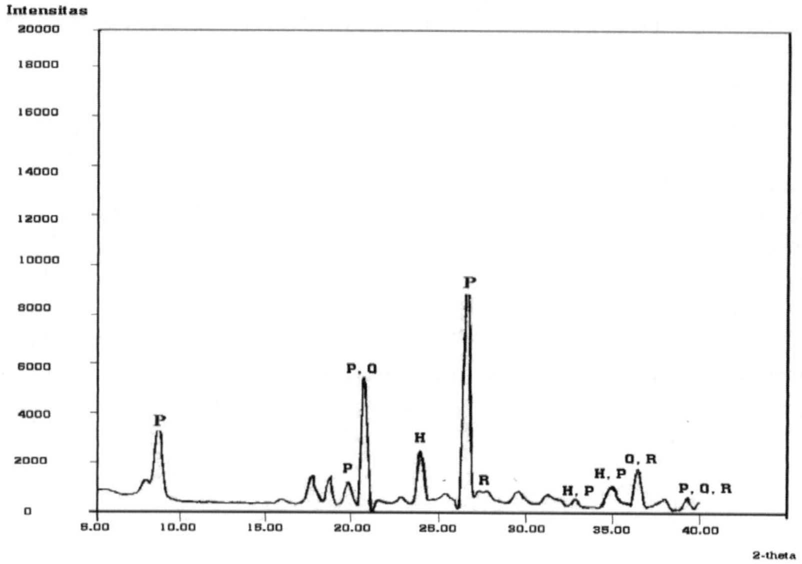

Figure 2. The difractogram of Nawangan-phyropyllite

Infrared study reveals that the characteristic peaks of phyropyllite are at 3630 cm^{-1}, 756 cm^{-1} and 938 cm^{-1} that are atributable for –OH bond to alumina. While the peak at 1021 cm^{-1} is for Si-O/Al-O which is compatible with the peaks at 538 cm^{-1} and 477 cm^{-1}. The spectrum of Nawangan- phyropyllite is displayed in Figure 3.

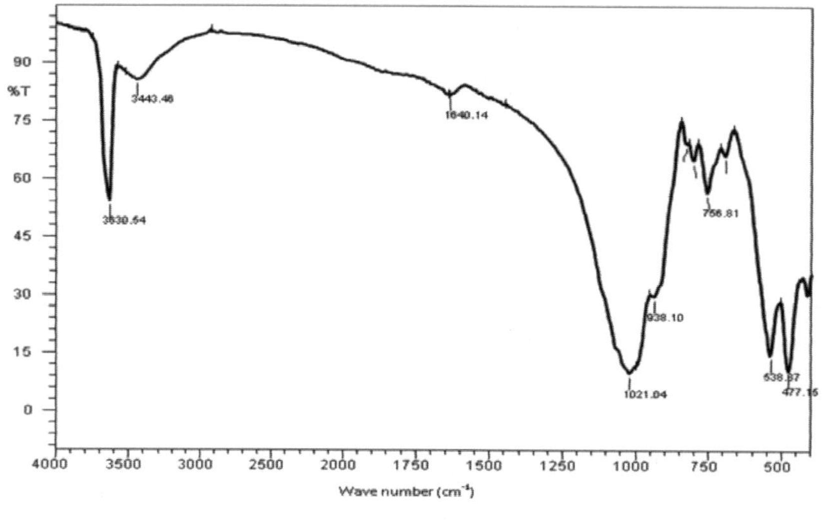

Figure 3. The spectrum of Nawangan-phyropyllite

The effect of heat treatment onto Nawangan-phyropyllite structure was studied by heating the mineral under various temperature, that is: 200 °C, 400 °C, 600 °C and 800 °C . It was heated for 5 hours. Hydroxyl group from water molecules which is shown by the peak at 3443 cm^{-1} disappears by heating at 200 °C. The peak at 3630 °C (Al-OH) still persist under heating up to 400 °C, but it disappears at 600 °C. At the temperature of 600 °C, the peak at 1021 cm^{-1} split into two peaks (1049 cm^{-1} and 992 cm^{-1}). It may be due to the displacive transition. The spectra of heated phyropyllite is shown in Figure 4.

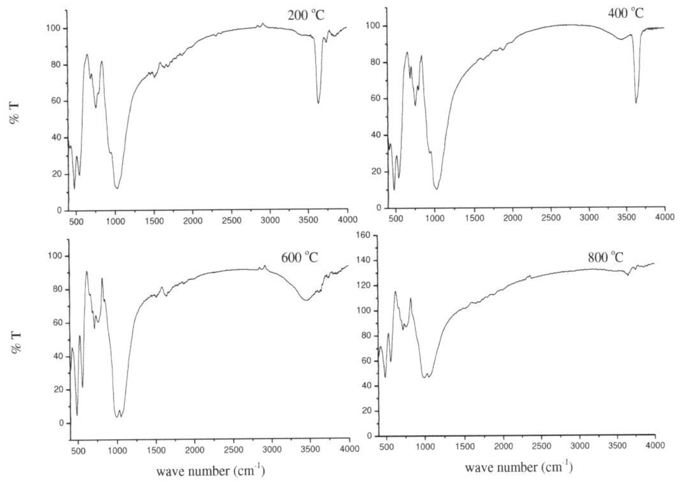

Figure 4. The spectra of phyropyllite treated under heating at 200°C, 400°C, 600°C and 800°C

To study the possibillity of Nawangan-phyropyllite as geopolymer material, three treatment was applied by adding aquades, NaOH 10 M and NaOH 15 M to the phyropyllite powder (120 mesh) with the ratio of 2:1 (phyropyllite : solvent). Hydrolysis and condensation reaction involves which can produce three proposed new structure of phyropyllite containing Al-OH, Al=O and Al-O-Al by virtue of inter- and intra-interaction of terminal OH. The reaction scheme is in Figure 5.

The infrared spectra of phyropyllite (as reference) denoted as (p-r), phyropyllite + aquades (p-a) , phyropyllite + NaOH 5 M (p-n5), phyropyllite + NaOH 10 M (p-n10)/15 M (p-n15) are shown in Figure 5. The infrared spectra of phyropyllite treated with KOH 5M (p-k5), KOH 10 M (p-k10) and KOH 15 M (p-k15) are displayed in Figure 6.

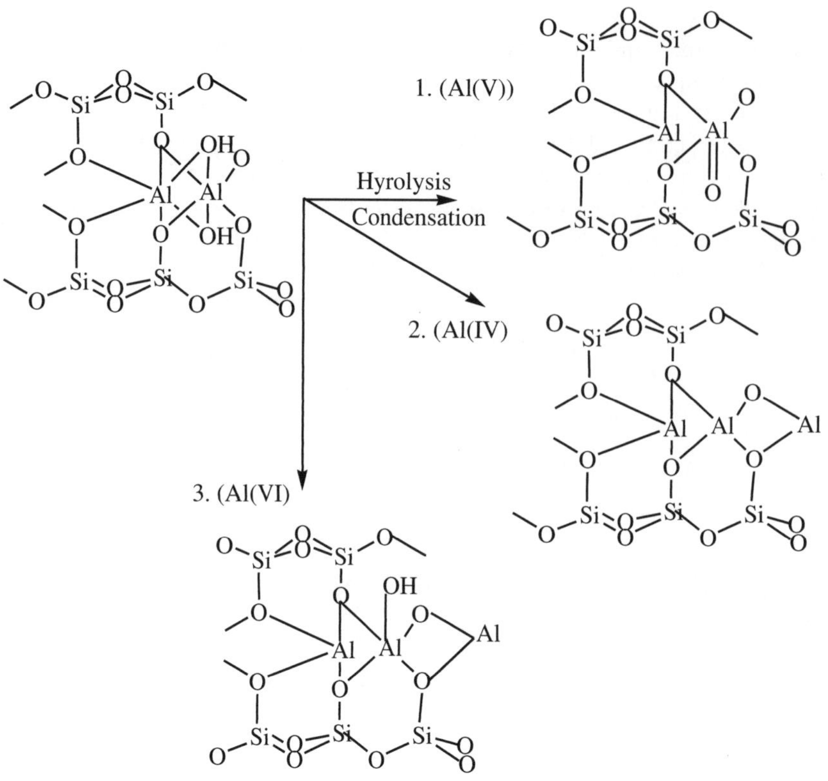

Figure 5. The hyrolysis and condensation reaction of phyropyllite

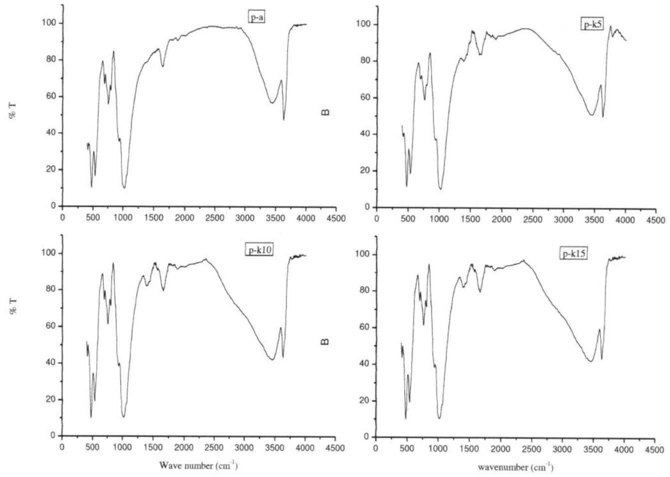

Figure 6. The spectra of phyropyllite+aquades (p-a), phyropyllite + NaOH 5 M (p-n5), phyropyllite + NaOH 10 M (p-n10) and phyropyllite + NaOH 15 M (p-n15)

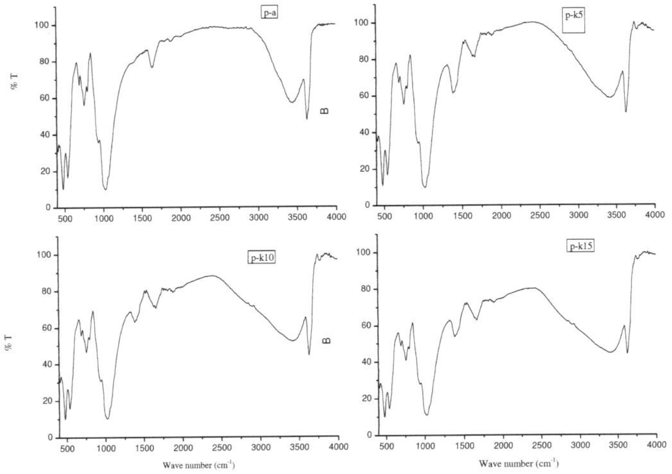

Figure 7. The spectra of phyropyllite+aquades (p-a), phyropyllite + KOH 5M (p-k5), phyropyllite + KOH 10 M (p-k10) and phyropyllite + KOH 15 M (p-k15)

There is a new peak at 1638 cm^{-1} for the spectrum of p-a which supposes the peak for Al=O (8% of intensity) while the peak intensity of Al-OH (3630 cm^{-1}) reduces to 24 % compared to p-r (33 %). It shows that hydrolysis-condensation reaction occurs reactively in phyropyllite only by adding water. Alkaline treatment by adding NaOH 10 M shows a new peak at 1387 cm^{-1} (5% of intensity) that is proposed for Al-O-Al vibration. The peak intensity of Al-OH reduces to 17 %, the peak intensity of Al=O rises to 14 % and the peak intensity for Al-O-Al is 5 %. Adding NaOH 15 M exhibits the peaks of Al-OH (22%), Al=O (5%) and Al-O-Al (30%). The most reactive Al is in oxidation state of V, it means that by adding NaOH 10 M, the most reactive product of geo-polymer is obtained. All the data are tabulated into Table 2. Based on the data tabulated in Table 3, the most reactive functional groups were obtained by adding KOH 5 M.

Table 2. The effect of adding water and sodium hydroxide on the infrared spectra of phyropyllite

Functional group	Intensity (%)				
	p-r (phyropyllite as reference)	p-a	p-n5	p-n10	p-n15
Al-OH	33	24	18	17	22
Al=O	3	8	14	14	5
O-Al-O	-	-	2	5	30

Table 3. The effect of adding water and potassium hydroxide on the infrared spectra of phyropyllite

Functional group	Intensity (%)				
	p-r (phyropyllite as reference)	p-a	p-k5	p-k10	p-k15
Al-OH	33	24	22	21	15
Al=O	3	8	8	9	9
O-Al-O	-	-	17	6	7

IOP Conf. Series: Materials Science and Engineering **285** (2017) 012026 doi:10.1088/1757-899X/285/1/012026

4. Conclusions

The physical chemistry of Indonesian's phyropyllite are: It is composed of Si (57.7%), Al (16,7%), K(20.6%), Fe (2.47%) ,Ti (2.33) and Cu (0.088%), amorphous that has characteristic peak for phyropyllite at $2\theta = 9^{o}, 20^{o}, 21^{o}, 26^{o}, 34^{o}, 36^{o}$ and 39°. Its characteristic peak of infrared are at 3630 cm^{-1}, 756 cm^{-1} and 938 cm^{-1}, attributed to –OH bonded to alumina. By heating at 600 ^{o}C for 5 hours, the displace transition occur. It is just by adding water, the functional groups of phyropyllite change to the reactive and adding NaOH 10 M and KOH 5 M producing the most reactive sides of phyropyllite.

References

[1] Anonymous 1996 *Memperkenalkan bahan Galian Golongan C di jawa Timur: Piropilit* Dinas Pertambangan Daerah propinsi Jawa Timur.
[2] Anonymous 2012 *Bumi Pacitan, Surga Tambang*, https://pacitanku.com/2012/12/06/bumi-pacitan-surga-tambang/
[3] Desai H, Biswal N R, Paria S 2010 *Ind. Eng. Chem. Res.* **49** 5400-06
[4] Mutrofin S, Setyaningsih T, Susilo A 2006 *Natural* **10** 1-7
[5] Kurniadi A, Lasmini A, Mutrofin S 2010 *International Journal of Academic Research*
[6] Davidovits J 2017 *State of the Geopolymer* R&D
[7] Miyah Y, lahrichi A, Idrissi M, Boujraf S, Taouda H, Zerrouq F 2017 *Journal of The Association of Arab Universities for Basic and Applied Sciences* **23** 20-8

IOP Conf. Series: Materials Science and Engineering **285** (2017) 012027 doi:10.1088/1757-899X/285/1/012027

Study on Sumbawa gold recovery using centrifuge

A D Ferdana[1], H T B M Petrus[1], I M Bendiyasa[1], I D Prijambada[2], F Hamada[3], and T Sachiko[3]

[1]Advanced Materials and Sustainable Mineral Processing Research Group, Dept. of Chemical Engineering, Universitas Gadjah Mada, Indonesia
[2]Faculty of Agriculture, Universitas Gadjah Mada, Indonesia
[3]Department of Regional studies and Humanities, Faculty of Education and Human Studies Course, The Education and Human Studies, Akita University, Japan

E-mail: bayupetrus@ugm.ac.id

Abstract. The Artisanal Small Gold Mining in Sumbawa has been processing gold with mercury (Hg), which poses a serious threat to the mining and global environment. One method of gold processing that does not use mercury is by gravity method. Before processing the ore first performed an analysis of Mineragraphy and analysis of compound with XRD. Mineragraphy results show that gold is associated with chalcopyrite and covelite and is a single particle (native) on size 58.8 m, 117 m up to 294 m. characterization with XRD shows that the Sumbawa Gold Ore is composed of quartz, pyrite, pyroxene, and sericite compounds. Sentrifugation is one of separation equipment of gravity method to increase concentrate based on difference of specific gravity. The optimum concentration result is influenced by several variables, such as water flow rate and particle size. In this present research, the range of flow rate is 5 lpm and 10 lpm, the particle size -100 + 200 mesh and -200 +300 mesh. Gold concentration in concentrate is measured by EDX. The result shows that the optimum condition is obtained at a separation with flow rate 5 lpm and a particle size of -100 + 200 mesh.

1. Introduction

Sumbawa is Indonesia's eastern region which rich in mineral resources. Potential of mineral resource Estimated form of gold at 180 thousand m^3, Copper at 1.575 million m^3, and Limestone at 274.29 million m^3, Several studies show that the amount of labor involved in ASGM in Indonesia is between 50,000 and 250,000 people [1].

The amalgamation process is used because of its simple method, The Artisanal Small Gold Mining in Sumbawa is processed with mercury, which poses a serious threat to the mining and global environment [2, 3]. Artisanal Small gold mining uses amalgamation techniques with mercury as its metal catcher. In this practice, mercury is used with excess amount and thus caused significant environmental pollution. But, the process itself is difficult to abolish because of its effectiveness. From local government data of Sumbawa Barat regency there are about 800 kg mercury released per week. Ministry of Environment republic Indonesia has made a plan of action for free mercury gold processing in 2018. This regulation is stemmed from the responsibility to protect the environment while on the other hand also paying attention to the mining rights of locals.

The alternative method for gold processing in artisanal small gold miners is by using the gravity method, the gravity separation method depends on the specific gravity of the mineral, the relative movement of the object, and the magnitude of the frictional force [4]. Separation by gravity technique is the separation of particles from the impurities based on the differences in specific gravity utilizing fluid flow. In the fluid flow there is force resultant between the gravity, friction and buoyancy force which make up terminal velocity. The terminal velocity becomes the base for separation process by gravity separation technique using centrifugation because the magnitude of each mineral's terminal velocity will increase with greater density. The equation of terminal velocity for the laminar flow follows the following Stokes equation.

Not all gold ores can be processed using gravity separation techniques, only native gold and electrum can be processed effectively. Therefore, the characterization of gold ore is crucial to be done before gravity separation process. Table 1 shows the value of the terminals velocity of several compounds of Gold Ore Sumbawa with size -100 mesh.

Table 1. Terminals velocity of several compounds of sumbawa gold ore

Single Partikel	Density (gr/cm3)	Terminal Velocity (cm/s)
Gold	19.3	0.22
Silicate mineral (SiO2)	2.65	0.02
Sulfide compounds	7.60	0.08

2. Experimental procedure

There were three main steps in this research, the first was characterization, Analysis of gold characteristic of Sumbawa was done in two stages, the first is X-ray diffraction analysis (XRD) to observe the composition of gold ore from Sumbawa, and the second was making a thin slice of section rock for mineragraphy analysis.

In this study used the optimal conditions in milling process by grinding in a rodmill consisted of 7 pieces of rod with rotational speed of 85 rpm for 120 minutes [5]. Afterward, separation optimization test was done by centrifugation with variable sizes of -100 + 200 mesh and -200 + 300 mesh with response is gold concentration.

3. Results and discussion

3.1 X-Ray diffraction (XRD) Sumbawa gold ore

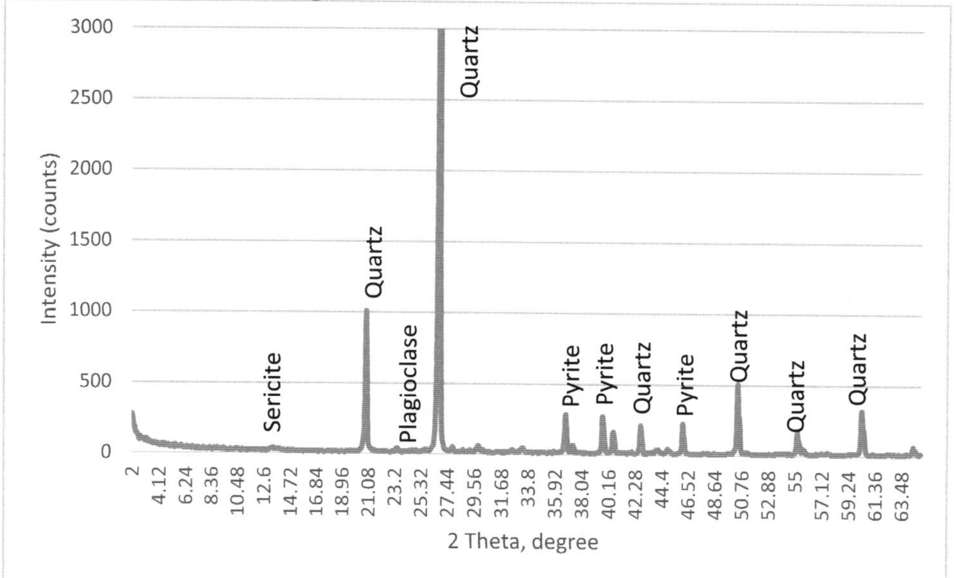

Figure 1. XRD Analysis Result

IOP Conf. Series: Materials Science and Engineering **285** (2017) 012027 doi:10.1088/1757-899X/285/1/012027

XRD analysis was performed to analyze the compounds of Sumbawa Gold Ore. The XRD gold ore has a mineral composition dominated by quartz, pyrite, sericite, pyroxene. XRD analysis results show that quartz minerals had the largest portion and were dominant in the ore, whereas pyrite and sericite minerals have far less quantities with nearly equal distribution. The analysis outcome shows that gold is mostly associated with quartz. So in gold ore indicated associated with the above minerals.

3.2 Mineragraphy of gold ore Sumbawa

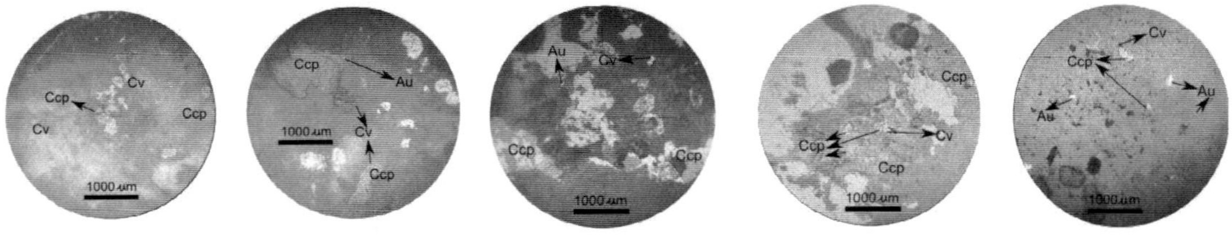

Figure 2. Mineragraphy analysis of gold ore sumbawa

Result of Mineragraphy analysis shows that gold ore from Sumbawa contained two main minerals, covelite and chalcopyrite, which are associated with gold with Covellite as the dominant phase. Native gold from this ore was present as individual mineral with shapes ranged from euhedra to subhedra and maximum size of 3000 m. Native gold in this experiment were in smaller sizes with particle sizes of 294 m, 882 m, and 2352.8 m. Ore minerals were bright yellow with shapes ranged from euhedra to subhedra. These minerals can be sized up to 64 m, but more dominant on the size 58.8 m, 117 m up to 294 m [6]. Where: Ccp = chalcopyrite, Cv = covellite, Au = gold. These results serve as a reference for the liberation process using a rod mill

3.3 Gold liberation

To get particle size that was smaller and more homogeneous, the particles was subjected through the stages of comminution with grinding. Comminution in the world of mining was instrumental in liberating precious mineral (gold) from the impurities, and the liberation process is done using rodmill. The grinding was done to reach a size of -100 + 200 and -200 + 300 mesh, it was done based on the results of the Mineragraphy of native gold size. After the classification of size was analyzed, the content of gold ore were analyzed with EDXRF.

3.4 Separation with centrifugation

In this research, the effectiveness of separation of gold from gangue with centrifugation was tested. The process was influenced by three main variables which are particle size, flow rate, and the response variable in the value of Effectiveness of separation (newton efficiency).

Table 2. Experimental data on the increase of gold concentration by centrifugation

No	Flow rate (Lpm)	Particle size (mm)	Gold concentration (% EDX)	
			Before centrifugation	After centrifugation
1	5	0.1115	0.218	0.595
2	10	0.0590	0.382	0.377
3	5	0.0590	0.382	0.414
4	10	0.1115	0.218	0.409

In this process, apart from the increased concentration of gold ore, observations were also made about the decrease of Si content as the main impurity on Sumbawan gold ore. Table 3 shows decrease of Si content on every variable of centrifugation operation.

Table 3. Decrease of silica concentrate on centrifugation operations

No	Flow rate (Lpm)	Particle size (mm)	Gold concentration (% EDX)	
			Before centrifugation	After centrifugation
1	5	0.0590	48.297	9.225
2	10	0.0590	48.297	2.221
3	5	0.1115	35.068	11.893
4	10	0.1115	35.068	3.301

With three free variables which were slope of the table, shaking speed, and particle Size with each had only two variations of the upper limit and lower limit. By using minitab 17, the experimental was designed to be divided into eight data points, where each data point was determined to evaluate the value of the response variable which in this case is the separation effectiveness which would determine the gold concentration results in the experiment and in turn determine the value of Newton's efficiency.

Newton's efficiency is a parameter that identifies the effectiveness of a separation process. Newton's efficiency can be calculated based on the following equation:

$$N = Rc-(1-Rfa) \quad (1)$$

Where

N : Newton's efficiency;
Rc : Gold Recovery (%)
Rfa : Impurities (%)

Further recovery of gold minerals from ores in material balance is based on how large the gold mineral content is in the feed, tailings, and concentrates, as shown in the following equation:

$$Rc = \frac{CcWc}{CfWf} = \frac{(Cf-Cfa)Cc}{(Cc-Cfa)Cf} x\ 100\% \quad (2)$$

$$Rfa = \frac{Wfa(1-Cfa)}{Wf(1-C)} = \frac{(Cf-Cc)(1-Cfa)}{(Cfa-Cc)(1-Cf)} x\ 100\% \quad (3)$$

Where :

Wf = weight of feed (kg) (gram);
Wc = weight of gold concentrated (gram);
Wfa = weight of impurities concentrated (gram);
Cf = concentration of gold in feed (%)
Cc = concentration of gold in concentrate (%)
Cfa = concentration of gold in *tailing* (%)

Based on the above equation, the effectiveness of the separation of gold from impurities on each combination of variables is as shown in Table 4:

Table 4. Newton's efficiency value on each variable combination

IOP Conf. Series: Materials Science and Engineering **285** (2017) 012027 doi:10.1088/1757-899X/285/1/012027

Flow Rate	Particle Size	Feed Weight	Gold Concentration in Concentrate	newton efficiency
		Wf (gr)	Cc (%)	N
5	0.1115	34.18	0.377	94.15686
10	0.0590	34.18	0.414	89.60647
5	0.0590	282.58	0.409	99.46954
10	0.1115	282.58	0.595	97.33463

3.5 Main effect and interaction studies

The experimental result of the effectiveness of separation is shown by the increase of newton efficiency vs flow rate and particle size as shown in figures 3 and 4.

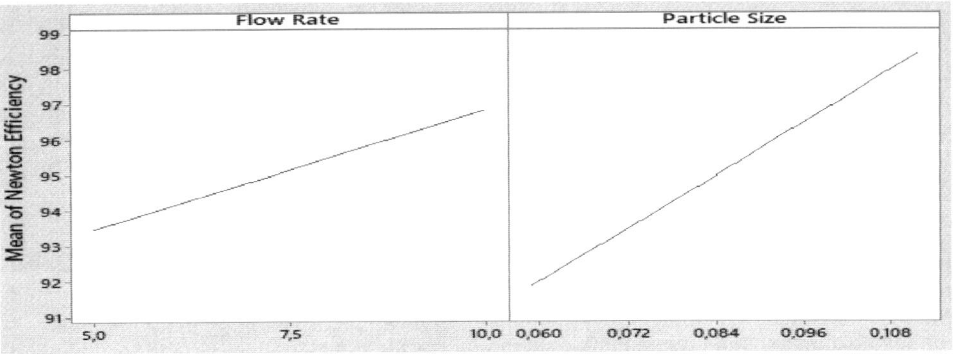

Figure 3. Main effect of each single variable

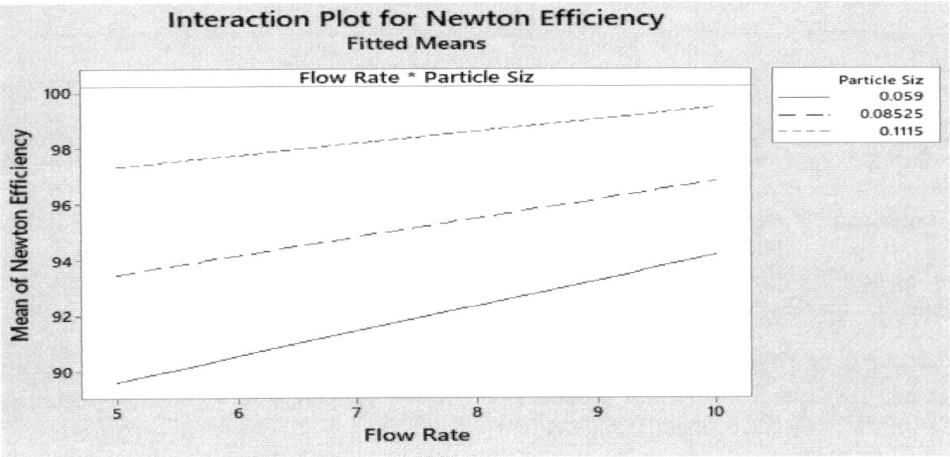

Figure 4. Interaction of variables and their effects on Newton efficiency improvements

In this experiment, there were two variable of the flow rate velocity observed which were 5 lpm and 10 lpm. 10 lpm was observed because it was adjacent to the value of the gold terminal velocity that had the highest density on the particle size of -100 + 200 mesh and 5 lpm was observed because it was adjacent to the value of the gold terminal velocity that had the density highest on the particle size of -200 + 300 mesh. For variable size, also two variables were observed which were -100 + 200 and -200 + 300 because they were in the range of gold size which was free from impurities based on mineragraphy analysis.

From figure 3 and 4, it is apparent that the effect of the flow rate on the effectiveness of the separation shows with higher flow rate, the greater the effectiveness of the separation, and with larger particle size, the effectiveness also increase. This is shown in the effectiveness of the variable flow rate of 5 lpm with 0.059 particle size compared to the effectiveness at flow rate of 10 lpm with particle size of 0.1115 in which the value of newton efficiency are 99.46954 and 97,33463 respectively. Because it is the basis of separation with the size of the velocity terminal of each particle. Comparison of the effectiveness gold separation with different amount gold concentration in each variable is shown in Figure 5:

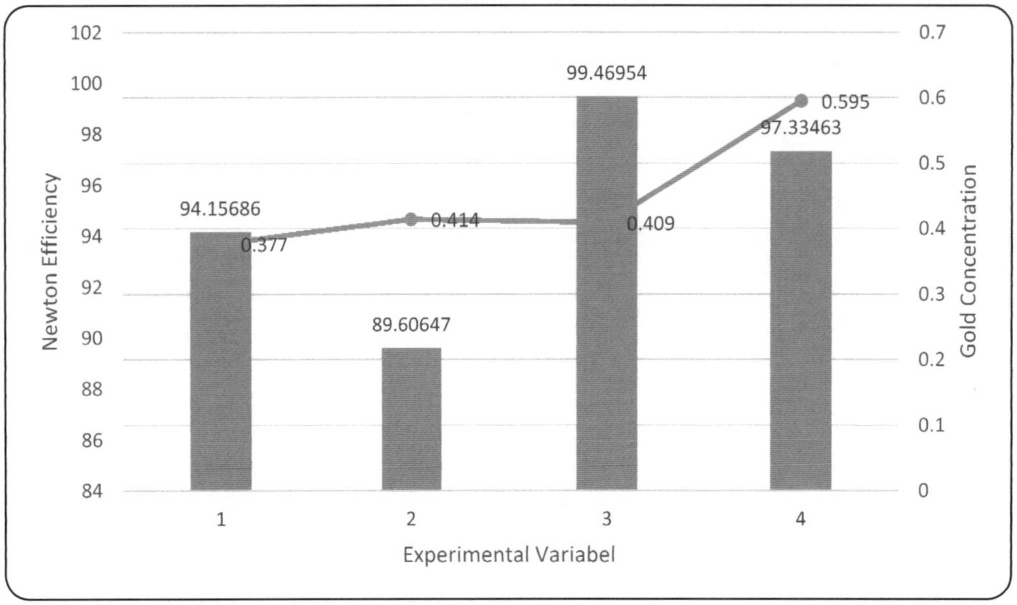

Figure 5. Newton efficiency and gold concentration of each variables

Figure 5 shows that the best separation effectiveness which indicated by the best newton efficiency values is obtained in the experiment with variables of flow rate of 5 lpm and particle size of 0.059 mm with Newton Efficiency value of 99. 46954%. If viewed from the amount of effectiveness of separation with the same particle size, Experiment with flowrate of 10 lpm resulted in Newton Efficiency of 89.60647%. These results differ greatly, as this result also shows that in determining the flow rate in separation by centrifugation, it must correspond to the speed of the nearest terminal velocity of the heaviest impurities so that the impurities will be carried away with the flow and only gold will be left behind. These result were also shown at the experiment with the particle size of 0.1115 mm, at a flow rate of 5 lpm, effectiveness of separation of 94.15686 were obtained while at flowrate 10 lpm the effectiveness was 97.33463. This shows that in the separation by gravity technique using centrifugation the main variable to watch for the success of the separation is the value of the terminals velocity needed for each particle to be separated.

4. Conclusion

Based on the results of the mineragraphy analysis, the gold was associated with chalcopyrite, and covellite. Native gold had size between 58.8 m, 117 m to 294 m. Based on the results of XRD analysis, gold ore from Sumbawa composed of quartz, pyrite, piroksen, and serisit. The optimal conditions in the gold ore separation using centrifugation were between flow rate 5 lpm with particle size 0.059 mm and flow rate 10 Lpm with particle size 0.1115 lpm.

Acknowledgments
The Authors would like to thank for the support of LPPM UGM which funded this research from appropriate technology program (425/DIT.PM/2017) and to the Japan Society For the Promotion of Science (JSPS) for their support in conducting this research.

References
[1] Sutrisno 2006 *CCOP-GSJ/AIST-GAI CASM* Pusat Sumber Daya Geologi-Bandung
[2] Polii B J, Sonya D N 2002 *Pendugaan Kandungan Merkuri dan Sianida di Daerah Aliran Sungai (DAS) Buyat Minahasa*
[3] Clarkson T 2006 *Critical Reviews in Toxicology* **36** 609-62
[4] Gaudin A M 1939 *Principles of Mineral Dressing* New York: McGraw Hill Book Company
[5] Himawan T B M P, Achmad D F, Hadhiansyah I, Arini M R P, Agus P 2017 *Unpar Prosiding* pp 2477-1694
[6] Aye M T, Pramumijoyo S, Idrus A, Setiadji L D, Imai A, Araki N, Arif J 2011 *The Mineralogy of Gold-Copper Skarn Related Porphyry at The Batu Hijau Deposit, Sumbawa, Indonesia* Yogyakarta: Department of Geological Engineering, Universitas Gadjah Mada

IOP Conf. Series: Materials Science and Engineering **285** (2017) 012028 doi:10.1088/1757-899X/285/1/012028

Additive manufacturing for steels: a review

A Zadi-Maad[1,2], R Rohib[3,4], A Irawan[5]

[1] Department of Metallurgy Engineering, University of Technology Sumbawa, Sumbawa, Indonesia
[2] Graduate Institute of Ferrous Technology, Pohang University of Science and Technology, Pohang, South Korea
[3] Department of Metallurgical and Materials Engineering, Kalimantan Institute of Technology, Balikpapan, Indonesia
[4] Department of Materials Science and Engineering, Yeungnam University, Gyeongsan, South Korea
[5] School of Business and Management, Bandung Institute of Technology, Bandung, Indonesia.

Email: ahmad2014@postech.ac.kr

Abstract. Additive manufacturing (AM) of steels involves the layer by layer consolidation of powder or wire feedstock using a heating beam to form near net shape products. For the past decades, the AM technique reaches the maturation of both research grade and commercial production due to significant research work from academic, government and industrial research organization worldwide. AM process has been implemented to replace the conventional process of steel fabrication due to its potentially lower cost and flexibility manufacturing. This paper provides a review of previous research related to the AM methods followed by current challenges issues. The relationship between microstructure, mechanical properties, and process parameters will be discussed. Future trends and recommendation for further works are also provided.

1. Introduction

Steel has been widely used in various applications, starting from the defense, petroleum, automotive, nuclear and chemical industries, due to its excellent mechanical properties and cost efficient [1]. Among them, stainless steels are getting more attention due to its excellent corrosion and oxidation resistance. Chromium additions impart passive protection layer when the amount of more than 11% [2]. However, there is a difficulty using conventional manufacturing to fabricate complex shape parts with cooling channels, mesh structure or inner cavities.

Additive manufacturing (AM) is a revolutionize technology which can manufacture solid parts from 3D image data through layer by layer processing. This technique melts the powder particle using electron or laser beam, while heat source is moving relative to base materials and then parts will solidify. AM process can be classified into two categories: powder bed (PB) and flow-based technique [3–5]. The PB process covers the electron beam melting (EBM) and selective laser melting (SLM), while flow-based method includes the laser-engineered net shaping (LENS), direct metal deposition (DMD) and direct metal laser sintering (DMLS).

PB technique starts from the bed of powder and the beam scanning the powder that melts the powder and solidified as final parts. The illustration of the process is shown in figure 1. These methods give parts with high density, excellent mechanical properties, and smooth surface, but it is limited for small scale product due to small beam size. While flow-based method begins with the injection of powder feedstock through the deposition head and heat from beam melts layer by layer until the desired part. This technique can build large scale products due to high deposition rate and volume, but it creates a rough surface and imprecision dimension. The methods are illustrated in figure 2.

IOP Conf. Series: Materials Science and Engineering **285** (2017) 012028 doi:10.1088/1757-899X/285/1/012028

Several types of steels have been processed through AM technique. Starting from pure iron, stainless steel (304, 316, 321, 347, 420, 17-4PH), tool steel (H13, M2 HSS), maraging steel (18Ni300), until low-alloyed steel (4140, 4340). The lists of previous work on the AM of steel are shown in table 1.

The focus of this paper is the development of the AM technique for steels materials, along with the current scientific, technical challenges and economic consideration that still need to be solved.

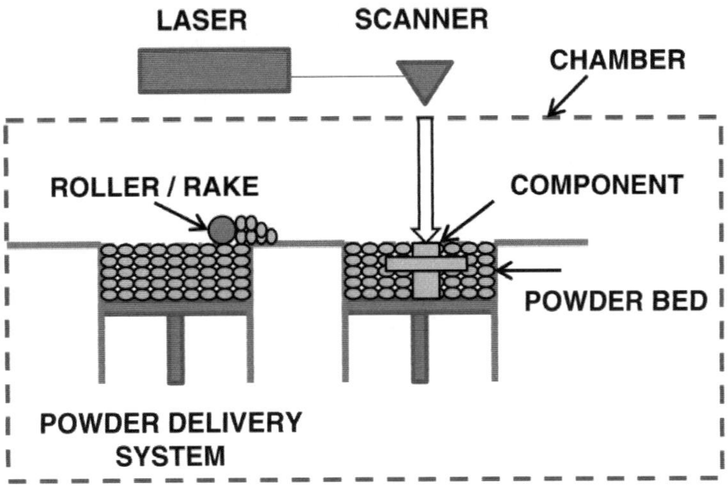

Figure 1. Powder bed fusion illustration scheme (adapted from [4])

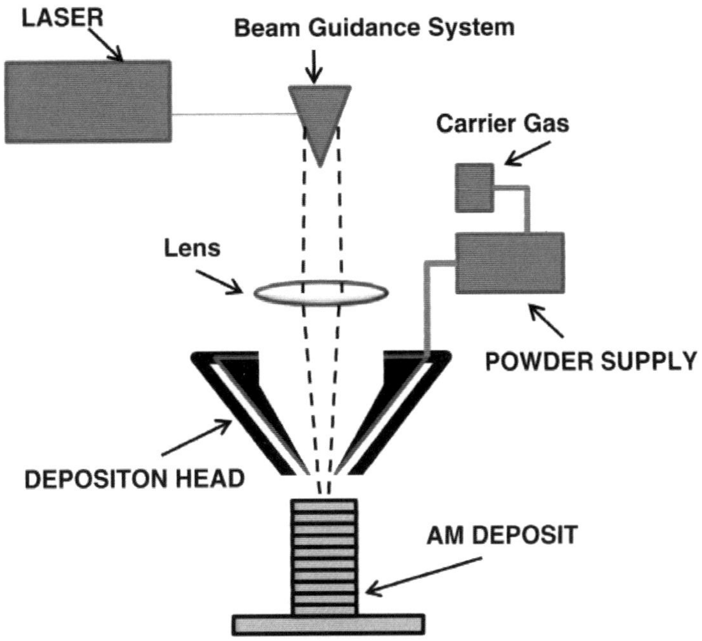

Figure 2. Flow-based illustration scheme (adapted from [4])

IOP Conf. Series: Materials Science and Engineering **285** (2017) 012028 doi:10.1088/1757-899X/285/1/012028

Table 1. List of publication

AM Technique	Machine Type	Alloy	Heat Source Power	Input Feedstock Characteristics	References
EBM	Arcam (Sweden)	H13 steel	Electron beam	Powder particle	[6]
EBM	Arcam S12	Pure iron	Electron beam	Gas atomized powder	[7]
EBM	Arcam AB	316	Electron beam	Powder particle	[8]
EBM	EBSM (China)	316L	Electron beam	Gas atomized powder	[9]
EBM	Sciaky (USA)	Stainless steel 321, 347	Electron beam	Wire filler	[10]
SLM	EOS (Germany)	Stainless steel (17-4, 15-5 PH)	Yb-fiber laser beam	Powder particle	[11,12]
SLM	Realizer (Germany)	316L	Laser beam	Powder particle	[13]
SLM	SLM-Solution (Germany)	316-L	Yttrium fiber laser	Powder particle	[14–16]
SLM	Concept Laser (Germany)	Maraging steel (18Ni300)	Continuous fiber laser	Powder particle	[17]
SLM	Concept Laser (Germany)	M2 HSS	Nd: YAG laser beam	Gas atomized powder	[18]
SLM	Concept Laser (Germany)	316L	Laser beam	Powder particle	[12,19–21]
SLM	Concept Laser (Germany)	17-4 PH stainless steel	Fiber laser beam	Water atomized powder	[22]
SLM	Concept Laser (Germany)	Low alloyed steel	Nd: YAG laser beam	Water and gas atomized powder	[23]
SLM	HRPM (China)	316L	Fiber laser	Gas atomized powder	[24]
SLM	HRPM (China)	AISI 420	Continuous fiber laser	Gas atomized powder	[25]
SLM	LSNF (China)	304 SS	Continuous fiber laser	Gas atomized	[1]
DMLS	EOS (Germany)	4340 steel	Ytterbium fiber laser	Powder particle	[26]
DMLS	Laser-based	316L	Continuous CO_2 laser	Powder particle	[27,28]
LENS	LENS (SNL-USA)	304L and 17-4PH	Laser beam	Atomized powder	[29]
LENS	LENS (SNL-USA)	316 and 316L SS	Nd: YAG laser beam	Powder particle	[30–33]
LENS	LENS	H-13 steel	Laser beam	Powder particle	[34–36]
LENS	LENS	AISI 4140	Laser beam	Pre-alloyed powder	[37]
DMD	Laser-based	AISI 4340 steel	Laser beam	Pre-alloyed powder	[38]
DMD	Laser-based	H13 steel	Laser beam	Powder particle	[39,40]

2. Microstructure and Mechanical Properties of AM Parts

The final properties of an AM product are depending on the process parameter such as laser power, scanning speed, beam diameter, layer thickness, beam scan pattern, build direction, and powder mass flow rate. The microstructures are also controlled by thermal history cause by repeating heating and cooling process during AM or also called thermal cycle. The thermal cycle induced the grain to growth on preferred crystallographic orientation, for example <100> direction for fcc alloys, which cause strong crystallographic texture, anisotropic tensile properties and evolve the microstructure [41]. However, Dehoff et al. show that the microstructure can vary from columnar grains with very strong texture to almost equiaxed grains with weaker texture using a modification of build parameters [42]. It makes the randomly crystallographic texture is possible in the AM process. On the other hand, Peter demonstrated that hot-isostatic processing (HIP) could also result in the equiaxed grain formation [43].

Table 2. Tensile properties of steel fabricated by AM technique

Machine Type	Alloy	Condition	Specimen Orientation	σ_y (MPa)	σ_{UTS} (MPa)	ε (%)	Hardness (HB)	Ref.
SLM 250	316L	As-built	Z	500	600.2	55	NA	[16]
		Heat-treated		475	617.9	54.1	NA	
		HIP		380	586.6	64.4	NA	
SLM M1	316L	As-built	NA	640	760	30	NA	[12]
LENS	316L	As-built	Z	405-415	620-660	34-40	NA	[33]
		Heat treated		325-355	600-620	42-43		
Wrought Alloy	316L	Anneal treated		235	560	55	146	a
EOS M270	15-5 PH	As-built	NA	1100	1470	15	NA	[12]
Wrought Alloy	15-5 PH	H900 condition		1275	1380	14	420	b
EOS M250	AISI 4340	Stress relieved	XY	1303	1372	16-17	430-468	[26]
DMD	4340	Stress relieved	XY	NA	1398	1.66	NA	[38]
Wrought Alloy	4340	Heat treated		1475	1595	12	NA	c

a Nominal wrought 316L data from Online Metals Corp. (www.onlinemetals.com)
b Nominal wrought 15-5 PH data from Online Metals Corp. (www.onlinemetals.com)
c Nominal wrought AISI 4340 data from Material Property Data (www.matweb.com)

The thermal cycles can also trigger a variety of metallurgical phenomena such as solid-state phase transformation and segregation behavior. El Kadiri et al. investigate the phase transformation of low-alloyed steel fabricated by LENS [37]. They found that delta ferrite is the primary phase of the solidification process. Due to very high cooling rates, they observed fine allotriomorphs ferrite in the microstructure which can lead very brittle behavior. In addition, Jagle et al. studied the microstructure evolution in the maraging steel fabricated by SLM [17]. They found that martensite is the main phase in the as-built condition also due to high cooling rates. Several post-heat treatments have been used to formed precipitate phase and reported that it could increase the hardness of the maraging steels. Therefore, these unconventional thermal history processes unleash possibility to control the microstructure of any kinds of steels materials.

Defects density can induce by the process parameters, including microvoids, inclusion, pores de-bonding and weak grain boundary. Xue et al. investigate the effect of microporosity on the tensile strength of 316L fabricated by LENS methods [32]. It is reported that an increasing porosity volume can significantly decrease the tensile properties. Therefore, the HIP treatment has been used to reduce the amount of porosity and proven to increase the fatigue limit on the 316L stainless steel [15].

There are only a few research works using wire filler. Qi et al. reported that high amount of MC carbides formed in the stainless steel fabrication [9]. This secondary phase can lead brittle properties and lower the ductility. Therefore, most of the previous works used powder as an input feedstock. Table 2 summarizes the tensile properties of steel from the previous research works. The table shows that most of the AM fabricated steel has an excellent mechanical properties compare to the conventional process product.

3. Technology Challenges

From the heat source, AM methods can be divided as two, using laser or electron beam. Due to its different nature energy carried out by photons and electron, it gives significant differences. The electron beam can provide higher scan rate up to 10^4 mm/s compared to the laser beam that only 1200 mm/s [44]. The electron beam can leap instantly from point to point and move inertia-free, but it has significant disadvantages which need a vacuum atmosphere to operate. Therefore, there are only a few studies using electron beam for steel fabrication, but it is popular for Ti alloy and Ni-base alloys manufacturing.

One of the major drawbacks on the AM technologies is the residual stresses that decrease the mechanical properties of the final product. The melting process creates thermal gradients between different layers which lead to significant residual stress. These residual stresses can be accurately measured using x-ray or neutron diffraction [20]. According to Rangaswamy et al. the magnitude of local residual stress reach up to 75% of the yield strength of the material, and it goes higher for superalloys parts [45]. As a result, these residual stresses also cause considerable inaccuracies on the dimension in the final product. Additional treatment has been proposed by Klingbeil et al and Shiomi et al which showed that substrate preheating and post-annealing treatment can be used to limit warping displacement induced by residual stresses [46,47].

4. Economic Consideration

According to the conclusion from Baumers, there are two AM advantages over conventional manufacturing techniques [48]. First, AM methods can efficiently fabricate complex geometry components and second, this technology can manufacture very small production quantities at relatively low average cost. This indicates that AM have future in economic perspective. The implementation of AM method can reduce cost budgeting, increase the efficiency of raw material used and reduce the waste of production in various industries.

The AM process can be replacing the production of tool dies in the injection moulding machine, which made of tool steel. The die requires a complex shape with a cooling channel and inner cavities. Kinsella investigated the flow-based AM method for Nickel superalloy (IN718) in forged engine case [49]. It is reported that more than 30% cost saving can be achieved using AM process compared to conventional process. However, only a few researches have been done investigating the prospective studies in steel materials.

The cost in manufacturing can be divided into (i) fixed cost such as tools, dies, buildings, etc, and (ii) recurring costs include raw materials, labour, etc. Figure 3 shows the illustration of total cost of manufacturing component via AM and conventional process. The total cost was calculated as a linear function of amount of parts being produced. The slope line represent the ratio of the recurring cost of AM divided by the recurring cost of conventional method. This cost model calculated that the budget of AM has a cheaper manufacturing cost than the conventional one for small-scale production, proven by interception in 175 parts for 2 recurring cost while 90 parts for 1.5 recurring cost. In summary, the

AM is currently favoured in small-production due to significant recurring cost driven by the high cost of raw material.

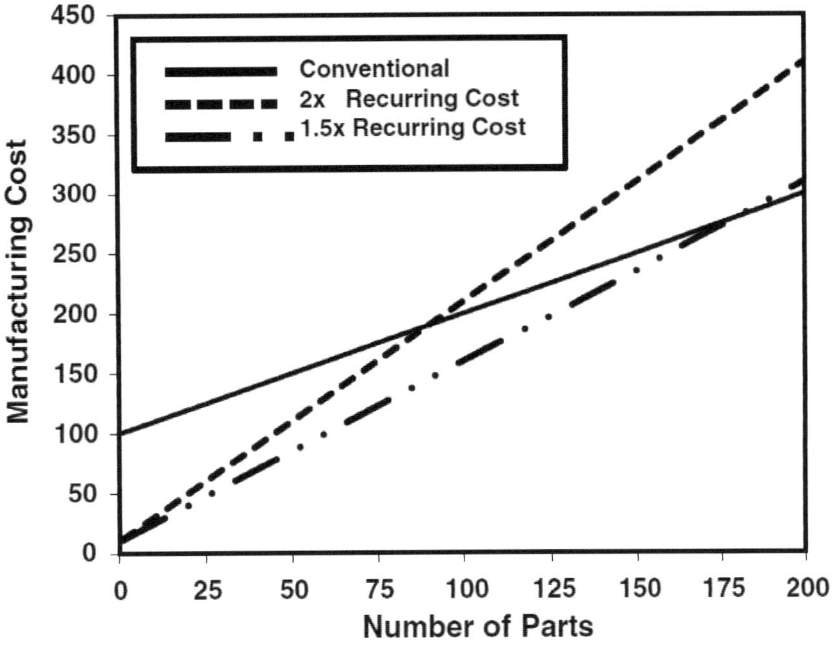

Figure 3. Cost comparison between AM and conventional process (adapted from [4]).

5. Conclusion and Further Works

Additive manufacturing revolutionize future industrial production by offering several advantages compare to conventional one, such as production of small quantities with complex geometry, design freedom, and reduction of development times. Along with optimum processing parameters and post-treatment, the resulting AM mechanical properties are comparable or even better than the conventional production methods. However, the AM processes with various type of machine, are presently far from being completely developed to manufacture the controlled-microstructure materials. Therefore, the future works should focus on the better understanding of process control, enhance the machine power and design new alloys. The computational approach also could be used to minimize the trial and error experiment. In addition, the cost reduction of raw materials should be continued.

Acknowledgements

The authors would like to thank Bandung Institute of Technology, Yogyakarta University of Technology and Kalimantan Institute of Technology for support of this work.

References

[1] Guan K, Wang Z, Gao M, Li X and Zeng X 2013 *Mater. Des.* **50** 581–6
[2] ASM 1993 *ASM Handbook Volume 1 Properties and Selection: Irons Steels and High Performance Alloys*
[3] Elahinia M, Shayesteh Moghaddam N, Taheri Andani M, Amerinatanzi A, Bimber B A and Hamilton R F 2016 *Prog. Mater. Sci.* **83** 630–63
[4] Frazier W E 2014 *J. Mater. Eng. Perform.* **23** 1917–28
[5] Froes F H and Dutta B 2014 *Adv. Mater. Res.* **1019** 19–25
[6] Cormier D, Harrysson O and West H 2004 *Rapid Prototyp. J.* **10** 35–41
[7] Murr L E, Martinez E, Pan X, Meng C, Yang J, Li S, Yang F, Xu Q, Hernandez J, Zhu W, Gaytan S M, Medina F and Wicker R B 2013 *J. Mater. Res. Technol.* **2** 376–85

[8] Hinojos A, Mireles J, Reichardt A, Frigola P, Hosemann P, Murr L E and Wicker R B 2016 *Mater. Des.* **94** 17–27
[9] Qi N B, Yan Y N, Lin F, He W and Zhang R J 2006 *Proc. Inst. Mech. Eng. Part B (Journal Eng. Manuf.* **220** 1845–53
[10] Wanjara P, Brochu M and Jahazi M 2007 *Mater. Des.* **28** 2278–86
[11] Abele E, Stoffregen H A, Kniepkamp M, Lang S and Hampe M 2015 *J. Mater. Process. Technol.* **215** 114–22
[12] Spierings A B, Starr T L and Wegener K 2013 *Rapid Prototyp. J.* **19** 88–94
[13] Tolosa I, Garciandía F, Zubiri F, Zapirain F and Esnaola A 2010 *Int. J. Adv. Manuf. Technol.* **51** 639–47
[14] Buican G R, Oancea G, Lancea C and Pop M A 2015 *Appl. Mech. Mater.* **760** 515–20
[15] Riemer A, Leuders S, Thöne M, Richard H A, Tröster T and Niendorf T 2014 *Eng. Fract. Mech.* **120** 15–25
[16] Leuders S, Lieneke T, Lammers S, Tröster T and Niendorf T 2014 *J. Mater. Res.* **29** 1911–9
[17] Jägle E A, Choi P, Humbeeck J Van and Raabe D 2014 *J. Mater. Res.* **29** 2072
[18] Kempen K, Vrancken B, Buls S, Thijs L, Van Humbeeck J and Kruth J-P 2014 *J. Manuf. Sci. Eng.* **136** 61026
[19] King W E, Barth H D, Castillo V M, Gallegos G F, Gibbs J W, Hahn D E, Kamath C and Rubenchik A M 2014 *J. Mater. Process. Technol.* **214** 2915–25
[20] Wu A S, Brown D W, Kumar M, Gallegos G F and King W E 2014 *Metall. Mater. Trans. A Phys. Metall. Mater. Sci.* **45** 6260–70
[21] Yusuf S, Chen Y, Boardman R, Yang S and Gao N 2017 *Metals (Basel).* **7** 64
[22] Lebrun T, Tanigaki K, Horikawa K and Kobayashi H 2014 *Mech. Eng. J.* **1** 1–13
[23] Rombouts M, Kruth J P, Froyen L and Mercelis P 2006 *CIRP Ann. - Manuf. Technol.* **55** 187–92
[24] Li R, Liu J, Shi Y, Wang L and Jiang W 2012 *Int. J. Adv. Manuf. Technol.* **59** 1025–35
[25] Zhao X, Wei Q, Song B, Liu Y, Luo X, Wen S and Shi Y 2015 *Mater. Manuf. Process.* **6914** 37–41
[26] Jelis E, Clemente M, Kerwien S, Ravindra N M and Hespos M R 2015 *JOM* **67** 582–9
[27] Gu D and Shen Y 2009 *Mater. Des.* **30** 2903–10
[28] Gu D and Shen Y 2008 *Appl. Surf. Sci.* **255** 1880–7
[29] Susan D F, Puskar J D, Brooks J A and Robino C V 2000 *Proceedings of the 11th Solid Freeform Fabrication Symposium*
[30] Griffith M L, Harwell L D, Romero J T, Schlienger E, Atwood C L and Smugeresky J E 1997 *Proc. 8th Solid Free. Fabr. Symp.* 387–94
[31] Lewis G K and Schlienger E 2000 *Mater. Des.* **21** 417–23
[32] Xue Y, Pascu A, Horstemeyer M F, Wang L and Wang P T 2010 *Acta Mater.* **58** 4029–38
[33] Yadollahi A, Shamsaei N, Thompson S M and Seely D W 2015 *Mater. Sci. Eng. A* **644** 171–83
[34] Maziasz P ., Payzant E ., Schlienger M . and McHugh K . 1998 *Scr. Mater.* **39** 1471–6
[35] Griffith M ., Schlienger M ., Harwell L ., Oliver M ., Baldwin M ., Ensz M ., Essien M, Brooks J, Robino C ., Smugeresky J ., Hofmeister W ., Wert M . and Nelson D . 1999 *Mater. Des.* **20** 107–13
[36] Brooks J, Robino C, Headley T, Goods S and Griffith M 1999 *Proc. Solid Free. Fabr. Symp.* 375–82
[37] El Kadiri H, Wang L, Horstemeyer M F, Yassar R S, Berry J T, Felicelli S and Wang P T 2008 *Mater. Sci. Eng. A* **494** 10–20
[38] Sun G, Zhou R, Lu J and Mazumder J 2015 *Acta Mater.* **84** 172–89
[39] Mazumder J, Choi J, Nagarathnam K, Koch J and Hetzner D 1997 *JOM* **49** 55–60
[40] Choi J and Chang Y 2005 *Int. J. Mach. Tools Manuf.* **45** 597–607
[41] Cakmak E, Kirka M M, Watkins T R, Cooper R C, An K, Choo H, Wu W, Dehoff R R and Babu S S 2016 *Acta Mater.* **108** 161–75

[42] Dehoff R R, Kirka M M, List F A, Unocic K A and Sames W J 2015 *Mater. Sci. Technol.* **31** 939–44
[43] Peter W H, Nandwana P, Kirka M M, Dehoff R R, Sames W, Erdman D, Eklund A and Howard R 2015 *Undestanding the Role of Hot Isostatic Pressing Parameters on the Microstructural Evolution of Ti-6Al-4V and Inconel 718 Fabricated by Electron Beam Melting*
[44] Murr L E, Gaytan S M, Ramirez D A, Martinez E, Hernandez J, Amato K N, Shindo P W, Medina F R and Wicker R B 2012 *J. Mater. Sci. Technol.* **28** 1–14
[45] Rangaswamy P, Griffith M L, Prime M B, Holden T M, Rogge R B, Edwards J M and Sebring R J 2005 *Mater. Sci. Eng. A* **399** 72–83
[46] Klingbeil N W, Beuth J L, Chin R K and Amon C H 2002 *Int. J. Mech. Sci.* **44** 57–77
[47] Shiomi M, Osakada K, Nakamura K, Yamashita T and Abe F 2004 *CIRP Ann. - Manuf. Technol.* **53** 195–8
[48] Baumers M 2012 *Economic aspects of additive manufacturing : benefits , costs and energy consumption* (Loughborough University)
[49] Kinsella M E 2008 *Additive Manufacturing of Superalloys for Aerospace Applications*

Simulation for grinding balls production using sand mold-gravity casting

F Nurjaman, A Shofi, U Herlina, N M Prilitasari, and Y Triapriani

Research Unit for Mineral Technology, Indonesian Institute of Sciences, Lampung-Indonesia

E-mail: fajar.nurjaman@lipi.go.id

Abstract. In this present work, the grinding balls from high chromium white cast iron (ASTM A-532) were produced by using sand mold-gravity casting. The simulation casting process was conducted before making these grinding balls by using SOLIDCast™ version 8.2.0. The gating system design and the pouring temperature of hot metal were investigated clearly to obtain grinding balls with no-defect. The sound casting of grinding balls was resulted by using the proper gating system with the addition of vent air on the top of each grinding ball's mold. The dimension of vent air was reduced by the increasing of pouring temperature, thus it resulted on the increasing of the yield production of grinding balls.

1. Introduction

Grinding ball is a spherical component that was rotated in ball mill unit. It is generally used in mining industries for grinding raw materials, such as coal, iron ore, quartz, etc [1-2]. High chromium white cast iron is one of material that commonly used as grinding ball, because it has high hardness and wear resistance, especially after being heat treated [3-5]. Grinding ball that was made from high chromium white cast iron, commonly produced by sand mold-gravity casting method [6-7]. Porosity is one of the casting defects that mostly found in this method [8]. This porosity will be the source of crack inside the grinding ball. It will rupture due to its crack propagation that has rapidly occurred with a high density of impact load on grinding ball movement in ball mill unit. Thus, gating system design should be calculated approximately to produce sound casting or grinding ball with no-defect. Furthermore, casting simulation is the best tool to help engineers to produce sound casting product by visualizing the mold filling, solidification, and cooling. It also can predict the location of internal defects, such as shrinkage and porosity [9]. *Gating system* is the inlet where the hot metal will flow and fill the mold. The appropriate gating system will contribute in preventing the casting defect, such as shrinkage, porosity, miss-run, cold-shut, etc. The outline of gating system in gravity casting is shown in Figure 1. The gating system design for grinding balls production in this experiment was investigated clearly. Pouring basin and sprue are the first entrance of hot metal into the mold. In Figure 2, there is some of the important dimension for pouring basin and sprue, they are A_1, A_2, H_1, dan A_2. The dimension of choked area (A_2) can be obtained from equation (1).

$$A_2 = \frac{w}{d.t.c.\sqrt{2.g.h}} \quad (1)$$

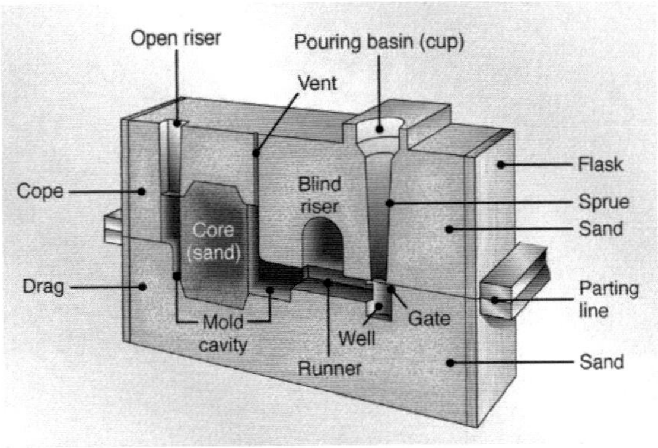

Figure 1. Gating system of sand mold casting [10]

where, w = weight of product (lb); d = density of hot metal (lb/in^3); C = efficiency process; H = height of product (in); t = pouring time (s); g = gravity acceleration (386.4 in/s^2).
The carbon content of high chromium white cast iron, that is 2.2% C, is similar to high carbon steel. Thus, the pouring time (t) is expressed by equation (2) [11].

$$t = k\sqrt{w} \tag{2}$$

For high chromium white cast iron with 2.2% C and less than 100 lb weight of a product, the value of k is 1.2.

The total height of sprue and pouring basin (H_2) is greatly determined by the flow type of hot metal when it is poured into the mold. The flow of hot metal must be in the mixed of laminar-turbulent with *Reynold Number* (Re) about 2000-20.000. The Reynolds number is expressed in equation (3).

$$v = \frac{\text{Re}.\mu}{\rho.D_2} \tag{3}$$

where, v = velocity of hot metal in A_2 (in/s); μ = dynamic viscosity (lb. in^{-1}. s^{-1}); D_2 = diameter of A_2 (in). By using the Bernouli's equation, thus H_2 can be obtained from equation (4).

$$V = \sqrt{2.g.H_2} \tag{4}$$

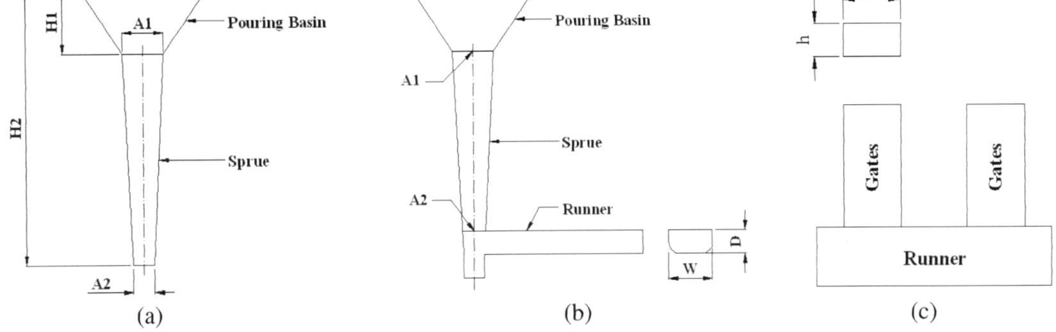

Figure 2. Dimension of gating system : (a) pouring basin and sprue, (b) runner, (c) gates

IOP Conf. Series: Materials Science and Engineering **285** (2017) 012029 doi:10.1088/1757-899X/285/1/012029

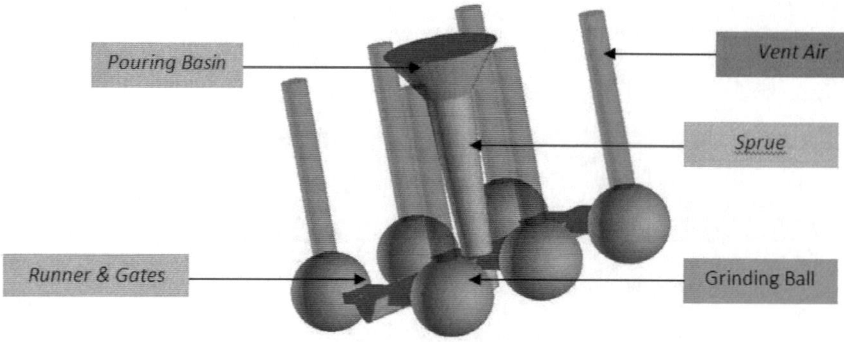

Figure 3. Gating system design for grinding ball casting simulation

Table 1. Parameters and constants for gating system design

Parameter/Constants	Index	Units	Value	Remark
weight of grinding balls	W	Kg	3.8705	6 pcs of *grinding ball*
density hot metal	□$_{(1450\ ^{o}C)}$	Kg/m^3	6840.76	cast iron
efficiency process	C	-	0.8	assumption
total height	H	m	0.16	-
gravity acceleration	g	m/s^2	9.8	-
reynold number	Re	-	7500	laminar-turbulent flow
dynamic viscosity	□	m^2/s	5.5×10^{-7}	cast iron

Table 2. Dimension of gating system design

Gating System	Index	Units	Value
pouring time	t	second	5
pouring basin and sprue	A_1	m^2	0.00068
	A_2	m^2	0.00014
	H_1	mm	30.226
	H_2	mm	120.9
runner	w	mm	9.82
	D	mm	11.18
gates	h	mm	8.31
	l	mm	25.4

The dimension of A_1 and H_1 can be obtained from the ratio of a total height of sand mold and diameter of sprue that was expressed in equation (5) [10].

$$\frac{A_1}{A_2} = \sqrt{\frac{H_2}{H_1}} = 2 \quad (5)$$

Runner and gates were shown in Figure. 2b. The hot metal that flowing from sprue will enter the runner and distributed into several gates, then it flows into grinding ball's mold. For high carbon steel, the ratio of *choke area* (A_2), *runner* (A_{runner}), and *gates* (A_{gates}) is 1: 2: 1.5 [11], where W (width) > D (diameter) for a *runner*, and L > h for *gates*. Air vent, as shown in Figure 1 was used for preventing the porosity that caused by gas entrapment in the sand mold after pouring process. In this study, the effect of a various dimension of an air vent in gating system on grinding ball's defect casting was investigated clearly. The influence of pouring temperature and yield production on grinding balls was also investigated to produce the sound casting product. In this study, the casting simulation of grinding balls production that was made from high chromium white cast iron was conducted by using SOLIDCast^{TM} verion 8.2.0 [12].

2. Material and Method

The gating system was designed for six grinding balls with 50 mm diameter for each grinding ball, as shown in Figure 3. The dimension of gating system design was obtained by input some parameters in table 1 into equation (1) to (5). Thus, the dimension of the grinding balls gating system was listed in table 2. Three kinds of gating system design were made, as shown in Figure 4. The influence of vent air will be investigated in this simulation and production of grinding balls. Gating system design-1 was without vent air. Gating system design-2 had diameter 6 mm of vent air, while gating system design-3 had a larger diameter of vent air that is 15.5 mm.The simulation for grinding ball casting process was conducted by using SOLIDCast^{TM} version 8.2.0 [12]. The casting simulation was started by selecting the parameter from the SOLIDCast^{TM}, as shown in Figure 5, such as grinding balls material, that was ASTM A-532, and the material for mold, that was silica sand.

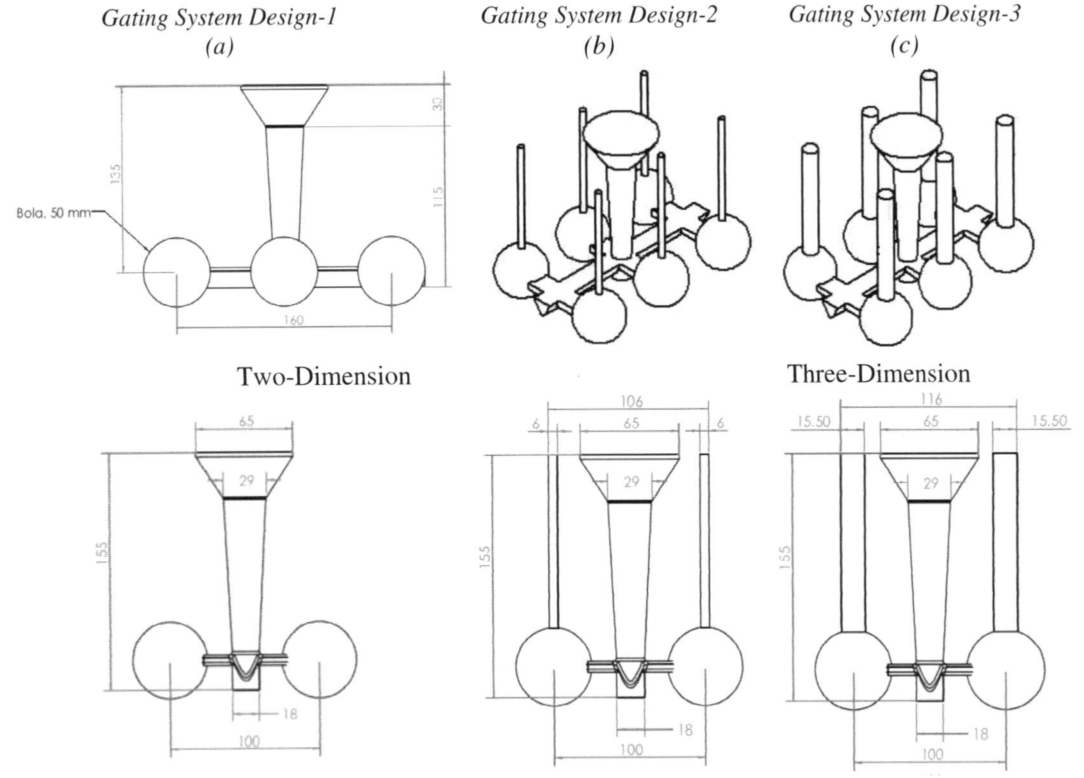

Figure 4. Gating system design for casting simulation of grinding balls

After the simulation data was obtained, the grinding ball casting process was conducted. Induction furnace, with capacity 500 Kg/heat, was used to melt scrap steel and ferroalloys, such as ferrochromium and ferromolybdenum. The hot metal then poured into sand molds. The composition of grinding balls was 2.18C - 0.72Si - 13Cr - 1.38Mo. The casting defects either on the surface or inside the grinding balls were examined carefully.

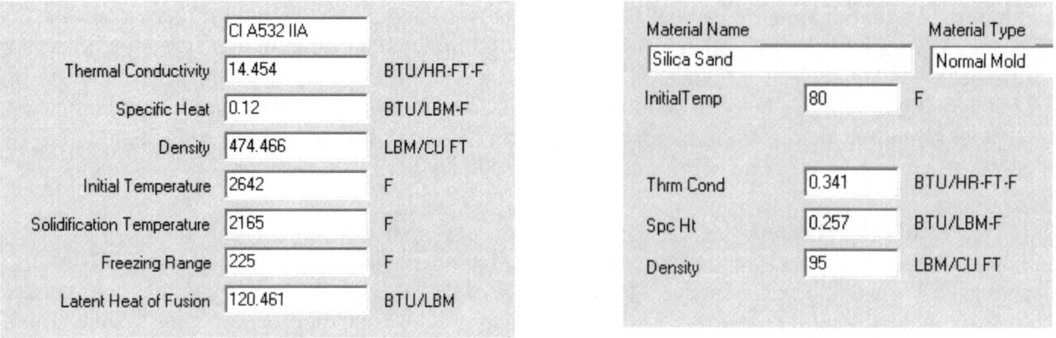

(a) (b)

Figure 5. Parameters for casting simulation: (a) material selection, (b) mold selection

3. Result and Discussion

The casting simulation of three kinds grinding ball's gating system would be emphasized on solidification time, gradient temperature, and niyama criterion. Solidification time would provide an overview of solidification stages for a specific time, while temperature gradient and niyama criterion were related to the formation of casting defects, such as shrinkage and porosity [13]. The simulation of casting process was shown in Figure 6 to 8. The simulation was conducted by using initial pouring temperature 2642 °F (1450 °C), as shown in Figure 4a. All of the gating system design in Figure 4 had a similar tendency of solidification stages, as shown in Figure 6. The solidification was started from the smallest volume of gating system part that was sprue and runner. The last solidification area was pouring basin. The addition of vent air on gating system (Figure 4b), and also the addition of larger volume vent air (Figure 4c) did not result in solidification time difference significantly. The solidification time of all design system was about 5-7 seconds.

The gating system area with high-temperature gradient was very susceptible to the formation of casting defects, such as shrinkage and porosity. From Figure 7, it was shown that the addition of vent air on gating system (Figure 4b) and larger volume of *vent air* (Figure 4c) had resulted in a positive impact on the reduction of high-temperature gradient area in gating system, which was indicated by blue and yellow color. The casting defects, such as shrinkage and porosity, were identified by using niyama criterion in this simulation casting process, as shown in Figure 8, where it was indicated by the blue color in the last row.

From this casting simulation, it was resulted that vent air would contribute in the sound casting of grinding balls production. Vent air volume as $1.89x10^{-5}$ m^3 (cross-sectional area of *vent air* with diameter 15.5 mm) in gating system design-3 resulted in grinding ball with no-defect.

The yield production for each gating system, as listed in table 3, was calculated by the ratio of the total weight of grinding ball (GB) product to the total weight of grinding balls (GB) and gating system (GS), as expressed in equation (6).

$$\% \, Yield = \frac{weight\ of\ GB}{weight\ of\ (GB+GS)} . 100\% \quad (6)$$

From table 3, the higher volume of vent air, the lower yield would obtain. Though gating system design-3 had no defect and it also had the lowest percentage of yield production. Thus, the optimization casting simulation of grinding balls should be enhanced by changing the casting simulation parameter, such as pouring temperature.

IOP Conf. Series: Materials Science and Engineering **285** (2017) 012029 doi:10.1088/1757-899X/285/1/012029

Time (s)	Solidification Time		
	Gating system design-1	*Gating system design-2*	*Gating system design-3*
1			
2	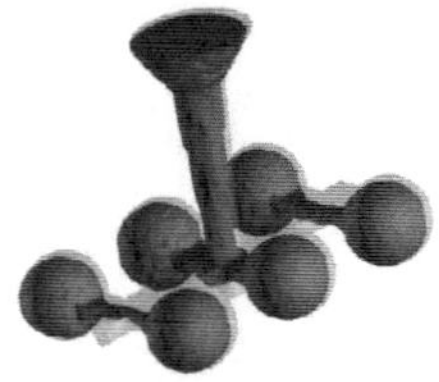		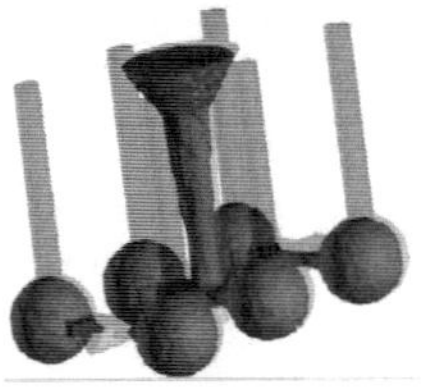
3	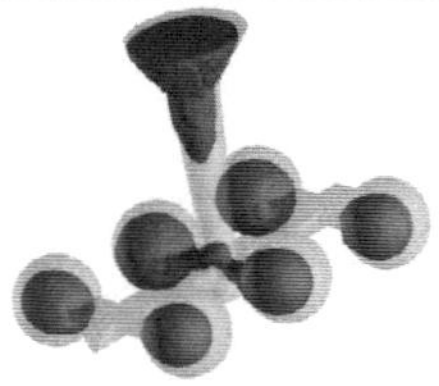	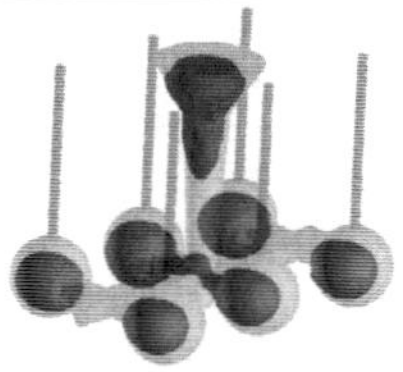	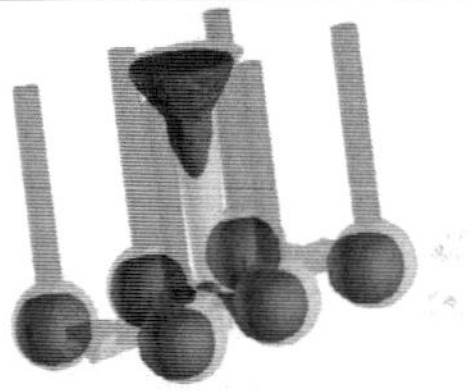
4			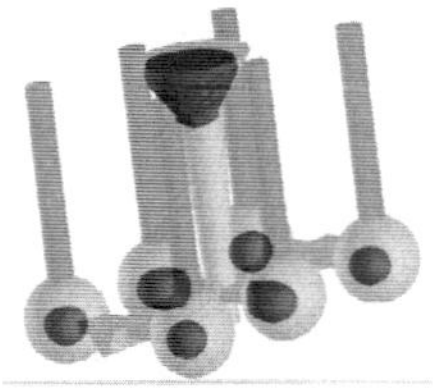
5	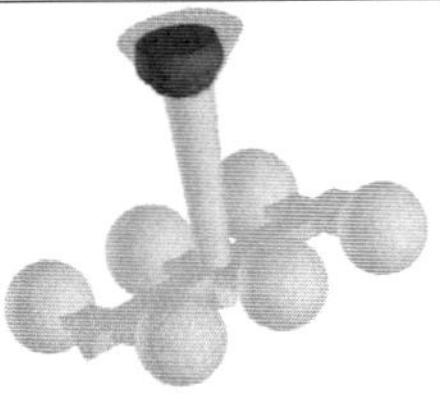	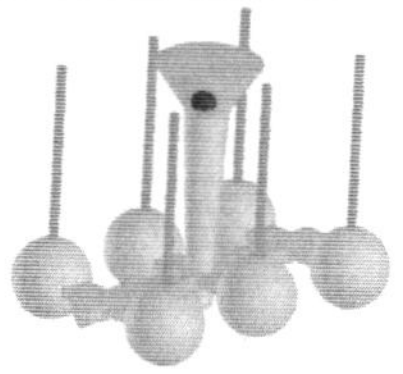	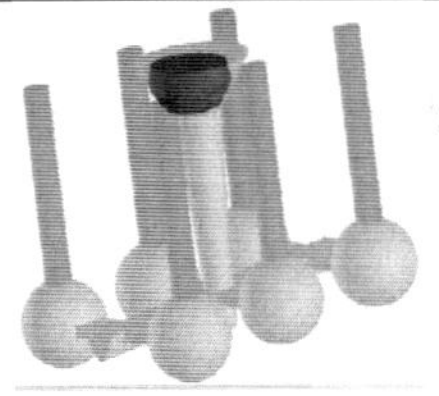

Figure 6. Casting simulation for solidification time

IOP Conf. Series: Materials Science and Engineering **285** (2017) 012029 doi:10.1088/1757-899X/285/1/012029

Percentage (%)	*Temperature Gradient*		
	Gating system desain-1	*Gating system desain-2*	*Gating system desain-3*
0	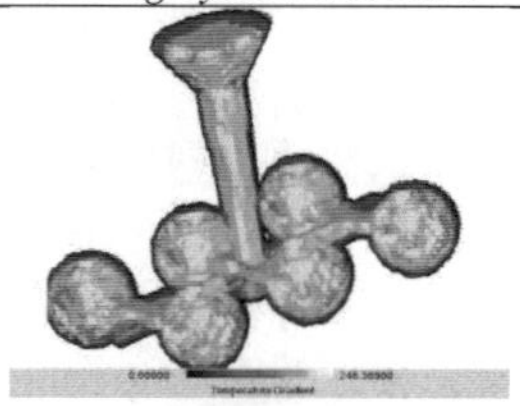	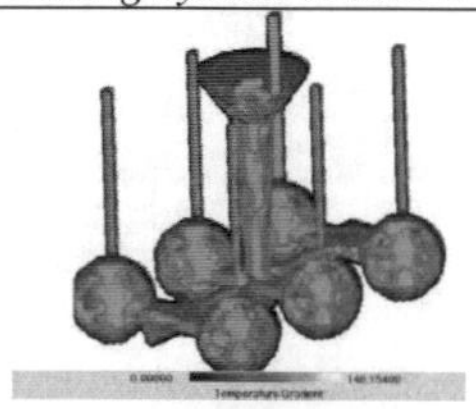	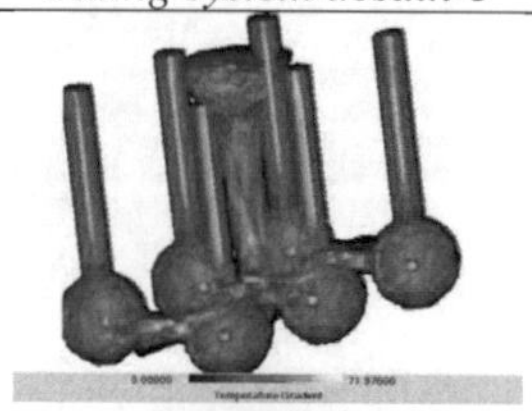
16.7	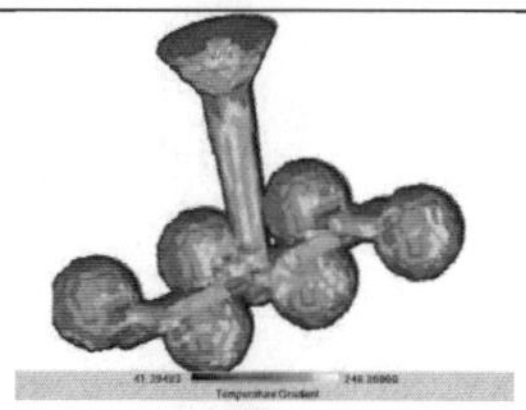	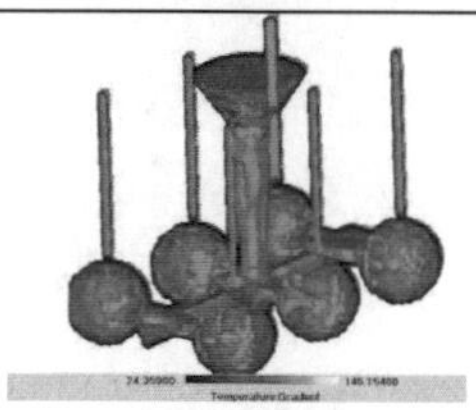	
58.5	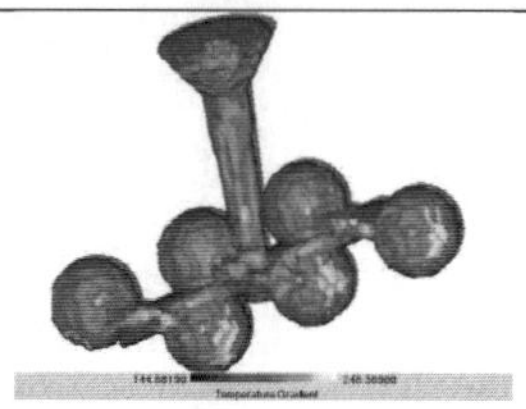	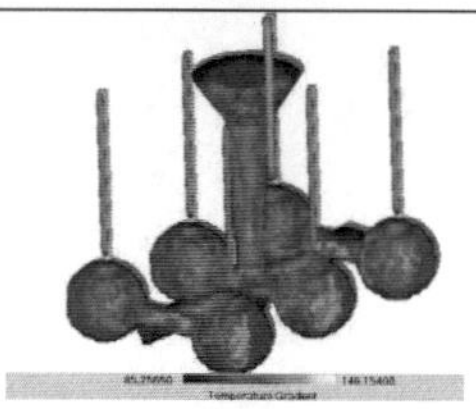	
91.7			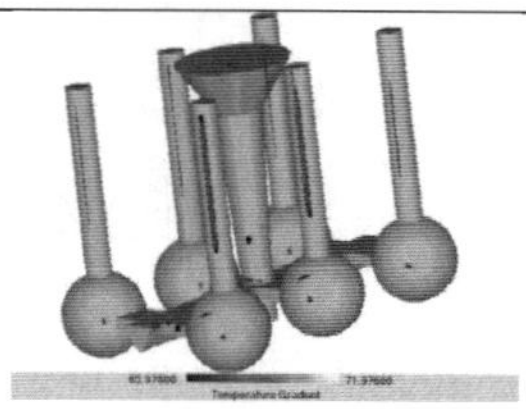
99.6	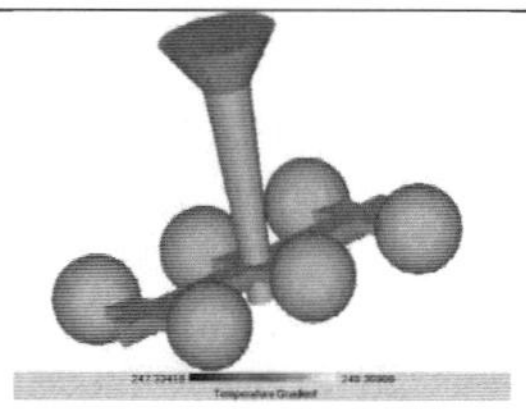	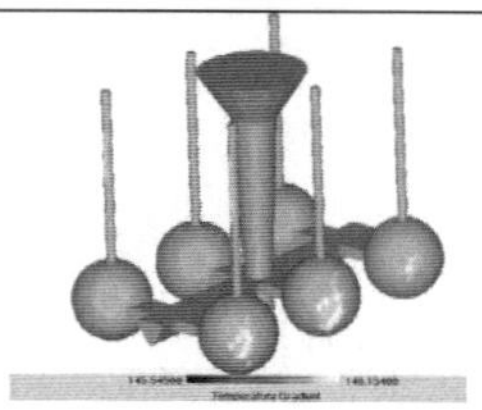	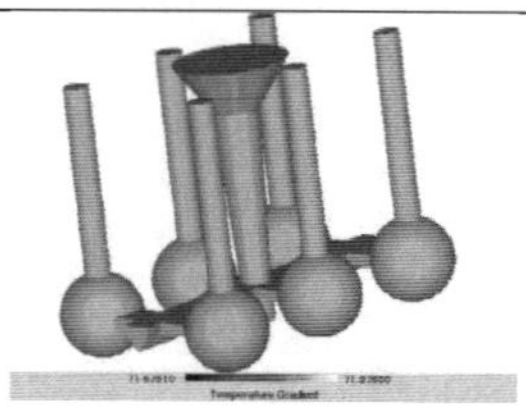

Figure 7. Casting simulation for temperature gradient

IOP Conf. Series: Materials Science and Engineering **285** (2017) 012029 doi:10.1088/1757-899X/285/1/012029

Percentage (%)	Niyama Criterion		
	Gating system desain-1	*Gating system desain-2*	*Gating system desain-3*
0			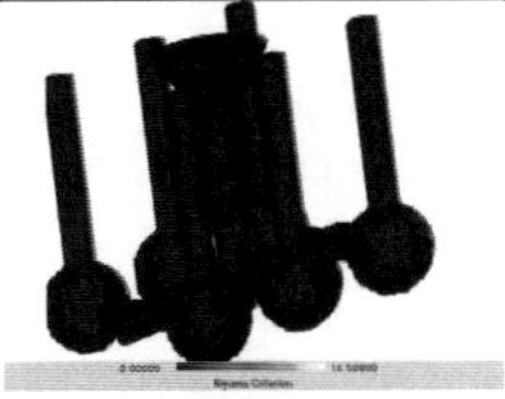
6.7		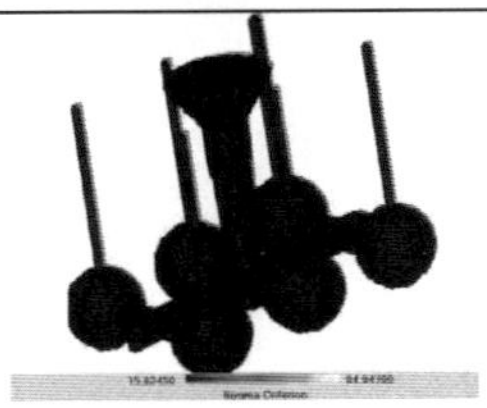	
58.3	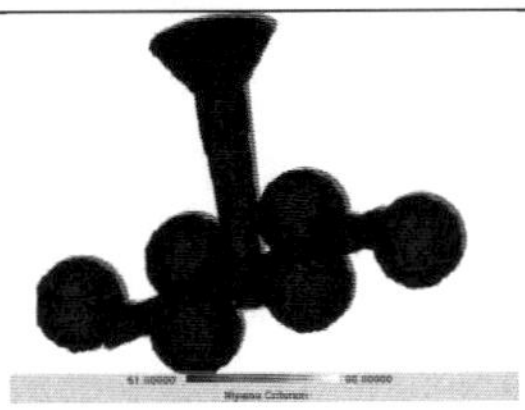	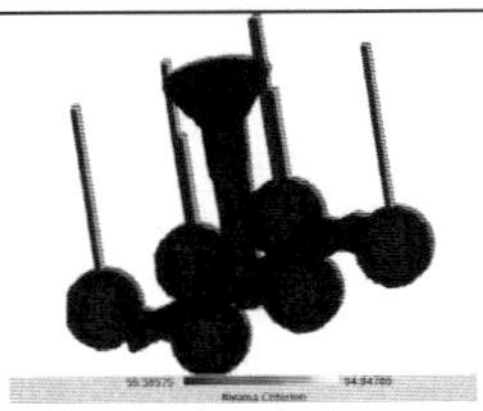	
95.8		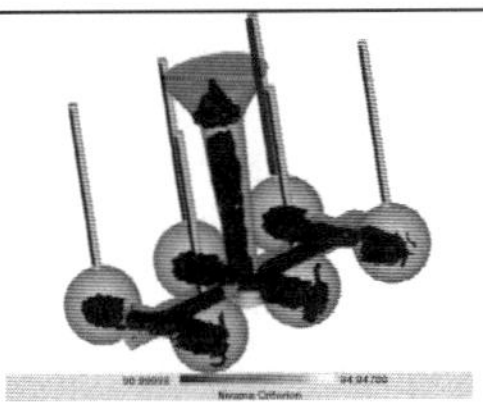	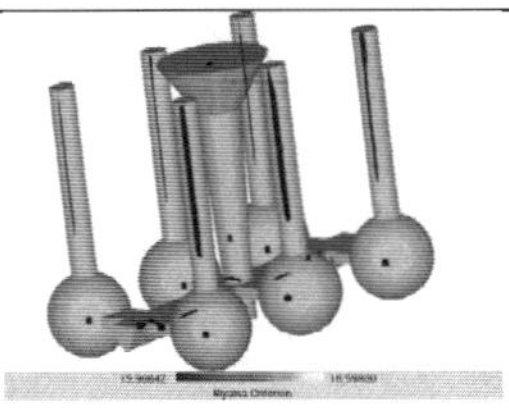
99.6	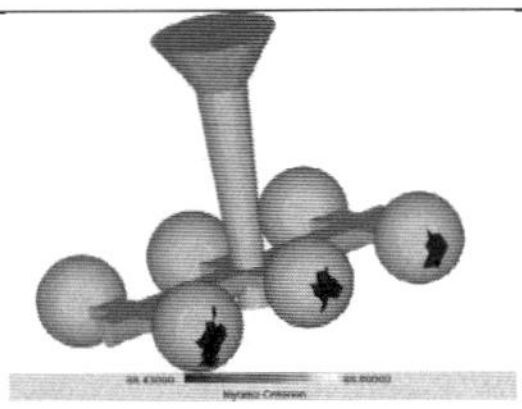	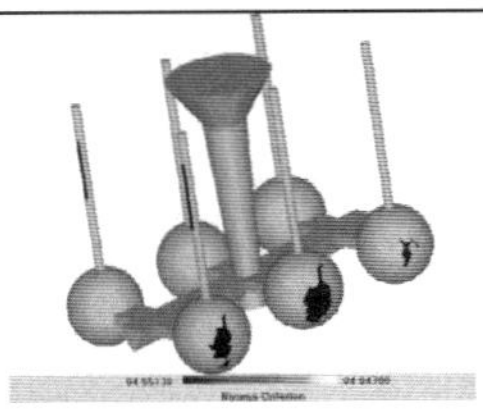	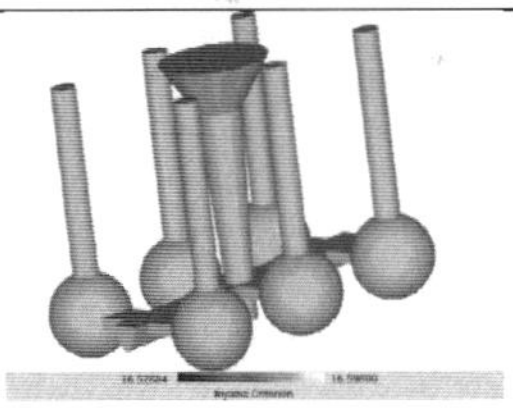

Figure 8. Casting simulation for niyama criterion

IOP Conf. Series: Materials Science and Engineering **285** (2017) 012029 doi:10.1088/1757-899X/285/1/012029

s / (%)	Gating system design-2		
	Solidification Time	*Temperature Gradient*	*Niyama Criterion*
1/0			
2/16.7	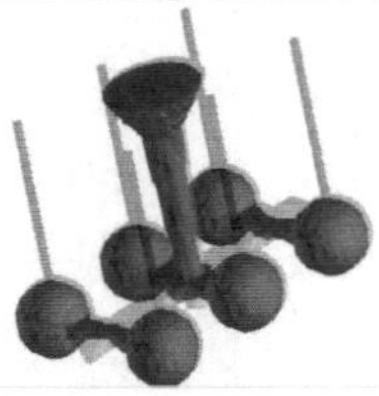		
3/58.3	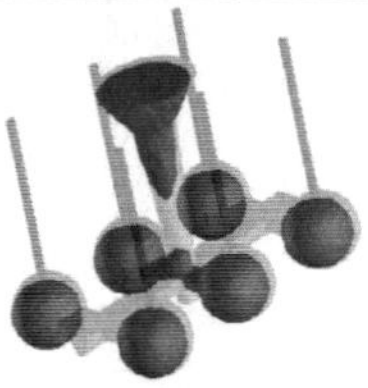		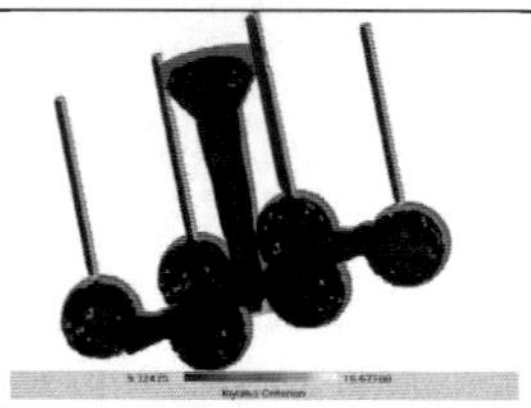
4/91.7	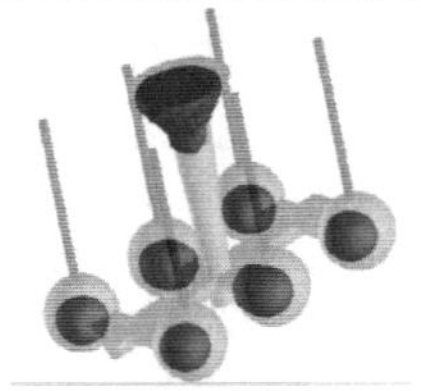	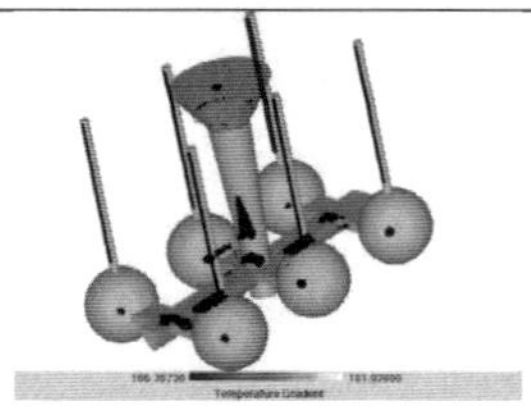	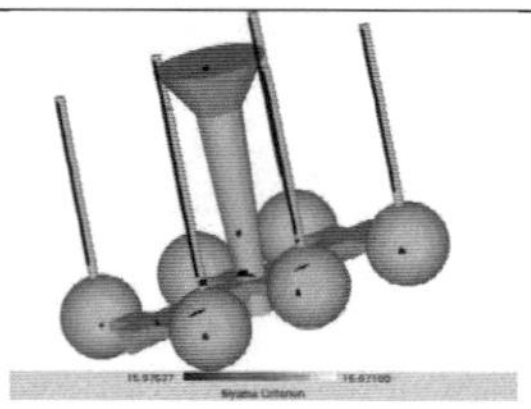
5/99.6		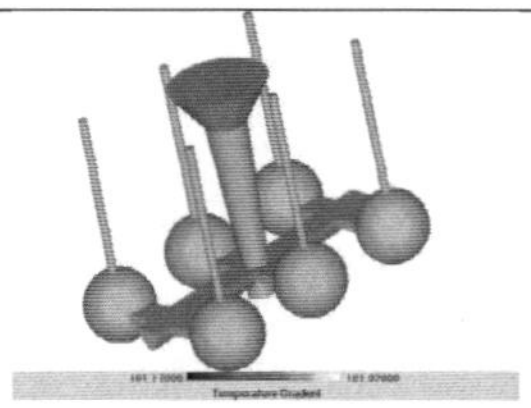	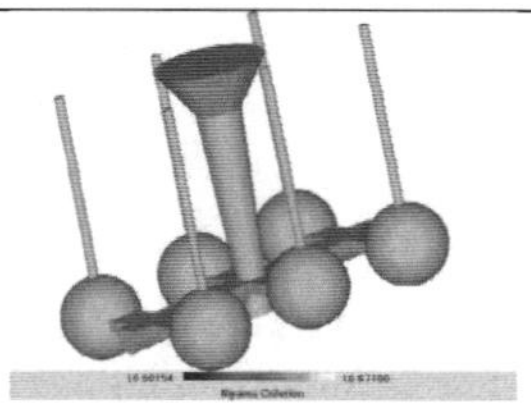

Figure 9. Optimization casting simulation for gating system design-2 with pouring temperature of hot metal at 2696 F (1480 °C)

Table 3. Percentage of yield production of grinding balls

Gating System Design	Weight of GB (Kg)	Weight of GS (Kg)	Total Weight (GB+GS) (Kg)	Yield Production (%)
1	2.97	1.31	4.28	69.39
2	2.97	1.425	4.395	67.58
3	2.97	2.37	5.34	55.65

High chromium white cast iron (ASTM A532) is one of cast iron that could solidify rapidly from a liquid phase. Thus, the pouring temperature will be a very important parameter in this casting process. Thus for the next simulation, the pouring temperature was changed to 2696 F (1480 °C).

As previously described that vent air would prevent the formation of casting defects. Therefore, the optimization of gating system design-2, that had a better in yield production than gating system design-3, had been conducted. The new pouring temperature (1480 °C) had been applied on gating system design-2. The casting simulation was included solidification time, temperature gradient, and niyama criterion, as shown in Figure 9.

From the result, it was shown that there was no defect (no-blue color) on simulation of niyama criterion for the gating system design-2. It was found that by changing the pouring temperature from 2642 F (1450 °C) to 2696 F (1480 °C) had improved the quality of grinding ball product significantly.

By using all parameters of previous casting simulation, the grinding balls were produced by smelting a scrap of cast iron and steel, and ferroalloys in an induction furnace. The hot metal was poured into a sand mold in two difference of pouring temperature that was 1450 °C to 1480 °C. The grinding ball casting product was shown in Figure 10. The hot metal that was poured into grinding ball's mold with gating system design-1 at 1450 °C showed defect/porosity inside the grinding ball, as shown in Figure 10a. Meanwhile, the hot metal that was poured into grinding ball's mold with gating system design-2 at 1480 °C showed no-porosity inside the grinding ball, as shown in Figure 10b. Therefore, the casting simulation of grinding balls was agreed with the actual casting process in this experiment.

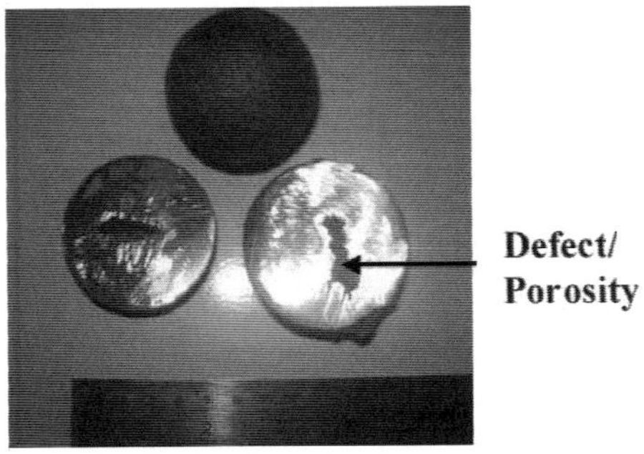

(a)

(b)

Figure 10. Grinding ball casting product, (a) Gating system design-1 at pouring temperature 1450 °C, (b) Gating system design-2 at pouring temperature 1480 °C

4. Conclusion

Grinding balls with 50 mm diameter, that had 2.18C - 0.72Si - 13Cr - 1.38Mo, were made with no-defect using the sand mold-gravity casting process. From the casting simulation using SOLIDCast^{TM} version 8.2.0, the vent air in gating system could prevent the internal defect on grinding balls. The vent air with 6 mm diameter resulted in the sound casting of grinding ball by applying pouring temperature at 1480 °C. Thus it was resulted in better yield production rather than by using pouring temperature at 1450 °C.

Acknowledgments

We would like thanks to Mr. Jeff Meredith as Trainee Staff from SOLIDCast^{TM} version 8.2.0 who gave this software license for a month to authors for completing this study.

References

[1] Gundewar C S, Natarajan K A, Nayak U B and Satyanarayana K 1990 *Minerals Engineering* **3** 207
[2] Albertin E and Moraes S L 2007 *Wear* **263** 43
[3] Albertin E and Sinatora A 2001 *Wear* **250** 492
[4] Reda R, Nofal A, Ibrahim K H and Hussien A 2010 *China Foundry* **7** 438.
[5] Wang J, Xiong J, Fan H, Yang, H S and Liu H H 2009 *Journal of Materials Processing Technology* **209** 3236.
[6] Chenje T W, Simbi D J and Navara E 2004 *Materials and Design* **25** 11.
[7] Amland H, Barnard L, Bates C, Birkey H, Dunbar D, Duncan R, Dutcher D, Ferguson G, Heger J, Hollander D, Houghton K, Kerwin T, Monroe R, Nariman R, Nelson R, Rudd P, Sailors D, Scott D, Schleg F, Steel C, Stevens T and Voigt R 1995 *Steel Casting Handbook-Sixth Edition* ed Blair M and Stevens T L (United States of America: Steel Founders' Society of America and ASM International) chapter 13 4-6.
[8] Naro R L 1999 *AFS Transactions* **839**
[9] Ravi B 2008 Casting Simulation: Benefits, Bottlenecks, and Best Practices. Indian Foundry Journal 1
[10] Kalpakjian S and Schmid S R 2006 *Manufacturing, Engineering, and Technology-Fifth Edition* (Singapore: Prentice Hall, Pearson Education South Asia Pte Ltd) 285-321.
[11] Heine R W, Loper C R and Rosenthal P C 1967 *Principle of Metal Casting-Second Edition* (Newyork: Mc. Graw Hill Book Company)
[12] SOLIDCast^{TM} version 8.2.0 (June 20) 2012 Finite Solution Incorporated
[13] Carlson K D and Beckerman C 2008 *Proc. Int. Conf. on the 62nd SFSA Technical and Operating Conference (Chicago: Steel Founders' Society of America) 1*

Effect of SiC addition to the characteristics of Al-11Zn-6.7Mg composite produced by squeeze casting for ballistic application

R F Adiputra, R Wijanarko, I Angela, and B T Sofyan[1]

[1]Department of Metallurgy and Materials Engineering, Faculty of Engineering, Universitas Indonesia, Kampus UI Depok 16424, Indonesia

Email: bondan@eng.ui.ac.id

Abstract. Aluminium composite material as an alternative to steel used in body of tactical vehicles has been studied. Addition of SiC was expected to have strengthening effect on the composite matrix therefore improving its ballistic performance. Composites of Al-11Zn-6.7Mg matrix and SiC strengthening particles with the fraction of 0, 10, and 15 vol. % were fabricated through squeeze casting process. Composite samples were then precipitation strengthened at 130 °C for 102 h to further improve their toughness. Final products were characterized by using chemical composition testing, optical microscopy, Scanning Electron Microscope – Energy Dispersive Spectroscopy (SEM-EDS) and quantitative metallography to calculate porosity, hardness test, impact test, and type III ballistic test in accordance with NIJ 0108.04 standard. The results showed that increase in SiC volume fraction from 0 to 10 and 15 vol. % managed to improve the hardness from 73 to 85 and 87 HRB, respectively, while on the other hand reduced the impact values from 12,278.69 to 11,290.35 and 9,924.54 J/m^2. SEM-EDS observation confirmed the presence of $Mg_3Zn_3Al_2$ intermetallic compound which formed during solidification and indicated the precipitation of $MgZn_2$ precipitates during ageing. The ballistic testing demonstrated a promising result of the potential of Al-11Zn-6.7Mg composite strengthened by 15 vol. % SiC to withstand penetration of type III bullet (7.62 mm).

1. Introduction

Aluminium 7xxx is a type of heat-treatable Al-alloys with superior strength compared to other series. It has been found that the alloys exhibited higher strength with addition of 9 wt. % Zn [1]. The high mechanical properties of these alloys are also influenced by the presence of $MgZn_2$ precipitates that form during heat treatment, be it either natural or artificial aging processes [2]. In a research conducted by Demir et al.[3] on comparison of ballistic impact resistance between Al 7075, Al 5083, and HSLA AISI 4040 steel, it was found that Al 7075 possessed the best ballistic resistance of all three materials. This shows that Al7xxx alloys are promising to be used as materials for tactical vehicles.

The mechanical properties of Al 7xxx alloys can be further enhanced by using it as the matrix (Metal Matrix Composites/MMC). These composites are engineered combination of the metal matrix and hard ceramic particle reinforcement in order to get tailored properties. Aluminium matrix composites are recently considered as alternatives to metals used in military applications due to their low density, high mechanical properties, and good resistance to corrosion. Common reinforcements for the Al-matrix composites are Al_2O_3, SiC, and MgO particles, which are utilized due to their high specific strength that remains steady at ambient temperatures [4]. Kheder et al. concluded that SiC acted as the most potential reinforcement in terms of increasing the strength and hardness of Al-MMC [5]. However, Zhou and Xu [6] found that SiC particles tended to float on the surface of molten metal during mixing due to their high surface tension and poor wetting mechanism. Nishchev et al. [7] also showed that SiC microstructure was chemically separated from the aluminium matrix which was well observed on the matrix-reinforcement interface. A research by Sigit and Sofyan [8] on solution treated and aged Al-7Si-Mg-5Zn composites with SiC reinforcement revealed that Mg acted as a wetting agent that lowered the surface tension between the matrix and the reinforcement, hence improving the matrix-reinforcement bond in the composites. Behera et al. [9] added that higher SiC content leads to higher Al-MMC composites hardness. Addition of SiC also contributed in grain refining mechanism of the composite, as the reinforcement acted as nucleating agents during solidification.

Squeeze casting is a casting method in which an external pressure is applied to the molten metal during solidification [10]. The pressure will minimize the occurrence of porosities and shrinkage defects. Rahmalina et al. [11] performed an experiment on fabricating Al-7Si-6Mg-5Zn composite with SiC reinforcement through squeeze casting method which demonstrated a good distribution of SiC particles of 5, 10, and 20 vol.% in the fabricated composite. In addition, Rahmalina et al.[12] also found that precipitation strengthening mechanism was needed in order to improve the toughness of the composites matrix to withstand the ballistic impact. Therefore, this research focused on studying the Al-11Zn-6.7Mg composites with variation of SiC reinforcement of 10 and 15 vol. % for ballistic application. The samples were fabricated using squeeze casting. A plate of the corresponding Al-alloy without strengthening particles was also fabricated and viewed as the reference matrix.

2. Materials and Method

Experimental matrix material of Al-11Zn-6.7Mg alloy was prepared by commercial pure Al (99.99 wt. %), Zn (99.99 wt. %), and Mg (99.99 wt. %) ingots. Particulates of #400 SiC from Sigma-Aldrich were prepared as the reinforcement with the amount of 0, 10, and 15 vol. %. The SiC particles were preheated in a muffle furnace at 1000 °C for 1 h to increase their wettability with the aluminium molten.

Melting was inducted in a graphite crucible at 870 °C and degassed with argon gas for 1 minute. Preheated SiC particles were then added to the melt and stirred at 5000 rpm for 2 minutes. The melts were poured into a 170 x 170 x 15 mm^3 preheated metal mould and then squeeze casted at 76 MPa for 10 minutes. To improve hardness and toughness, the composites were subsequently solution treated at 450 °C for 1 h, water-quenched, and aged at 130 °C for 102 h. The actual composition of the composites samples are shown in table 1.

The chemical composition of the composites was analyzed using ThermoARL Optical Emission Spectroscopy (OES). Hardness testing was conducted in accordance to ASTM E18 with 4 indentations using Rockwell B method, while impact testing was conducted by using Charpy method. Standard metallographic preparation was conducted with 0.5 % HF etchant. The microstructures were observed by Scanning Electron Microscope (SEM) – Energy Dispersive Spectroscopy (EDS) and Zeiss Primotech Optical Microscope (OM). Calculation of porosities was done using Image Pro-Plus Software. Type III ballistic testing was performed to three layers of 10 x 10 x 1.5 $cm^3$15 vol. %. SiC composite plates in accordance to NIJ 0108.01 (Ballistic Resistant Protective Material) standard using 7.62 mm caliber bullets, shot twice by an SPR-1 shotgun from the distance of 15 m.

Table 1. The actual composition of the samples.

Sample	Zn	Mg	Si	Cu	Fe	Mn	Ti	Cr	Ni	Al
A (0 vol.% SiC)	10.318	6.741	0.036	0.001	0.172	0.017	0.004	0.010	0.011	Rem
B (10 vol.% SiC)	11.546	7.052	1.039	0.001	0.209	0.001	0.001	0.012	0.011	Rem
C (15 vol.% SiC)	11.156	6.290	2.643	0.001	0.254	0.007	0.005	0.008	0.042	Rem

3. Results and discussion

3.1. Microstructural characterization

Figure 1 show the microstructures of the composites with different amounts of SiC particulates which were observed as gray particles. The microstructures revealed dendritic grain structure in which intermetallic phases $Mg_3Zn_3Al_2$ could be observed (*see arrows*). The intermetallic phases were found in a net shape with dark contrast. From Figure 1 (b-c), it is clear that SiC particles are uniformly distributed in the 10 and 15 vol. % SiC composites. This was proven by the hardness value examined of the top, middle, and bottom of the composite plate as shown in Figure 2. It can be seen that the hardness value are similar for each position in both composites.

Figure 1. Microstructures of Al-11Zn-6.7Mg composites with additions of (a) 0, (b) 10, and (c) 15 vol.% SiC. Arrows showing intermetallic $Mg_3Zn_3Al_2$ phase

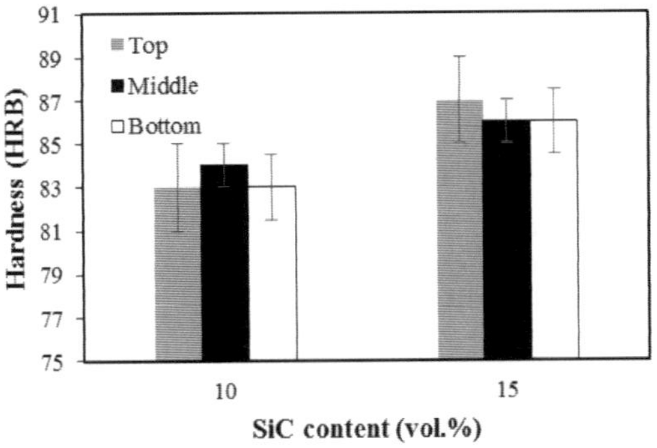

Figure 2. Amount of porosities in Al-11Zn-6.7Mg with different vol. % SiC addition

Figure 3. Microstructure of Al-11Zn-6.7Mg with 15 vol. % SiC as 20 µm at 130 °C for 102 h

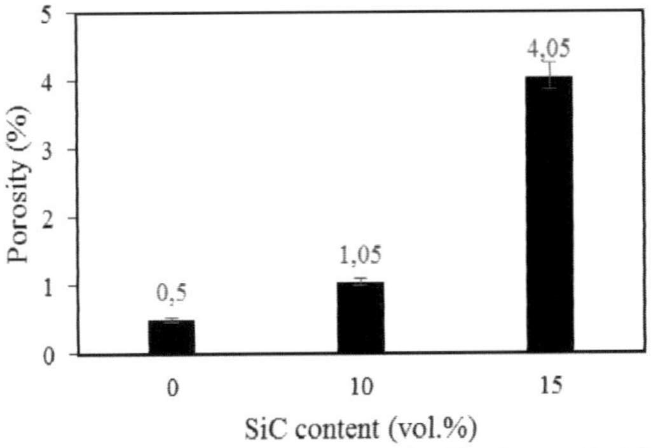

Figure 4. Hardness distribution of Al-11Zn-6.7Mg with 10 and 15 vol. % SiC composites

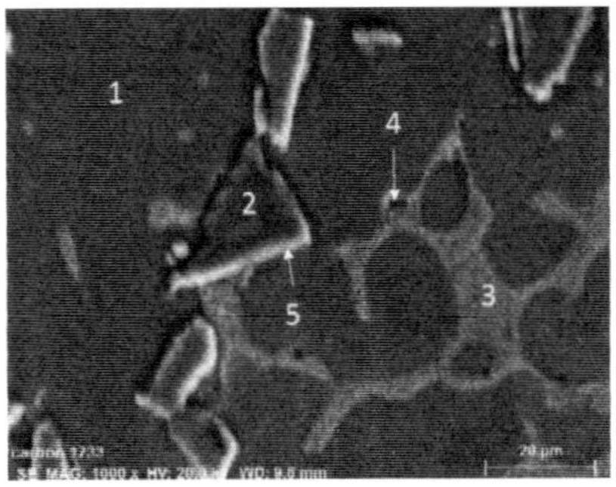

Figure 5. SEM micrograph of Al-11Zn-6.7Mg with 15 vol. % SiC as aged at 130 °C for 102 h

Addition of Mg is expected to reduce the contact angle between matrix and reinforcement, thus increasing the composite wettability. Mg has lower superficial stress than Al and SiC does, that it can support the infiltration process of Al – SiC and prevent the formation of Al_4C_3 interface which is hard and brittle [13]. However, in Figure 3, we remain see visible voids around the interface of 15 vol.% SiC composite due to the lack of wettability. The amount of porosities increased with more SiC content. Figure 4 shows that the porosities of samples with addition of 0, 10, and 15 vol.% SiC were 0.5 %, 1.05 %, and 4.05 % in average, respectively.

SEM-EDS micrograph of the composite with 15 vol. % SiC was presented in Figure 5. The results of EDS characterization at corresponding positions are presented in table 2. Furthermore, in Figure 5 α-Al was detected to be the matrix as pointed out by point 1. In addition, with Zn:Mg ratio at that point which is approaching 2:1, $MgZn_2$ is predicted to form [14]. At point 2, detected contents of Si and C atoms were 47.25 and 46.40 at. %, respectively, which was nearly 1:1 in ratio. This proved that SiC particles reinforcement did exist at this point. Meanwhile at point 3, it was noticed that the phases in the dendrites was mainly composed of $Mg_3Zn_3Al_2$. Point 4 indicated intermetallic phase as well as $Mg_3Zn_3Al_2$ due to the percentage of Mg:Zn are almost the same, 1:1. While point 5 in which the atomic percentage of Mg is higher than Zn and exists between SiC and the matrix, was identified as the interface area of the composite that consists of high Mg content. This indicates occurrence of segregation of Mg to the matrix-SiC interface that improves the overall wettability.

Table 2. EDS data of Al-11Zn-6.7 Mg with 15 vol. %SiC on as-aged condition at 130 °C for 102 h at positions as shown in Figure 5

Position	Element content (at.%)						Phase
	Zn	Mg	Si	C	O	Al	
1	4.59	1.96	-	18.77	2.62	72.06	□-Al, $MgZn_2$
2	0,29	0.24	47.25	46.40	4.38	1.43	SiC
3	25.16	24.98	-	22.21	3.64	24.01	$Mg_3Zn_3Al_2$
4	12.73	11.53	-	3.95	19.36	52.43	$Mg_3Zn_3Al_2$
5	1.79	4.52	-	18.58	2.87	72.24	Interface, □-Al

3.2. Mechanical Properties

Hardness of Al-11Zn-6.7Mg composites with 0, 10, and 15 vol. % SiC reinforcement is shown in Figure 6, which indicates an increase in overall composite hardness along with more SiC added into the matrix. Composite hardness increased from 73 to 85 and 87 HRB on addition of 10 and 15 vol. % SiC, respectively. This finding compromises results of the previous works [5, 9, 15].

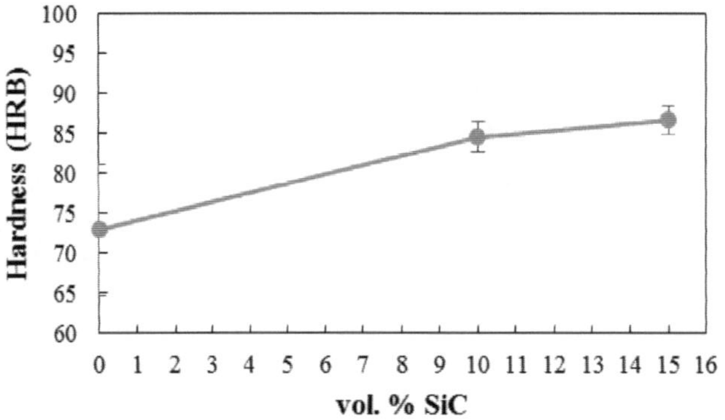

Figure 6. Effect of vol. % SiC on hardness of Al-11Zn-6.7Mg-SiC reinforced composites

Table 3. Calculated result of theoretical and actual composite hardness

SiC content (vol. %)	Theoritical Hardness (HRB)	Actual Hardness (HRB)
10	86	85
15	87	87

With help from image analysis, it was found that samples with targeted 10 and 15 vol. % SiC contained only 4.50 and 6.98 vol. % SiC. Using the rule of mixture theory [16], we can obtain the theoretical hardness of the composite plate as well as the amount of porosity and fraction of SiC contained within, that can be used further in comparison to the reference composite sample. Table 3 contains the calculated result of theoretical and actual composite hardness attained using the rule of mixture. Actual composite hardness was close to its theoretical hardness due to shortage in acquired final vol.% SiC inside the composite, which was lower than 10 and 15 vol.%. It is suspected that the mixing process was not done well that caused more material loss than presumed.

Figure 7 shows the effects of vol. % SiC addition on the toughness of Al-11Zn-6.7Mg composites. The figure has a decreasing trend of toughness with more SiC addition. This phenomenon might happen due to the increase of composite brittleness and the presence of porosities within the plate with the amount of 0.5, 1.05, and 4.05% on addition of 0, 10, and 15 vol. % SiC, respectively. It can be inferred from the fracture surface in Figure 8 that porosities exist around SiC reinforcement (marked with red arrows). These porosities may initiate crack propagation when composite is subjected to impact loading, making it more vulnerable to cracking, thus absorbing less energy before failure. Observed fracture area showed a brittle fracture due to the absence of plastic deformation and fibrous fracture morphology in the samples.

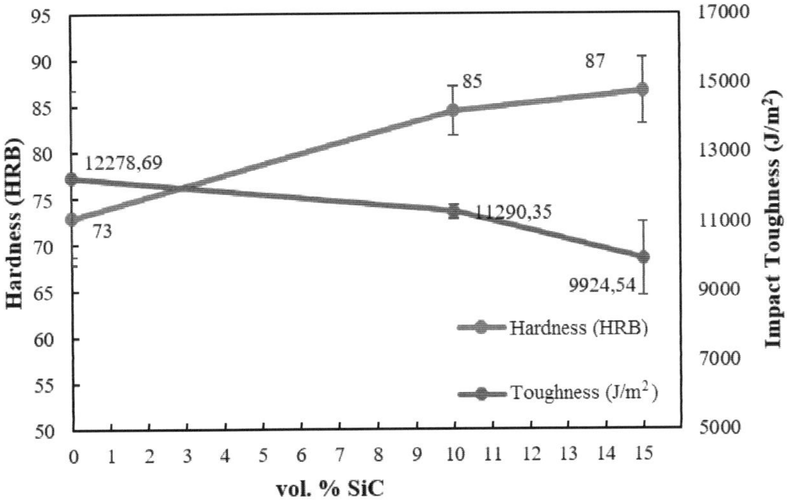

Figure 7. Effect of SiC reinforcement on impact toughness of the composite

3.3. Ballistic Properties

Ballistic testing was done on three composite plates with 15 vol. % SiC reinforcement. The surface of each plate was coated using High Velocity Oxygen Fuel method (HVOF) with 80 wt.% WC – 20 wt.% Co powders. The hardness of Plate 1, 2, and 3 are 87, 86, and 84 HRB, while toughness for each plate is 9,534.58, 10,387.76, and 9,851.29 J/m^2, respectively. Plate 3 with the lowest overall hardness was put up in front, followed by Plate 2 and 1 with higher hardness. Ballistic testing were done using SPR-1 shotgun with calibre 7.62 mm bullet. Shooting was done on 15 meter range which corresponded to the type III ballistic testing standard. Two shots were taken in the actual testing. Figure 9 showed the result of ballistic testing.

0 vol. % SiC 10 vol. % SiC 15 vol. % SiC

Figure 8. Fracture surface of Al-11Zn-6.7Mg-SiC reinforced composites with (a) 0, (b) 10, and (c) 15 vol. % SiC aged at 130 °C for 102 h

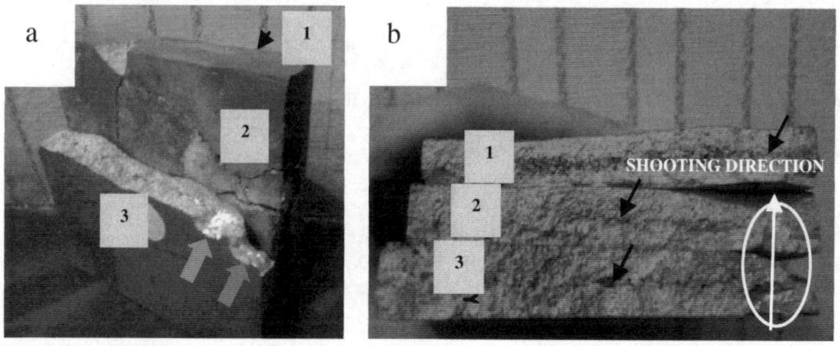

Figure 9. Result of ballistic testing: (a) propagation of crack (b) fracture surface morphology

Visual observation on macroscopic scale indicated a dominant brittle fracture mechanism due to the high hardness of each composite plate. It can also be concluded that porosities located around SiC reinforcement inside the composite were the source of crack propagation (Figure 9b, marked with black arrows). Plate 2, however, still showed traces of plastic deformation after the ballistic testing, which confirmed the occurrence of ductile fracture mechanism in the composite plate.

Experimental layer of three Al-10.6Zn-6.7Mg composite plates with 15 vol. % SiC reinforcement each shown an ability to resist bullet penetration in type III ballistic testing. With the occurrence of ductile fracture characteristic in Plate 2, it is safe to say that the fabricated composite plate is able to withstand and absorb the impact energy from the bullet shot. Previous ballistic testing on three Al-8Zn composite plates with 15 vol. % SiC reinforcement and variation of Mg addition by 3, 4, and 5 wt. % done by Septiani [17] showed that the produced composite could not withstand the ballistic penetration. Thus, this research has managed to improve the ballistic performance of composite plate fabricated using Al-11Zn-6.7Mg matrix with 15 vol. % SiC reinforcement.

4. Conclusion

Investigation on Al-11Zn-6.7Mg composite plates with variation of 0, 10, and 15 vol. % SiC reinforcement particles resulted in the following conclusions:

(1) The addition of SiC reinforcement of 10 and 15 vol. % to Al-11Zn-6Mg alloys showed an increase in strength and lower the toughness of the composites. The strength of Al-11Zn-6.7Mg composites with variation of 0, 10, and 15 vol. % SiC are 73, 85, and 87 HRB, respectively, whereas the toughness are 12,278.69, 11,290.35, and 9,924.54 J/m^2.

(2) Overall composite porosity increased from 0.5% to 1.05% and 4.05% with the addition of 10 and 15 vol. % SiC, respectively.

(3) $MgZn_2$ precipitate and $Mg_3Zn_3Al_2$ intermetallic phase which formed and distributed in the matrix were found to influence the final mechanical properties of the composite matrix.

(4) Layer constructed of three Al-11Zn-6.7Mg plates with 15 vol. % SiC reinforcement is potential to be used as materials for armor manufacturing due to its ability to withstand penetration of type III bullet despite of the cracking failure experienced.

Acknowledgements
The authors wish to acknowledge the financial support acquired from Ministry of Research Technology and Higher Education through Hibah Penelitian Unggulan Perguruan Tinggi (PUPT) 2016.

References

[1] Rahmalina D 2012 *Pengembangan Komposit Aluminium sebagai Material Armor dengan Keunggulan Kinerja Balistik* (Depok: Fakultas Teknik Universitas Indonesia)
[2] Sofyan B T, Askarningsih N, and Garjati V N 2014 *Prosiding Seminar Material Metalurgi* 95-102
[3] Demir T, Ubeyli, Mustafa, and Yildrim R 2008 *Mat. and Design* **29** 2009-2016
[4] Yilmaz H 2004 *Characterization of Silicon Carbide Particulate Reinforced Squeeze Cast Aluminium 7075 Matrix Composite* (Grad. School of Natural and App. Sci. Middle East Technical University)
[5] Kheder A, Marahleh G, and Al-Jamea D 2011 *Jordan J. Mech. and Indust. Eng.* **5**
[6] Zhou W and Xu Z 1997 *J. Mat. Sci.* **63** 358-383
[7] Nischev K, Novopoltsev M, Mishkin V, and Shchetanov B 2013 *Bulleting of the Russian Academy of Sci. Phy.* **77** 981-985
[8] Sigit S R and Sofyan B T 2013 *Adv. Mat. Res.* **789** 198-203
[9] Behera R, Mohanta, Nilhar R, and Sutradhar G 2012 *Int. J. of Emerging Trends in Eng. and Development*
[10] ASM International Committee 1988 *ASM Metals Handbook: Casting* **15** (Ohio: ASM International)
[11] Rahmalina D, Sofyan B T, Suharno B, and Siradj E S 2012 *M.P.I.* **6** 51-56
[12] Rahmalina D, Sofyan B T, Suharno B, and Siradj E S 2012 *Prosiding Annual Eng. Seminar* A88-A92
[13] Vijarayam T 2006 *J. Mat. Process. Tech.* **178** 34
[14] Mondolfo L 1976 *Alluminum Alloys: Structure and Properties* (London: William Clowes & Sons Limited)
[15] Rahmalina D, Sofyan B T, Suharno B, and Siradj E 2012 *Majalah Pengkajian Industri* **6** 51-56
[16] Matthews F and Rawling R 1994 *Composite Materials: Engineering and Science* (London: Imperial College of Science)
[17] Septiani A 2013 *Studi Pengaruh Penambahan Mg terhadap Karakteristik Komposit Al-8Zn berpenguarr SiC Hasil Squeeze Casting untuk Aplikasi Balistik* (Depok: Universitas Indonesia)

IOP Conf. Series: Materials Science and Engineering **285** (2017) 012031 doi:10.1088/1757-899X/285/1/012031

Effects of aluminum addition on the characteristics of Cu-28Zn brass produced by gravity casting

A Arisgraha, I Angela, N A Arandana, and B T Sofyan

Department of Metallurgy and Materials Engineering, Faculty of Engineering, Universitas Indonesia, Kampus UI Depok 16424, Indonesia

Email: bondan@eng.ui.ac.id

Abstract. Brass has an attractive combination of strength, ductility, and good corrosion resistance, which are required in many components, such as cartridge cases, valves, and fittings. In all of these components, brasses are usually fabricated through casting, rolling, deep drawing, and other mechanical deformation processes. Alloying elements were added in order to improve strength without sacrificing the ductility of brass. In this research, aluminium was added to Cu-28Zn brass of 1.9, 5.7, and 6.2 wt. % addition. Alloys were fabricated by gravity casting followed by homogenization at 800 ^{o}C for 2 h. Material characterizations were done through microstructural analysis using optical microscope and Scanning Electron Microscope – Energy Dispersive X-Ray (SEM-EDX), Rockwell B and Micro Vickers hardness testing methods, and tensile testing. The results showed that Al addition improved the mechanical properties of Cu-28Zn alloy. Hardness, tensile strength, and yield strength increased, but the elongation decreased.

1. Introduction

Cartridge brass is a 70Cu-30Zn wt. % alloy that has an exceptional mechanical strength and ductility. It is widely utilized in applications requiring high mechanical properties such as heat exchangers, pipelines, and cartridge cases. In the previous research, it was found that the alloy exhibited optimal strength and ductility properties at the composition of 28 wt. % Zn [1]. Further attempts in optimizing the mechanical properties of Cu-28Zn were then done through alloying. Basori, et al. [2] concluded that the addition of 0.22, 0.41, and 0.80 wt. % Bi did not contribute significantly in improving hardness of the alloy, instead, the presence of 0.41 and 0.80 wt. % Bi in the alloy will result in lenticular and network morphology, respectively. Both of the morphologies had negative impacts on elongation, tensile, and yield strength of the cartridge brass alloy. Pratiwi [3] then showed that Mn addition of 1.26, 3.23, and 5.83 wt. % could increase the hardness of Cu-28Zn alloy through solid solution strengthening and formation of β' second phase in Cu-28Zn-5.83Mn alloy composition. A research by Ovat, et al. [4] suggested that addition of up to 10 wt. % aluminium could enhance the hardness, strength, and elongation of Cu-15Zn brass. Optimum combination of ultimate tensile strength, elongation, and hardness was found in samples with 5 wt. % Al addition, while optimum toughness resulted from addition of 2 wt. % Al. However, the research did not provide any results of microstrucural observation.

In mathematical Al-Cu-Zn thermodynamic systems presented by Liang and Fetzer [5] that included a vertical phase diagram sections of the alloy at constant 2, 4, and 6 wt. % Al addition, it was stated that different phases will form with ranging aluminium addition. The α + β phases will form in 2 wt. %

aluminium addition, while single β-phase will result in addition of 4 and 6 wt. % aluminium, all occurring at temperatures below 950 °C. Martunis [6] attempted adding 6.4 wt. % Al to 79.8Cu-13.4Zn wt. % alloy of which final product revealed the presence of α-phase and eutectic α + γ microstructure with toughness and elongation of 0.355 $J.mm^{-2}$ and 10.65%, respectively, in contrary to the thermodynamic system by Liang [5]. The research added that the presence of Al in brass will promote the formation of β-phase, whilst further addition will initiate the hard γ-phase to form. Suprihadi [7] then observed the effect of 6 wt. % Al addition to Cu-30Zn alloy. The Cu-30Zn-6Al alloy was found to have dominant β-phase content along with some γ-phase of with overall toughness and elongation of 0.202 $J.mm^{-2}$ and 7.63%, respectively. In the present work, microstructural analysis and mechanical properties investigation were done on as-homogenized cast samples of Cu-28Zn alloy with 1.9, 5.7, and 6.2 wt. % Al addition. Effects of Al addition to the cartridge brass alloy was studied.

2. Experimental Method

Pure Cu, Zn, and Al ingots were used as starting materials. Cu and Al ingots were first melted in an electric furnace at 1150 °C. Zn ingots were added last to the molten metal. Borax was added to the molten metal as flux to reduce the formation of inclusions while no degassing process was conducted. Prepared molten alloys were casted into a metal mold preheated at 650 °C. The as-cast alloys were then homogenized in muffle furnace at 800 °C for 2 h followed by air-cooling. Nominal compositions of as-homogenized alloys were acquired by Optical Emission Spectroscopy (OES) as listed in Table 1.

Table 1. The chemical composition of the Cu-28Zn-Al (wt. %) alloy

Alloys	Zn	Al	Pb	Fe	P	Mg	S	Sn	Cr	Cu
A	29.516	1.934	0.009	0.015	<0.001	0.001	0.002	0.000	-	Bal.
B	28.369	5.744	0.006	0.015	<0.002	0.000	0.002	0.000	-	Bal.
C	25.1	6.24	<0.001	0.012	<0.003	<0.001	0.007	0.015	-	Bal.

Microstructures of the samples were observed by using optical microscope and Scanning Electron Microscope (SEM) EDX Oxford Quanta 650 instrument. Standard metallographic preparation was conducted based on ASM E3-11 using 10 gr $FeCl_3$ + 100 ml distilled water as etchant. *Image Pro Analysis* software was used to determine the actual volume fraction of each phase. Rockwell and Micro Vickers hardness testing were performed to each sample in accordance to ASTM E 18. Tensile testing was done on dog bone specimens according to ASTM E8 standard.

3. Results and Discussion

3.1. Microstructures

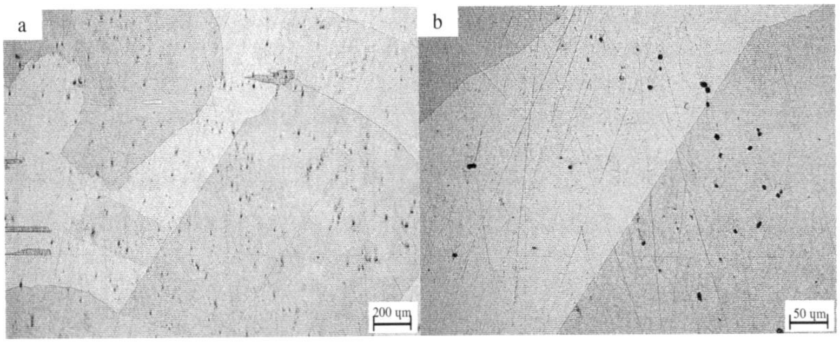

Figure 1. Microstructures of as-homogenized Cu-28Zn alloy of (a) 50 and (b) 200 x magnification

Figure 1 (a,b) shows large equiaxed grains of the Cu-28Zn alloy as a result of homogenization. However, such large and equiaxed grains were not found in alloys with 1.9 wt. % Al addition as seen in figure 2 (a.b). Microstructure of alloy A with Cu-29.5Zn-1.9Al alloy composition was consisted of irregular shapes in different colors that were identified as α + β phases [8]. According to Sampath [8], the more stable α-phase were formed from metastable β phase that transformed during homogenization and air cooling. Figure 2 (c,d) shows the microstructure of alloy B with Cu-28Zn-5.7Al wt. % composition, consisting of equiaxed grain identified as β phase [9,10]. This is supported with a research by Vilarinho [10], stating that the addition of Al in brass alloy will deplenish and further eliminate the existence of α phase, resulting in a single β phase morphology. In other words, Al was said to promote the formation of β phase in Cu-Zn alloy [10]. Subsequently, a second phase in four leaves clover shape was found in the microstructure of plate C with Cu-29.5Zn-6.2Al composition as shown in figure 2 (e,f). The corresponding shape was identified as γ phase [11].

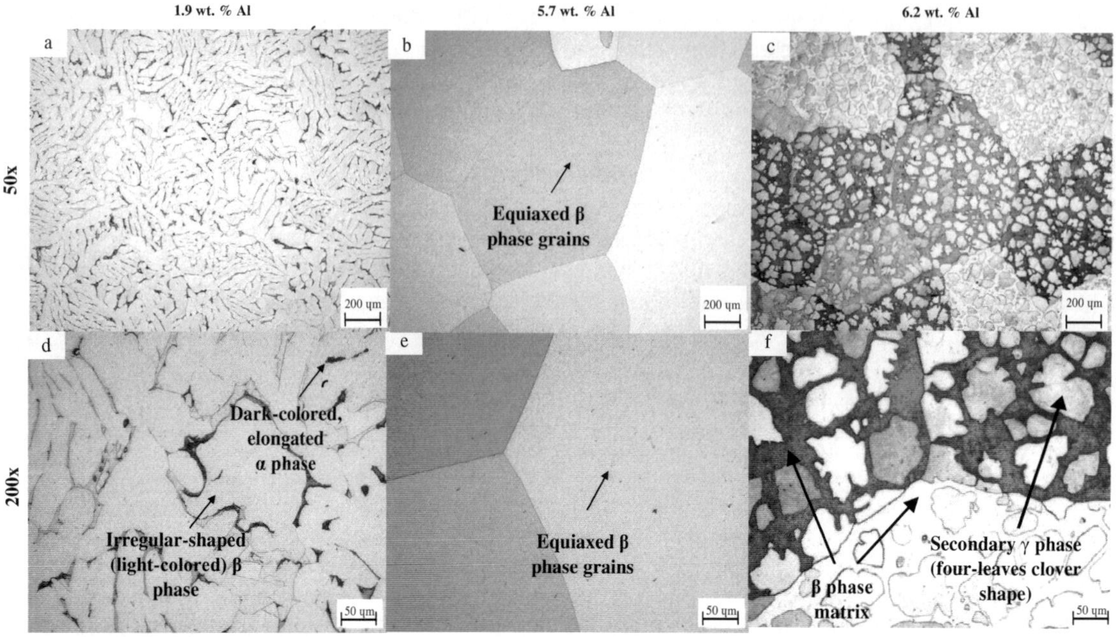

Figure 2. Microstructures of as-homogenized Cu-28Zn alloy with addition of: (a,b) 1.9, (c,d) 5.7, and (e,f) 6.2 wt. % Al

Microstructures of as-homogenized alloys A, B, and C were further inspected using SEM photography as presented in figure 3. Microstructure image of alloy A showed the presence of bright α phase and dark β phase. Equiaxed grain microstructures of β phase were also revealed from EDX testing of plate B, while plate C exhibited four leaves clover morphology of secondary γ phase with relatively uniform size. The EDX analysis also proved that the γ phase wielded greater Al content compared to those found in α and β phase as seen in table 2. From this result it can be reffered that the addition of aluminium in Cu-28Zn alloy could promote the formation of β phase and γ phase at further addition.

Volume fractions of each formed α, β, and γ- phase were calculated using Lever Rule principal on each sample compositions plotted on the vertical ternary diagram with constant 2, 4, and 6 wt. % aluminium content by Liang and Fetzer [5]. Both calculation and quantitative analysis using *Image Pro Analysis* results shows correspondence to each other. The Cu-30.36Zn-1.9Al wt. % sample was found to have more α phase to β phase as seen in figure 2 (a,b) where the light colored matrix was identified as α phase. Whereas the Cu-25.1-6.2Al wt. % sample was found to have greater amount of γ phase than β phase with volume fraction of 50.8 % and 49.15 %, respectively.

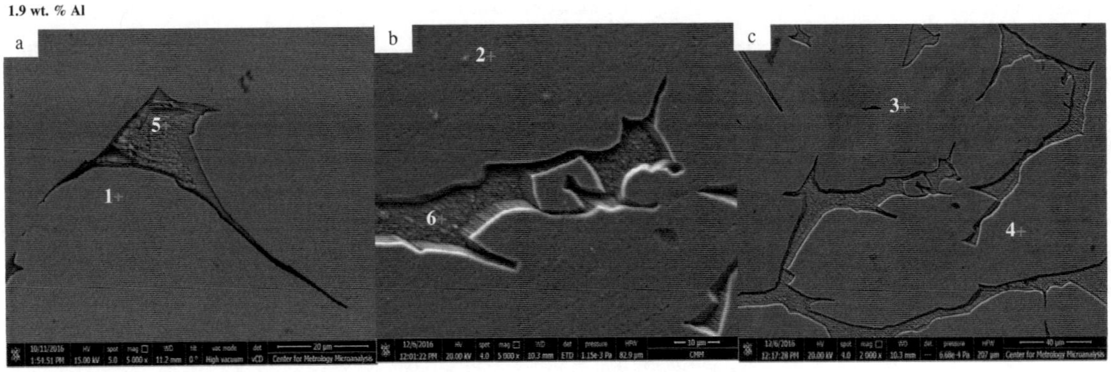

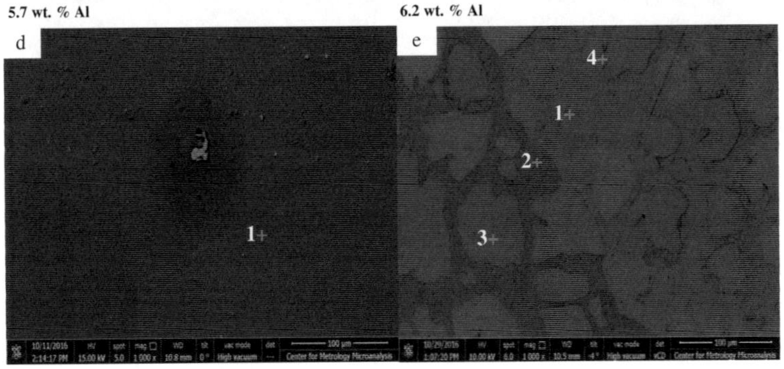

Figure 3. SEM-EDX investigation results on alloys (a,b,c) A; (d) B; and (e) C. Testing locations are indicated by numbers.

Table 2. Results of EDX analysis on alloys (a) A, (b) B, and (c) C

(a)

Loc. No.	Composition (at. %) Cu	Zn	Al	O	C	Possible Phase
1	63.64	30.25	5.59	1.12	2.65	α
2	69.70	25.60	4.80	-	-	α
3	70.00	25.80	4.30	-	-	α
4	69.80	25.60	4.60	-	-	α
5	60.49	28.41	11.11	3.56	2.12	β
6	60.00	31.90	8.10	-	-	β

(b)

Loc. No.	Composition (at. %) Cu	Zn	Al	O	C	Possible Phase
1	63.11	27.86	10.03	-	2.96	β

(c)

Loc. No.	Composition (at. %) Cu	Zn	Al	O	C	Possible Phase
1	63.15	25.40	11.45	0.21	2.96	β
2	61.05	24.15	10.82	-	-	β
3	60.67	23.59	15.74	0.76	3.02	γ
4	61.33	23.31	15.36	-	-	γ

3.2. Mechanical properties

Hardness of as-homogenized alloys A, B, and C with addition of 1.9, 5.7, and 6.2 wt. % Al were 40.3, 83.43, and 103.58 HRB, respectively, as seen in Figure 4. Increasing hardness was caused by the presence of aluminium, which significantly promoted the formation of β and γ phase at certain extent of addition. Figure 4 (a,b) shows the microhardness of plate A, B, and C. The number of microhardness of each plate A, B, and C were 77.64, 191.43, and 261.70 VHN, respectively. Both micro and macrohardness showed an increase along with more addition of aluminium to the alloy.

IOP Conf. Series: Materials Science and Engineering **285** (2017) 012031 doi:10.1088/1757-899X/285/1/012031

Plate B with single β phase was found to have higher strength compared to α + β phase in plate A due to the lattice strain caused by solute Al in the matrix, which prevented dislocation movement and initiated a solid solution strengthening mechanism. While further strength increase in plate C is assumed to happen due to the presence of γ phase which exhibited greater strength than β phase. Furthermore, it was supposed that the well-distributed γ phase contributed in the dispersion strengthening mechanism to occur.

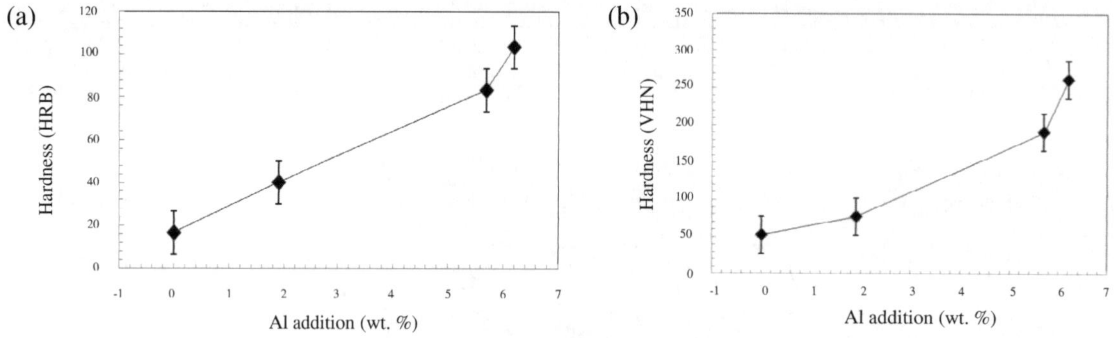

Figure 4. Increasing hardness properties of as-homogenized Cu-28Zn alloy with adddition of 1.9, 5.7, and 6.2 wt. % Al: (a) macrohardness and (b) microhardness

Figure 5 shows the mechanical properties of alloys A, B, and C with addition of 1.9, 5.7, and 6.2 wt. % Al, respectively. The tensile strength of alloys A, B, and C were 415.7, 545.25, and 412.65 MPa, while the yield strength were 21.39, 465.66, and 381.38 MPa, both respectively. From this data, it is known that both tensile and yield strength increased linearly with the hardness increase until the addition of 5.7 wt. % Al. However, contradicting result was found in plate C, where increasing hardness was followed by decreasing tensile and yield strength. This phenomenon was assumed to happen due to the presence of porosity that initiated microcracking before the specimen reached its actual yield strength and UTS. Elongation of plate A, B, and C were 43.32, 9.5, and 1%, respectively, while reference Cu-28Zn alloy showed elongation of 58% as shown in figure 5. It can be inferred that more Al content in the cartridge brass matrix could decrease the elongation of the alloy due to the presence of lattice strain caused by the addition of aluminum which further inhibited the dislocation movement through solid strengthening mechanism.

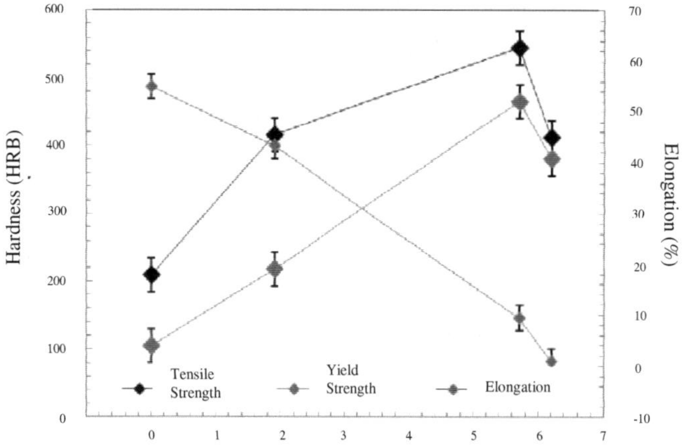

Figure 5. Effect of 1.9, 5.7, and 6.2 wt. % Al addition to several mechanical properties of as-homogenized Cu-28Zn brass alloy

IOP Conf. Series: Materials Science and Engineering **285** (2017) 012031 doi:10.1088/1757-899X/285/1/012031

Fracture surfaces of tensile samples of alloys A, B, and C are shown in figure 6. Figure 6 (a) shows the fracture surface of alloy A which appeared to have concave fracture surface (blue area), fibrous fracture surface (white area), and area of plastic deformation. Similarly, fracture surface of plate B in figure 6 (b) were having conformable fracture characteristics of plate A, with greater area of plastic deformation observed. Both plate A and B possessed area of plastic deformation or necking area due to tensile testing, which indicated that the two plates had a ductile fracture characteristic. On the other hand, fracture surface of plate C shown in figure 6 (c) was flat with a very small amount of plastic deformation which indicated a brittle fracture characteristic. This brittle characteristic is caused by the inhibition of dislocation movement when tensile force was applied. It is commonly found that materials with brittle fracture characteristics have a high tensile and yield strength, but on the contrary, plate C showed a decrease in both tensile and yield strength due to the presence of porosities in the specimen that allow cracks to propagate more swiftly.

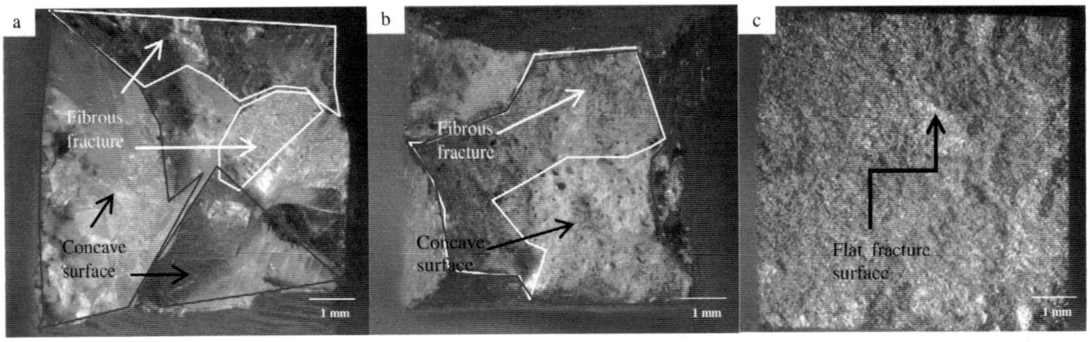

Figure 6. Fracture surfaces of as-homogenized Cu-28Zn tensile testing specimen with addition of (a) 1.9, (b) 5.7, and (c) 6.2 wt. % Al

4. Conclusions

Microstructure and mechanical properties of Cu-28Zn alloys with various wt. % Al addition were investigated. The following conclusions are drawn:

1) As homogenized Cu-28Zn alloy with 1.9 wt. % Al addition showed elongated α phase grain and β phase matrix morphology. The corresponding alloy with 5.7 wt. % Al content exhibited equiaxed grain morphology consisting of β phase while microstructure of Cu-28Zn-6.2Al wt. % alloy indicated the presence of γ phase in four leaves clover shapes dispersed on β matrix.
2) Addition of 1.9 and 5.7 wt. % Al increased the tensile strength of Cu-28Zn alloy from 209 MPa to 415.7 and 545.25 MPa and the yield strength from 105 MPa to 218.39 and 456.66 MPa. However a decrease in both tensile and yield strength of the alloy to 412.65 and 381.38 MPa were found in addition of 6.2 wt. % Al due to the existence of porosity.
3) Hardness of Cu-28Zn alloy increased along with more Al content from 53 VHN to 77.46, 191.43, and 356.23 VHN on the addition of 1.9, 5.7, and 6.2 wt. % Al, respectively. Strenghtening in alloys with 1.9 and 5.7 wt. % Al addition were due to the formation of β-phase and solid solution strengthening mechanism, while dispersion strengthening mechanism by the presence of well dispersed γ-phase was responsible for the improved hardness of alloy with 6.2 wt. % Al addition.
4) Elongation of Cu-28Zn alloy decreased from 58% to 43.32, 9, and 1% with the addition of 1.9, 5.7, and 6.2 wt. % Al, respectively. This result was confirmed by observing the fracture behaviors of each sample, where samples of 1.9 and 5.7 wt. % Al addition showed characteristics of ductile fracture while on the contrary, sample with 6.2 wt. % Al exhibited brittle fracture surfaces.

Acknowledgement
The authors are grateful with the provision of research funding through *Hibah Publikasi Internasional Terindeks untuk Tugas Akhir* (*PITTA*) 2017 from Universitas Indonesia.

References
[1] Basori I and Sofyan B T 2015 *Prosiding Seminar Nasional Mesin dan Teknologi Kejuruan* 351-356
[2] Basori I, Jendra D, and Sofyan B T 2016 *Prosiding Seminar Nasional Mesin dan Teknologi Kejuruan* 274-277
[3] Pratiwi H I 2015 *Pengaruh Penambahan Mn terhadap Karakteristik Cartridge Brass Hasil Pengecoran Gravitasi untuk Aplikasi Selongsong Peluru* Undergraduate Thesis (Depok: Universitas Indonesia)
[4] Ovat F A, Asuquo L O and Abam F I 2012 *Res. Journal in Eng. and Appl. Sci.* **4** 214-218
[5] Liang S M and Fetzer R S 2015 *CALPHAD* **52** 21-37
[6] Martunis 1993 *Studi Sifat Mekanis dan Struktur Mikro Aluminium Brass 79.8Cu-13.4Zn-6.4Al Sebelum dan Sesudah Perlakuan Panas Celup Oli dan Tempering* Undergraduate Thesis (Depok: Universitas Indonesia)
[7] Suprihadi N 1993 *Studi Karakteristik Paduan 64Cu-30Zn-6Al Hasil Pengecoran dan Perlakuan Panas terhadap Kekuatan Tarik, Impak, Kekerasan, dan Struktur Mikro* Undergraduate Thesis (Depok: Universitas Indonesia)
[8] Sampath V 2007 *Material and Manufacturing Proc.* **22** 9-14
[9] Asanovic V, Delijic K, and Jaukovic N 2008 *Scripta Materialia* **58** 599-601
[10] Vilarinho C, Davin J P, Soares D, Castro F, and Barbosa J 2005 *J. Materials Processing Tech.* **170** 441-447
[11] Castro M L and Romero R 1998 *Mat. Sci. and Eng.* **A 255** 1-6

IOP Conf. Series: Materials Science and Engineering **285** (2017) 012032 doi:10.1088/1757-899X/285/1/012032

The structure heterogeneity of silica mesopores of Sba-15 in respect to the pluronic 123 template concentration

D Dhaneswara, J F Fatriansyah, D A Putranto, S A A Utami and F Delayori

Metallurgical and Materials Engineering, Faculty of Engineering, Universitas Indonesia, Kampus UI Depok, Jawa Barat, 16424, Indonesia

Email: fajar@metal.ui.ac.id

Abstract. The analysis of structure heterogeneity factor of silica mesoporous SBA-15 has been conducted. The structure factor has been found to be different for low and high concentration of Pluronic-123 template. The structure heterogeneity of high concentration of Pluronic-123 has been found less than 1 while for low concentration, the structure heterogeneity was found to be larger than 1. This indicates the dissimilarity of the structure and can be used as a probe to detect the formation of large mesopores. It also was found that the system exhibits type IV and H1 adsorption type which indicates the capillary condensation and interconnected pores.

1. Introduction

Silica mesoporous is one of promising nanotechnology materials that has many potential applications in wide area. Recently, it has been used as catalysis [1], drug delivery [2-4], adsorption of biomolecules [5], biosensor [6] and membrane separation [7]. According to IUPAC [8], silica mesoporous material, is a silica based materials with mesoporous with pore of diameters in the range of 2-500 nm. Due to highly ordered mesostructures, this material yields high surface area which is extermely suitable for catalysis and sorption. This advantage allows diffusion and adsorption of larger molecules compared to zeolite or other micropores materials. No wonder that mesoporous silica can be applicable in wide applications.

Silica Mesoporous was invented in 1970 but this went largerly unnoticed [9]. However in 1990, Yanagisawa et. al [10], synthesized mesoporous silica and was named as M41S/MCM. Some of typical MCM mesoporous silica are MCM-41, MCM-48 and MCM-50. MCM-41 has hexagonal packed rod-shaped structure, MCM-48 has cubic structure and MCM-50 has lamellar structure. Again in 1998, Zhao et. al [11] succesfully sythesized mesoporous silica named as Santa Barbara Amorphous-15/SBA-15 (named after University of California, Santa Barbara). SBA-15 has the larger pores and thicker walls than older mesoporous silica. In addition it has good thermal, mechanical stability and chemical resistance properties which make SBA-15 a promising material for wide applications.

SBA-15 has a ordered hexagonal structure. It has diameter of pores of 4.6 nm to 30 nm and was synthesized using triblock copolymer [12]. Surface are of pure SBA-15 can be as high as 650 m^2/gram. The synthesis of SBA-15 using triblock coplymer as template has an advantage which is much easier to synthesize. In this paper we used Pluronic 123, a triblock copolymer surfactant [13]. The pluronic 123 has capability to form long cylindrical micelles which is spesifically suitable to synthesis SBA-15 with precursor silica Tetraethyorthosilicate/TEOS. The use of SBA-15 in sorption application is related to its adsorption capacity. The higher adsorption capacity/ adsorbate uptake towards the selective adsorbate is preferred. In order to study the possibility the use of silica mesoporous SBA-15 to be used as adsorbant to adsorb heavy metals, the isotherm adsorption study is important. To guarantee that the adsorption occurs homogenously in the surface, the structure heteregoneity factor knowledge is important in addition of adsorption capacity. In this paper we calculate of SBA-15 by varying the concentration of surfactant/Pluronic 123. Jaroniec [14] described the structure heterogeneity as a heterogeneity resulted from interconnected pores of different sizes and shapes. To calculate the structure heteregoneity factor Dubinin- Astakhov (D-A) isotherm model was used. This model actually was developed to explain the multilayer adsorption phenomenon. In the results and discussion subsection, the value of structure homogeneity as a function of the concentration of surfactant/Pluronic 123 will be discussed as a recommendation of the use SBA-15 as adsorbant materials

2. Experiment

2.1Materials

In order to synthesize silica mesoporous SBA-15, silica precursor Tetraethyorthosilicate/TEOS (Merck) and Surfactant Pluronic 123 (BASF) as template were used. HCl 2 M (Merck), ethanol and distilled water were used as solvent. All of materials were used without further purification.

2.2 Synthesis

To synthesize SBA-15, wet sol-gel process was conducted. The basic ingredients are building block of mesoporous materials which is silica and the wet template. The source of silica is TEOS and the template is surfactant Pluronic 123. The sol gel process should be conditioned in acidic condition to ensure the self-assembly of silica occurs on template. First step, 5 ml et.OH was added to 31.25 gram TEOS and then the mixing was stirred for 30 minutes under room temperature. Then, 5 ml et.OH was mixed again with 10 ml HCl to make acid environment. This solution then was mixed again with et.OH (10 ml) and addition of 50 ml distilled water solution and they were stirred for 30 minutes under room temperature. The mixed solution was then refluxed for 2 hours at temperature of 70 C. The TEOS solution was ready to be used in further step.

Next the surfactant template solution was synthesized. The concentration of surfactant was varied at 7 mM, 50 mM, 55 mM, 60 mM and 66 mM. This surfactant then was added into mixture of HCl (10 ml) and ethanol (25 ml) in acidic condition. This surfactant solution was then added drop wise into already made TEOS solution and was continuously stirred until the solution turned into gel. The gel then was dried at 100oC for an hour and then was calcined for the addition of 5 hours. The product was then ready for adsorption measurement.

2.3 Characterization

In order to measure the structure heterogeneity factor, the adsorption isotherm experiment was conducted using Quantachrome adsorption-desorption equipment at 77 K. N2 gas was used as adsorbate. The specific surface area was determined by the Brunauer-Emmer-Teller (BET). In order to check the crystallinity of the sample, the addition small angle X-ray scattering (SAXS) measurements (Shimadzu XRD-6000) were carried out.

3. Dubinin-Astakhov isotherm model of adsorption

Dubinin [15] proposed the micropore volume filling theory which is related to the Polanyi characteristic curve. Dubinin acknowlegded that heterogeneity of surface solid can affect the adsorption capacity/ gas uptake amounts. This is called as surface and homogeneity factor which can be originated from interconnected pores and crytall irregularities on the surface respectively. In this paper, those two heterogeneities are called structure heterogeneity factor. Dubinin assumed that the surface contains numerous distribution of pore size distributions related to their characteristics energy. This means that the mathematical form of generalized D-A model can be written as follows

$$C = C_0 \sum_{i=1}^{m} f_i \exp\left[-\left(\frac{A}{E} \right)^n \right] \tag{1}$$

IOP Conf. Series: Materials Science and Engineering **285** (2017) 012032 doi:10.1088/1757-899X/285/1/012032

where f_i is the local fraction of adsorption sites, $A = -\Delta G = RT \ln (P_0/P)$ is the adsorption potential of the Polanyi theory which is related to the change of Gibbs free energy ΔG and the change of relative pressure P_0/P, E is the characteristic energy, n is the structural heterogeneity parameter which is linked to the adsorption of structurally heterogeneous surface, $P_s = (T/T_c)^2 P_c$ (for $T > T_c$) is the saturated pressure, P_c and T_c are critical pressure and temperature respectively. This model is suitable to describe adsorption in highly activated carbon [16]. However in silica mesopore system, we assume that only one site homogenous energy which is related to only one distribution of pore size in order to make calculation easier. By taking this assumption, the simplified D-A model reads

$$C = C_0 \exp\left[-\left(\frac{A}{\beta E_0}\right)^n\right]. \tag{2}$$

The fitting can be conducted by linearizing equation (1) which can be written as follows:

$$\ln C = \ln C_0 - \left[\frac{RT}{\beta E_0}\ln\left(\frac{P_0}{P}\right)\right]^n. \tag{3}$$

D-A model is valid over a range of N_2 relative partial pressure up to 0.03.

4. Results and Discussion

In order to confirm that our sample is in the crystal form, we checked small angle X-ray scattering (SAXS) results which are shown in Figure 1. All of our SBA-15 measured samples show three main peaks (100), (110) and (200). These results are in good agreement with the results obtained by Su et. al [17]. These three peaks show that synthesized SBA-15 has 2D hexagonal structure. As we increase the concentration of surfactant, the peaks shift to the lower angle. This shifting is due to the expansion of lattice parameters. This increase yields the expansion of pore sizes. At concentration 60 mM, the (100) peak is much broader than other concentration which indicates that more non-uniform mesopores were produced.

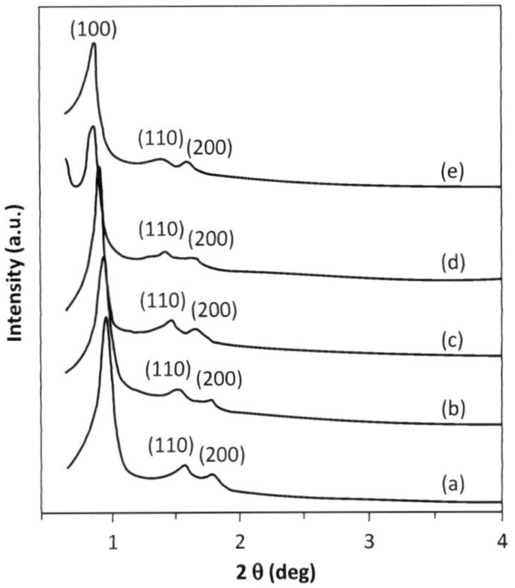

Figure 1. SAXS patterns of SBA-15 with different concentration of Pluronic 123 (a) 7 mM (b) 50 mM (c) 54 mM (d) 60 mM and (e) 66 mM.

The N_2 adsorption-desorption isotherms (bold-dash lines) of SBA-15 with various concentration of Pluronic 123 (7 mM, 50 mM, 54 mM, 60 mM and 66 mM) are shown in figure 2. BET results are given in table 1 including spesific surface area, average pore diameter. Wall size is pedicted using AFM image (which is not shown here) which is around 2-3 nm. In order to fit with the D-A model in equation 3, we took low relative pressure of adsorption isotherms and redrawn the plot. The fitting is shown in figure 3. From figure 2 we can conclude that there are two kinds of adsorption type. At low concentration of Pluronic 123 (7mM), the curve exhibits type I isotherm without hysterisis loop. This curve indicates that at low concentration of surfactant, SBA-15 has small external surface. The uptake is determined only by the accessible pores instead of surface area. At high concentration of Pluronic 123 (50 mM, 54 mM, 60 mM and 66 mM) curves exhibit type 4 isotherm. Their curves reveal a H1 type hysteresis loops. This shows a cillyndirically structure of pores, capillary condensation phenomenon and the possibility of interconneted networks. The relation between the surfactant concentration and adsorption capacity is given in figure 4. The lowest adsorption capacity is at 7 mM concentration of surfactant which stands at 175 cm^3/gram and the highest adsorption capacity is at 60 mM which stands at 421 cm^3/gram.

Table 1. BET results for various P123 concentrations.

No	P123 concentration (M)	Spesific surface area (m^2/g)	Average pore diameter (nm)
1	7 mM	482.2	3.440
2	50 mM	534.6	3.044
3	54 mM	598.5	3.040
4	60 mM	702.1	3.397
5	66 mM	746.7	3.045

Now let discuss the structure heterogeneity parameter. In this paper we do not differentiate between structure heterogeneity and surface heterogeneity. We just called both heterogeneity in single term called structure heterogeneity. This parameter reflects degree of heterogeneity of surface (pore size) as well as interconnectivity of pores. Take a look figure 5. At low concentration of surfactant, the value of structure heterogeneity is larger than 1 at high concentration of surfactant the value is less than 1. This may indicate the fundamental difference of structure. it can be inferred at low concentration, the pore sizes are much larger and more homogenous. There is minimal interconnectivity effect and no capillary condensation. This is in agreement with figure 3 and 4 results which tell us that at low concentration the type I isotherm adsorption occurs and it has lowest gas uptake/adsorption capacity. On the other hand, at high concentration of surfactant, it can be inferred from the results in figure 5, the pore sizes are smaller and thus more heterogeneous. There is interconnectivity effect and capillari condensation. This argument is reflected from figure 3 and 4 results which tell us that at high concentration the type IV isotherm adsorption (with H1 type of hysterisis) occurs and it has highest gas uptake/adsorption capacity. However, at concentration of surfactant of 66 mM however the n value bit higher while for other high concentation much less similar. Our analysis suggests that the structure heterogeneity parameter can be used to figure out indirectly the pore structure of micro/mesopores materials. The heterogeneity of surface information is crucial for mesoporous in order to be applied in adsorption application, heavy metal adsorption for example. This structure may determine the effectivity of adsorption.

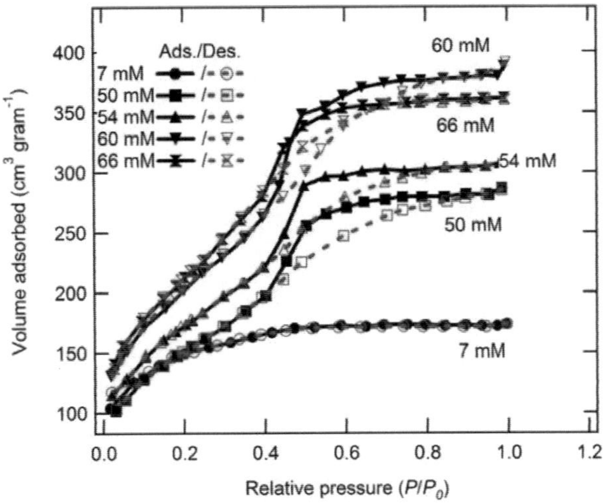

Figure 2 The N_2 adsorption-desorption isotherm (bold-dash lines) of SBA-15 with various concentration of Pluronic 123 (7 mM, 50 mM, 54 mM, 60 mM and 66 mM). The hysterisis curve observed in all samples except at Pluronic 123 concentration of 7 mM.

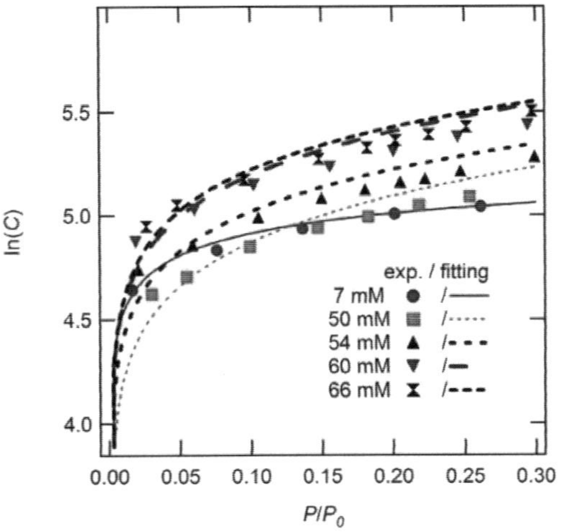

Figure 3 D-A isotherm model fitting for various concentration of Pluronic 123 (7 mM, 50 mM, 54 mM, 60 mM and 66 mM). All curves were fitted up to relative pressure up to 0.3.

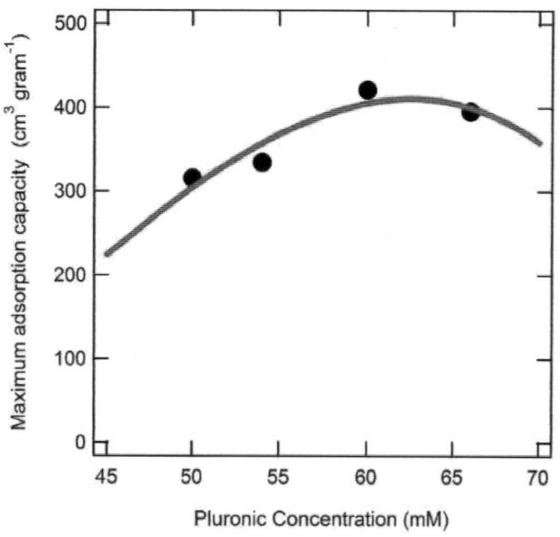

Figure 4 The fitting results of maximum filling capacity/maximum adsorption capacity (C_0) for various concentration of Pluronic 123 (7 mM, 50 mM, 54 mM, 60 mM and 66 mM).

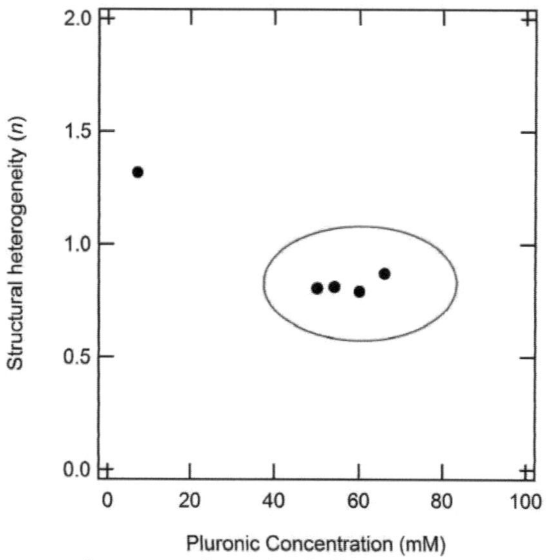

Figure 5. The structure heterogeneity (n) of different Pluronic 123 concentrations. Circled points show lower structural heterogeneity values at high concentration of surfactants.

5. Conclusions

SBA-15 has been succesfully synthesized with various concentration of Pluronic 123 (7 mM, 50 mM, 54 mM, 60 mM and 66 mM). D-A model was found to be fitted well up to relative pressure of 0.3 and can be used to estimate the structure heterogeneity factor of mesporous. It was found that at concentration 60 mM, SBA-15 has highest adsorption capacity. At low concentration of surfactant, the value of structure heterogeneity is larger than 1 at high concentration of surfactant the value is less than 1. This means that at high concentration of surfactant, SBA-15 much more heterogeneous than at low concentration. This results suggest that the structure heterogeneity factor is important to obtain the information of surface of mesoporous. This results can be applied for other applications such as study of adsorption for other gases or adsorption of heavy metals.

IOP Conf. Series: Materials Science and Engineering **285** (2017) 012032 doi:10.1088/1757-899X/285/1/012032

Acknowledgement
This writing is supported by DRPM (Universitas Indonesia Research Directorate) under PITTA 2017 grant.

References
[1] Abdalla Z E A and Baoshan L 2012 *Chemical Engineering Journal* **200** 113 – 21
[2] Tozuka Y, Wongmekiat A, Kimura K, Moribe K, Yamamura S and Yamamoto K 2005 *Chemical & Pharmaceutical Bulletin* **53(8)** 974 – 7
[3] Mellaerts R, Aerts C A, Humbeeck J V, Augustijns P, Den Mooter G V and Martens J A 2007 *Chemical Communications* **13** 1375 – 7
[4] Mellaerts R, Mols R, Kayaert P, Annaert P, Van Humbeeck J, Van Den M G, Martens J A and Augustijns P 2008 *International Journal of Pharmaceutics* **357 (1–2)** 169 – 79
[5] Katiyar A, Yadav S, Panagiotis G, Neville S and Pinto G 2006 *Journal of Chromatography* **1122 (1–2)** 13 – 20
[6] Radu D R, Lai C, Jeftinija K, Rowe E W, Jeftinija S and Lin Victor S Y 2004 *Journal of the Americal Chemical Society* **126(41)** 13216 – 17
[7] Zornoza B, Irusta S, Téllez C and Coronas J 2009 *Langmuir* **25(10)** 5903 – 9
[8] McCusker L B, Liebau F and Engelhardt G 2001 *Pure and Applied Chemistry* **73(2)** 381 – 94
[9] Chiola V, Ritsko J E and Vanderpool C D 1971 *Process for producing low-bulk density silica*, Application No. US 3556725D A filed on 26-Feb-1969, Publication No. US 3556725 A published on 19-Jan-1971
[10] Yanagisawa T, Shimizu T, Kuroda K and Kato C 1990 *Bulletin of the Chemical Society of Japan* **63(4)** 988 – 92
[11] Zhao D, Feng J, Huo Q, Melosh N, Fredrickson G H, Chmelka B F and Stucky G D 1998 *Science* **279.5350** 548 – 52
[12] Zhao D, Sun J, Li Q and Stucky G D 2000 *Chemistry of materials* **12.2** 275 – 9
[13] Dhaneswara D and Sofyan N 2016 *International Journal of Technology* **7(6)** 1009 – 15
[14] Jaroniec M 1995 *Characterization of Nanoporous Materials*, p. 259 in Access in Nanoporous Materials, Ed. T.J. Pinnavaia and M.F. Thorpe. Plenum Press, New York
[15] Dubinin M M 1975 *Physical Adsorption of Gases and Vapors in Micropores*, pp. 1-70 in Progress in Surface and Membrane Science, Ed. J.F. Danielli, M.D. Rosenberg and D.A. Cadenhead. Academic Press, New York
[16] Stoeckli F F, Jakubov T and Lavanchy T 1994 *J. Chem. Soc. Faraday Trans.* **90** 783 – 6
[17] Su F, Zeng J, Bao X, Yu Y, Lee J Y and Zhao X S 2005 *Chemistry of materials* **17(15)** 3960 – 7

Beneficiation and agglomeration of manganese ore fines (an area so important and yet so ignored)

R Sane

Consulting Metallurgist
704 / B, Rachana Sayantara-II, Hazari Pahad, Nagpur= 440007 (M.S.), India.

Email : rajendrasane1947@gmail.com

Abstract. Unpredictable changes in demand and prices varying from very attractive to depressing levels have thrown all Manganese ore mines out of normal operating gear. The supply has to be in time-bound fashion, of dependable quality and continuous. With setting-up of numerous small units and with existing ferro-alloy units, ore supply has become extremely sensitive issue. Due to unpredictable swing in price of Mn ore lumps , furnace operators found it economic and convenient to use fines, even at great risks to furnace equipment and operating persons and therefore risks & damages were conveniently & comfortably ignored. Beneficiation Cost(Operating) approx. – (ferruginous ore) – Roast reduction followed by magnetic separation route-particulars – Water 20/-, Power 490/-,Coal fines-675/-,OH-250/- totaling to Rs.1435/T.(Figures are based on actual data from investigations on Orissa & Karnataka sector ores).Feed Grade Mn- 28 to 32 %, Fe – 14 to 25 %, Concentrate (Beneficiated ore fines)- - Mn- 45 to 48 %, Fe – 6 to 8 %.,Recovery - 35 % , Price of 28-30 % Mn ore fines = Rs. 2400 /T, Cost of Concentrated fines (45/ 48% Mn grade) = Rs. 8300 /T, Price of 47-48 % Mn Lumpy ore = Rs.11,000 /T. Sintering Cost (Operating) – Approx- Rs.1195=00 / T Sinter. Therefore cost of Sinter produced from beneficiated concentrate is 9130+1195 = Rs. 10325. The difference in cost of 48%Mn ore Lumps & 48%Mn sintered concentrate = 11000-10325 = Rs.675/T. The main purpose of this paper is to show that establishment of beneficiation unit & Sintering unit is economically feasible. There are many misconcepts, still prevailing, about use of Mn ore sinters. Few of the main misconcepts are- 1)Sinters bring no benefit – technical or economical.2) Sinters are very friable and disintegrate easily into high fines during handling/transportation. 3) Fines below 100 mesh cannot be sintered. 4) Silica increases to high level during sintering, resulting in to high slag volume thereby higher power consumption. All are false. Sinters have already been proved to bring enormous technical benefits in smelting operation. However small mine owners find it practically impossible to set up a small beneficiation and/or sintering unit at mine site. Recent advances in dry or pneumatic separation which depends on density difference between valuable mineral and gangue/waste rock, are proving to be advantageous & effective because of its small capacity. Capacity is 5 T/hr or 10 T/hr . When applied to low grade high silica Iron ores Iron content has been found to be enhanced by 8 to 10 % from 48% to 58 % , with overall wt. recovery of 60 to 70 %. Recently this technique was applied to low grade siliceous Mn ore with very encouraging results. Mn content has been found to be enhanced from 25 +/- 2 % to 37 +/- 2 % and with 50 to 65% wt. recovery. Size of feed is restricted to (i) 6 to 25 mm and (ii) 2 to 6 mm. Due to Modular design it is possible to put up units from 50 TPD to 800 TPD capacity.

1. Introduction

Unpredictable changes in demand and prices, varying from very attractive to depressing levels have practically thrown all the Manganese ore mines out of normal operating gear. Although efforts are intensified in all the mines to cope up with the erratic demand, alternatives to supplement existing facilities have gained tremendous importance. The end users of Manganese ores are also experiencing the impact of such erratic & unpredictable changes which have ignited a competition in cost-effective measures to draw maximum benefit in present global market of Manganese based ferro alloys.

The ore consuming industry is exerting tremendous pressure on mining industry for supply of all grades of ores. The supply has to be in time-bound fashion, of dependable quality and continuous. With setting-up of numerous small smelting furnaces (capacity varying from 2.5 MVA to 5 MVA) and already existing ferro-alloy units, ore supply has become extremely difficult and very sensitive issue.
With such situation, setting up of beneficiation unit for lean ores and otherwise unsuitable ores has to be considered very seriously. Details of almost every known manganese ore deposit in India are studied and published, encompassing a wide range of its geological formations, mineral compositions, nature of ore bodies and chemical analysis with advantages & disadvantages. It is seen that, by and large, most of the Ferro-Alloy Producers are suffering due to one or more of the following technological gaps-

1) Existing technology unable to cope up with deterioration in Raw Material quality
2) Coupled with rising cost of power and its availability.
3) Continued use of outdated and inefficient Plant & Machinery.
4) Lack of automation and / or modernization in fairly old plants.

In fact end-users of manganese ores, prominently ferro-alloy units are required to put up intensive efforts in following areas for proper use and pragmatic conservation of this strategically important mineral-

1) Up gradation of lean ores by appropriate beneficiation methods.
2) Use of fines in Briquettes or Sintered or Pellet form
3) Charging of pre-heated and pre-reduced raw material in the furnace.
4) Installation of Gas Cleaning Plant and Waste Heat Recovery system.
5) Use of computer aided operations in optimization of process parameters.

If proper & consistent supply of our own resources is not ensured, end user is bound to embrace imported ore even if it turns out to be more expensive. He would not hesitate to make few business compromises even with his competitors because his shear existence would be at stake if proper raw material is not available within country.

Studies on utilization Of Indigenous Lean Grade Manganese Ores – India has done massive work on manganese ore beneficiation and agglomeration. The research work conducted in reputed laboratories like IBM, NML, NMDC, RRLS, CSIR labs and various engineering colleges is highly significant. In fact there is hardly any manganese ore deposit in India, which has not been investigated. Volumes and volumes of data / information is available on characteristics of various Mn ore deposits, different techniques applied in beneficiation, various parameters in briquetting /sintering / pelletization process, material balance and energy balance and other related technological data, and yet, beneficiation and sintering of manganese ores remained a grossly ignored area in India.

2. What could have gone wrong?

After beneficiation & sintering / pelletization of iron ore, beneficiation and sintering of manganese ore fines attracted technology experts due to massive expansion in manganese based ferro-alloys. Fast depleting high grade lumpy manganese ore deposits made the beneficiation and sintering all the more wanted solution. But the real problem started when price of lumpy ore was compared with beneficiated and/or sintered equivalent grade of fines. As the difference in prices of lumps and fines started fluctuating unpredictably with an element of uncertainty, chances of installing beneficiation

IOP Conf. Series: Materials Science and Engineering **285** (2017) 012033 doi:10.1088/1757-899X/285/1/012033

and sintering plant were dimmed. Rising power tariff in India, over-production in other countries, export-import restrictions in India complicated the matter further. Investors went in defensive shell and Ferro Alloy industry went on back-foot. Mining industry was not prepared for such situation and looked confused.

When situation started looking gloomy and immediate relief/solution was not in sight, many furnace operators opted to use fines without any treatment, due to probably following reasons -

1) Non-availability of economically viable lumpy ore.
2) To reduce cost of production even at the risk of equipment & human health, just to earn marginal profit or even just to cover fixed expenses to sustain the existence.(Few operators even had to bear losses just to keep the unit running and avoid retrenchment of people)

It was made amply clear by most of the furnace operators that fines were charged because there was no other option. So with all the risks of explosions, gas eruptions, slag boil, higher power, more slag and grade deterioration, fines continued to find its way to smelting furnaces. Due to unpredictable swing in price of lumpy Mn ore , furnace operators found it economic and convenient to use fines, even at great risks to furnace equipment and operating people. And that was how, unfortunately, a technological abuse or violation got glorified.

The population of manganese based ferro alloy units having furnace capacities up to 9MVA or less, is very large compared to big units with furnace capacity of 12MVA and above. Use of fines in small furnaces (2.5 MVA to 9 MVA) spread like a wild fire which is still continuing. Net result was a grossly wrong concept that fines work wonderfully in small furnace and lumps can be easily restricted to 10-15 %. Unfortunately small & medium sized private producers picked up wrong message– "Fines need not be beneficiated or sintered ". The apparent difference in prices of lumps & fines was too lucrative, the temptation was huge and therefore risks & damages were conveniently & comfortably ignored. The net outcome was that once again the proposals for Beneficiation and / or Sinter Plants received a massive set back .

Table 1. An illustration to indicate the viability of beneficiation and sintering process.

PARTICULARS	QTY	RATE Rs/T	TOTAL-Rs
Water	2 Cu M/T	10	20
Power	70kwh/T	7	490
Coal Fines	150 Kg/ T	4.5	675
Salary + Maintenance + other overheads			250
TOTAL			1435

(Figures are based on actual data from investigations on Orissa & Karnataka sector ores of India)

Feed Grade of ferruginous Manganese ore - - - - - - Mn- 28 to 32 %, Fe – 14 to 25 %

Grade of Concentrate (Beneficiated ore fines)- - Mn- 45 to 48 %, Fe – 6 to 8 %.

Recovery of concentrate by weight – (approx) - - - - - - - - - - - - - - - - - - 35 %

Price of 28-30 % Mn ore fines = Rs. 2400 /T

Cost of Concentrated fines (45 to 48% Mn grade) = Rs. 8300 /T

Price of 47-48 % Mn Lumpy ore = Rs.11,000 /T

Another barrier in establishment of a beneficiation plant is its size. Even a small capacity crushing-washing-screening-wet jigging circuit becomes uneconomical. Unlike Iron ore mines, manganese ore mines have small production capacity and have, many times, heterogeous nature of ore, making common beneficiation technique quite difficult. Therefore small mine owners find it practically impossible to set up a small beneficiation plant at mine site. Moreover any beneficiation technique which requires water becomes cumbersome due to expensive water recovery and recirculation system, apart from requirement of additional land for slurry pond.

Recent advances in gravity separation which depends on density difference between valuable mineral and gangue/waste rock by dry or pneumatic Jigging, are proving to be advantageous & effective. Presently small Iron ore mine owners are using this technique specially in size range of 6 to 25 mm as it is feed size for DRI units. The size is limited to due restriction caused by liberation of minerals at quite coarse size. Iron content has been found to be enhanced by 8 to 10 % from 48% to 58 % Fe with overall wt. recovery of 60 to 70 %. It has been proved to be economical. Recently this technique has been applied on low grade siliceous Mn ore with very encouraging results. Mn content has been found to be enhanced from 25 +/- 2% to 37 +/- 2% and with 50 % wt. recovery. Silica was reduced from 50 +/- 2 % to 18 +/- 2 %. Present price difference between 25 % Mn grade and 38 % Mn grade chips ore (i.e. size between 6 to 25 mm) justifies installation of this DRY or PNEUMATIC way of separation.

Table 2. Another ignored area is sintering of Mn ore fines sintering Cost (Operating) – Approx

PARTICULARS	QTY / T	RATE (Rs/ unit)	Rs/ T
Coke Fines	70 Kg	6.5	455
Water	100 Liters	L.S	10
Fuel (FO/LDO)	5 Liters	45	225
Power	65 Kw	7	455
Maintenance			50
TOTAL			1195

Therefore cost of Sinter produced from beneficiated concentrate is 8715+1195 = Rs. 9910. The difference in cost of 48%Mn ore Lumps & 48%Mn sintered concentrate = (11000-9910) = Rs.1090/T. A simple illustrative example to indicate possibility of sintering low grade fines and blending with high grade ore to arrive at economical blend, when compared with price of equivalent grade lumpy ore –(Based on data collected from various resources)) –

Hutch – 30 % Mn --Rs 2500 +/- 5%
Fines – 32 % Mn – Rs 2800 +/- 5%
Fines - 35% Mn – Rs 3300 +/- 5%
Ave. - 33 % Mn Rs 2850 +/- 5%

If these fines are sintered, then with requirement of 1.05 T of average grade fines and Rs.1195/T cost of Sintering, cost of 1 T of sintered fines will be Rs.4138 /T and grade will be approx. - 35% Mn. If Sinters are blended with high grade Mn ore (47% Mn) in ratio of 85:15, average grade will be 36.40 % Mn, with a cost of (0.85 x 4138 + 0.15 x 11000) = Rs 5167 /T of blend. Say – Rs 5170 / T. If prices of other grades of lumpy ore are compared to sintered ore, following picture would emerge -

Average cost of 30 % Mn lumpy ore ----- Rs 4750 +/--5%
Average cost of 32 % Mn Lumpy ore - - -Rs 5100 +/-- 5%
Average cost of 34 % Mn Lumpy ore - - -Rs 6000 +/-- 5%
Average cost of 37 % Mn Lumpy ore -- - Rs 7000 +/-- 5%
Average cost of 35/36 % Mn grade Lumpy ore - - - (approx) Rs.6200 / T +/-- 5%
(prices based on data from various resources)

The difference in Lumpy ore & sintered fines of 35%Mn grade = 6200-5170 = Rs. 1030 +/-- 5%. The purpose is indicate possibility of establishing a viable sinter plant unit. We need to work seriously in this direction.

3. MSP – (Mini Sinter Plant) – A New Concept in Sinter Plant Design.

The need for establishing manganese ore sintering units was realized long back but one of the major constraints was sizing of the plant. Most of the Iron ore sintering plants, as a part of integrated steel

plant had large capacities ranging from 10,000 TPD upwards. Attempts to miniaturize the conventional sinter plant did not prove to be a viable proposition in view of high capital investment and high cost of production. In India only 3 companies ventured into establishment of small capacity(100 TPD or less) Manganese ore sinter plants -1) Khandelwal Ferro Alloys Ltd , 2) Chandrapur Ferro Alloy Plant of SAIL (earlier known as Maharashtra Elektrosmelt Ltd) and 3) Navbharat Ferro Alloys Ltd. There were no known efforts since last 30 – 35 yrs. Successful establishment of MBFs (Mini Blast Furnaces) to cater to the MINI and MIDI steel plants necessitated production of Iron ore sinter on smaller scale. Mini Sinter Plant was conceived and established. In place of heavy, cumbersome and expensive conventional straight line travelling grate type sinter machine, a simple circular sinter machine comprising of number of small pots/pans or continuous bed type was designed successfully. MSPs gained enormous importance since last 10 to 15 years as MBFs were proving to be economic in steel industry. However unfortunately such efforts were missing in Ferro Alloy industry.
These modifications will help Manganese ore sintering to be less expensive due to reduced capital expenditure and ease to operate.

There are many misconcepts, still prevailing, about use of Mn ore sinters. Few of the main misconcepts & facts are as follows –

Misconcept 1	Fines cause no harm to furnace or process
Fact	Fines cause formation of crust, reduce gas permeability, choke the passage, increased dust losses in flue gas, which ultimately result in higher specific power consumption
Misconcept 2	Sinters bring no benefit – technical or economical
Fact	Sinters ensure higher permeability, higher resistivity of charge, higher power factor, more thermal stability, lesser flue dust losses, lesser power consumption-resulting into reduced cost of production
Misconcept 3	Sinters are very friable and disintegrate easily into high fines during handling/transportation
Fact	Sinters are as strong as Iron ore sinters and have T.I of +75 % and A.I. of less than 5 % . As Iron ore sinters pose no problems in Blast furnace charging or operation, Mn ore sinters are equally beneficial in Sub merged Arc Furnace operation
Misconcept 4	Fines below 100 mesh can not be sintered.
Fact	Fines up to 80 % below 1 mm and overall up to 50 % of fraction in fines below 100 mesh can be sintered without any problem, with special Intensive Mixers
Misconcept 5	Silica increases to high level during sintering, resulting in to high slag volume thereby higher power consumption
Fact	With addition of coke to max 8 % of sinter mix, silica increase in finished sinter is to the tune of 0.5 to 0.75 % only. This drawback is more than compensated by other major technical advantages by use of sinter.

End users of manganese ores, mostly ferro-alloy smelting units, are also required to put up intensive efforts in following areas for proper use and pragmatic conservation of this strategically important mineral, to ease pressure on Mining sector.

1) Use of Briquettes / Sinters / Pellets from fines – natural or beneficiated.
2) Charging of pre-heated and pre-reduced raw material in the furnace.
3) Replacement of coke by other carbonaceous reductants.
4) Installation of Waste Heat Recovery system.
5) Use of computer aided operations in optimization of process parameters.

The reserves of Manganese ore are limited, the constraints are well known, the demands are fairly well projected and there is a continuous pressure from consumer sector to provide better and consistent quality products which are not expensive. Global markets are opening to Indian end users with a wide variety of ore choice. The prices may not be very economic but availability and consistent supply are of prime importance which bring solace & consolation to Ferro Alloy smelting units. It is now entirely left to the Mining Engineers, Mineral Processing Engineers and Metallurgical Engineers to come together for a national cause, to sharpen their skills and rise to the revival of Mineral Industry in general and Manganese Ore Industry in particular, towards a promising future.

References

[1] Indian Bureau of Mines 2014 *Manganese ore vision 2020 and beyond* Ministry of Mines, Govt. of India
[2] Khitrik S I *Electro-Metallurgy of Manganese ferroalloys*
[3] Sane R Y 2004 *Improvisational efforts in ferromanganese production at MOIL* National conference by MEL (SAIL), Nagpur, India
[4] Sane R Y, Bothra S M, Reddy P M 2005 *Manganese ore sector-Problems, Challenges and Prospects at MOIL* M P T
[5] Sane R Y 2011 *Manganese ore sinters-misconcepts and facts* International Conference Indian Institute of Metals, Delhi